T0290357

Modern Quantum Mechanics and Quantum Information

Modern Quantum Mechanics and Quantum Information

J S Faulkner

Department of Physics, Florida Atlantic University, Boca Raton, Florida, FL, USA

IOP Publishing, Bristol, UK

ISBN   978-0-7503-2167-9 (ebook)
ISBN   978-0-7503-2165-5 (print)
ISBN   978-0-7503-2168-6 (myPrint)
ISBN   978-0-7503-2166-2 (mobi)

DOI   10.1088/978-0-7503-2167-9

Version: 20211201

IOP ebooks

British Library Cataloguing-in-Publication Data: A catalogue record for this book is available from the British Library.

Published by IOP Publishing, wholly owned by The Institute of Physics, London

IOP Publishing, Temple Circus, Temple Way, Bristol, BS1 6HG, UK

US Office: IOP Publishing, Inc., 190 North Independence Mall West, Suite 601, Philadelphia, PA 19106, USA

I would like to express my gratitude to the ladies in my life. My wife Dora and my daughters Emilia and Lee Anne.

Contents

Preface

The teaching of graduate quantum mechanics in the twenty-first century.

The choice of topics and emphasis in *Modern Quantum Mechanics and Quantum Information* stem from the demonstrated interests of the physics community. Those interests are best illustrated by the list of the divisions of the American Physical Society (APS) that represent physicists who use quantum mechanics. The divisions established between 1943 and 1950 cover the fields of atomic physics, condensed matter physics, and chemical physics. The interest in applying the developments in those fields to the understanding of commercially interesting materials instigated the establishment of the Division of Materials Physics in 1984, and the fact that these applications require the development of techniques to carry out quantum calculations on large supercomputers prompted the foundation of the Division of Computational Physics in 1986. In 2017, the rapidly growing applications of quantum mechanics into such new areas as quantum computing and quantum cryptography led to the establishment of the Division of Quantum Information. There are many topics in *Modern Quantum Mechanics and Quantum Information* that do not appear in older texts.

Acknowledgement

My quantum mechanics students over the years showed patience and gave me useful feedback as the ideas in this book were developed. I had useful conversations with my colleagues at the Oak Ridge National Laboratory and Florida Atlantic University.

Author biography

J S Faulkner

Prof. John Samuel (Sam) Faulkner was born in Memphis, Tennessee. He obtained BS and MS degrees in physics from Auburn University. He was awarded a PhD in physics by The Ohio State University. He has published over 86 journal articles in the area of theoretical condensed matter physics. He wrote a book on the theory of alloys, *The Modern Theory of Alloys*, J S Faulkner, 1982, published as Vol. 27 of the series Progress in Materials Science, edited J W Christian, P Haasen and T B Massalski (Oxford: Pergamon Press). He edited a book in the same field, *Metallic Alloys: Experimental and Theoretical Perspectives*, 1982, NATO ASI Series, Series E: Applied Sciences, Vol. 256, edited by J S Faulkner and R G Jordan (Dordrecht: Kluwer Academic Publishers).

Prof. Faulkner was a research scientist and head of the theory group at the Oak Ridge National Laboratory and is now an Emeritus Professor of Physics at Florida Atlantic University. He is the Associate Director of the Center for Biomedical and Materials Physics (CBAMP).

Prof. Faulkner's PhD adviser at The Ohio State University was the famous Dutch theoretical physicist Jan Korringa, who is best known for conceiving the multiple-scattering method for calculating the electronic structure of condensed matter. This technique was proposed in 1947, but the first realistic calculations using were done by Faulkner and his colleagues in 1967. During the ensuing years he and his collaborators developed the method further and applied it to many studies of metals and alloys. All of this progress has been summarized in the book *Multiple Scattering Theory: Electronic structure of solids*, J S Faulkner, G M Stocks, and Y Wang, 2019 (Bristol: IOP Publishing). His novel, written under the name "Sam Faulkner" and entitled *The Sparrow*, is available as an ebook from Kindle Bookstore (Amazon). It is an action-adventure novel, but the plot-line contains many scientifically accurate references from the fields of physics, cryptology and quantum computers. Prof. Faulkner is a Fellow of the American Physical Society and the American Association for the Advancement of Science. He is profiled in Who's Who in America. He is married with two children.

IOP Publishing

Modern Quantum Mechanics and Quantum Information
J S Faulkner

Chapter 1

Review of basics

1.1 About quantum mechanics

The metadata on physics publications shows that the number of papers on quantum mechanics has increased exponentially every year since the early nineteen hundreds. This is surprising because the interest most topics in physics that started so long ago has saturated, and they are no longer considered to be on the forefront of the field. The reason that quantum mechanics is different is that when one aspect of the field is worked out, another appears. The new applications are uniformly unpredictable. A way to illustrate this is to consider the sub divisions of the American Physical Society (APS) that rely on quantum mechanics and are listed in table 1.1.

It is not surprising that the earliest application of quantum mechanics was to explain the properties of atoms and simple molecules, and this is reflected in the establishment of DAMOP in 1943. The invention of the transistor in 1947 was completely unexpected, and the development of electronic devices based on semi-conductors led to an explosion of interest in quantum mechanics. The effort to understand superconductivity was also a major goal of physics at this time, as well as the study of the Fermi surfaces of metals. All of this led to the establishment of DCMP in 1947. It is still by far the largest Division of the APS. One of the most important electronic device is the modern high-speed computer. As computers became more powerful, it became feasible to do quantum mechanical calculations on larger molecules and complex multi-phase solids. This led to the founding of the DCP in 1950 and the DMP in 1984, as well as DCOMP in 1986.

All of these exciting developments are still progressing, but with the coming of the new millennium another explosive development arrived on the horizon. The Division of Quantum Information (DQI) was founded in 2017 to focus attention on the underlying theory and the construction of devices such as quantum computers and quantum key distributors that incorporate quantum principles in their structure. Although only a few quantum electronic devices have been made, it has been shown theoretically and experimentally that this new class of devices has the potential to do

Table 1.1. The names and dates for six of the Divisions of the American Physical Society.

Year founded	Name
1943	Division of Atomic, Molecular & Optical Physics (DAMOP)
1947	Division of Condensed Matter Physics (DCMP)
1950	Division of Chemical Physics (DCP)
1984	Division of Materials Physics (DMP)
1986	Division of Computational Physics (DCOMP)
2017	Division of Quantum Information (DQI)

calculations that are impossible with conventional computers and also revolutionize the important field of encryption. All of these applications will be discussed in the following chapters. In the remainder of this chapter, several concepts that are normally introduced in more elementary courses will be defined more precisely.

1.2 Hilbert space

The mathematical language of quantum mechanics is based on linear algebra and linear vector spaces. A vector $\mathbf{v}$ is a mathematical object that has two properties, commonly called a length and a direction. The inner product of two vectors is a number that may be written $a = \mathbf{u} \cdot \mathbf{v}$. The length of a vector is called the norm, and may be written $|\mathbf{v}| = \sqrt{\mathbf{v} \cdot \mathbf{v}}$. A vector with a zero norm $|\mathbf{v}| = 0$ is called a null vector. If the vector $a\mathbf{u} + b\mathbf{v}$ is in the space when $\mathbf{u}$ and $\mathbf{v}$ are, the vector space is said to be linear.

A two-dimensional vector space in which every vector may be written

$$\mathbf{v} = a_1\mathbf{e}_1 + a_2\mathbf{e}_2, \tag{1.1}$$

where the a_i are numbers and $\mathbf{e}_i$ are basis vectors is called complete. Any vector in a complete three-dimensional vector space may be written

$$\mathbf{v} = a_1\mathbf{e}_1 + a_2\mathbf{e}_2 + a_3\mathbf{e}_3. \tag{1.2}$$

It is easy to imagine a complete N-dimensional vector space, in which

$$\mathbf{v} = a_1\mathbf{e}_1 + a_2\mathbf{e}_2 + \ldots + a_N\mathbf{e}_N, \tag{1.3}$$

or even a complete vector space whose dimension is countably infinite

$$\mathbf{v} = \sum_{i=1}^{\infty} a_i\mathbf{e}_i. \tag{1.4}$$

All of these expressions can be written

$$\mathbf{v} = \sum_{i} a_i\mathbf{e}_i, \tag{1.5}$$

where the sum is from 1 to N or ∞. It is convenient, although not necessary, to require the basis vectors to be orthonormal

$$\mathbf{e}_i \cdot \mathbf{e}_j = \delta_{ij}, \tag{1.6}$$

where δ_{ij} is the Kronecker delta. There is no requirement that the numbers a_i must be real. Using these conventions, the inner product of two vectors is

$$\mathbf{u} \cdot \mathbf{v} = \sum_i b_i^* a_i, \tag{1.7}$$

where $\mathbf{u} = \sum_i b_i \mathbf{e}_i$ and b_i^* is the complex conjugate of b_i. The norm of a vector is

$$|\mathbf{v}| = \sqrt{\sum_i a_i^* a_i} = \sqrt{\sum_i |a_i|^2}. \tag{1.8}$$

Definition: a vector space in which inner products are defined for all vectors, vectors have a real positive norm, the space is linear and also complete, is called a Hilbert space. In a Hilbert space, the norm of every vector is positive or zero. There is no requirement that the numbers that appear in the definitions be real.

Operators differ from scalars (numbers) in that they can change the direction of a vector as well as its magnitude (norm). A way to write them is

$$\mathbf{u} = \mathbf{A}\mathbf{v}. \tag{1.9}$$

The operators of interest are linear in the sense that, if $\mathbf{A}$ and $\mathbf{B}$ are operators, $\mathbf{C} = a\mathbf{A} + b\mathbf{B}$ is also an operator. The effect of a linear operator on a vector can best be seen by considering the components of the vectors

$$b_i = \sum A_{ij} a_j. \tag{1.10}$$

The connection between linear algebra, linear operators on Hilbert spaces, and matrix theory is obvious. The relation in equation (1.10) is that a column vector with elements b_i is obtained by multiplying the one with elements a_i by a square matrix with elements A_{ij}. Just as matrices do not necessarily commute with each other, operators do not commute in general,

$$\mathbf{A}\mathbf{B} \neq \mathbf{B}\mathbf{A}. \tag{1.11}$$

The mathematical language of Hilbert space and linear algebra is the natural one for Heisenberg's matrix formulation of quantum mechanics. However, Schrödinger's wave formulation requires a different kind of mathematics.

1.3 Elementary quantum mechanics

Let us consider some of the simple examples that were worked out in elementary quantum mechanics:

One-dimensional particle in a box. The Schrödinger equation is

$$-\frac{\hbar^2}{2m}\frac{d^2\psi}{dx^2} + V(x)\psi = E\psi, \tag{1.12}$$

with $V(x) = 0$ for $0 \leqslant x \leqslant L$ and $V(x) = \infty$ for all other x. It has a countable infinity of solutions

$$\psi_n(x) = A_n \sin\left(\frac{n\pi x}{L}\right), \tag{1.13}$$

and $E_n = \frac{\sqrt{2m}}{\hbar}\sqrt{\frac{n\pi}{L}}$.

One-dimensional harmonic oscillator. For this case, the potential inserted into the Schrödinger equation in (1.12) is $V(x) = 1/2kx^2$ for all x. It has a countable infinity of solutions $\psi_n(x) = c_n H_n(\alpha x)e^{-\frac{\alpha^2 x^2}{2}}$ where H_n is a Hermite polynomial, $\alpha = \sqrt{m\omega}/\hbar$, and $\omega = \sqrt{k/m}$. These solutions correspond to the countable infinity of energies $E_n = \hbar\omega(n + 1/2)$.

Free particle. The potential in equation (1.12) is set equal to zero for this case,

$$-\frac{\hbar^2}{2m}\frac{d^2\psi}{dx^2} = E\psi. \tag{1.14}$$

There is an uncountable infinity of solutions $\psi_p(x) = ce^{\frac{ipx}{\hbar}}$ corresponding to an uncountable infinity of energies $E = p^2/2m$.

Hydrogen atom. This is a three-dimensional problem so the Schrödinger equation is written

$$-\frac{\hbar^2}{2m}\nabla^2\psi - \frac{e^2}{r}\psi = E\psi. \tag{1.15}$$

There are a countable infinity of negative bound state energies $E_{nlm} = -R/n^2$, where $R = me^4/4\pi c\hbar^3$ is the Rydberg constant. The bound state wave functions corresponding to these energies are normalizable. The subscripts l and m describe the different angular momenta that the electron can have for a given n, so the degeneracy of each energy is n^2. Equation (1.15) has a solution for every positive energy. These non-normalizable solutions describe the scattering of an electron from a proton.

It is not clear that the Hilbert space described in the preceding section is helpful in connection with these functions of x, particularly the non-normalizable free-electron functions and scattering functions. There are two additions to the formalism that will make it a proper mathematical foundation for quantum mechanics.

1.4 Dirac and von Neumann

The tools for the axiomatic formulation of quantum mechanics were put forward by P A M Dirac [1] and John von Neumann. [2] This formulation is called the Dirac and von Neumann (DvN) theory.

Von Neumann based his axioms on the mathematical theory of linear operators and linear vector spaces. In addition to couching quantum mechanics in this language, he made original contributions to the development of the mathematics of linear algebra. He actually introduced the term 'Hilbert space', naming it after the mathematician David Hilbert.

Dirac introduced a notation in which the vector $\mathbf{v}$ is written as a ket $|v\rangle$. For a complete finite or countably infinite dimensional vector space, any vector in the space can be written

$$|v\rangle = \sum_i a_i \, | \, i\rangle. \tag{1.16}$$

In order to discuss inner products, a conjugate space must be introduced. The inner product of $|u\rangle$ and $|v\rangle$ is $\langle u|v\rangle$, where $\langle u|$ is the conjugate of $|u\rangle$. If $|u\rangle$ is a constant times $|v\rangle$, $|u\rangle = c|v\rangle$, the conjugate of $|u\rangle$ is $\langle u|=c^*\langle v|$, where c^* is the complex conjugate of c. The basis vectors are orthonormal if

$$\langle i|j\rangle = \delta_{ij}. \tag{1.17}$$

If $|u\rangle = \sum_i b_i|i\rangle$ and $|v\rangle$ is given by equation (1.16), the inner product is a number

$$\langle u|v\rangle = \sum b_i^* a_i. \tag{1.18}$$

From this equation, it is easy to show that the complex conjugate of an inner product is the inner product of the same vectors in the opposite order

$$\langle u|v\rangle^* = \langle v|u\rangle. \tag{1.19}$$

A vector in the conjugate space is called a bra. Dirac's notation is intended to be a mnemonic device. Consider the nomenclature

$$
\begin{array}{c}
\langle b \parallel k\rangle \\
\text{bra}\parallel\text{ket} \\
\langle b|k\rangle \\
\text{bracket}
\end{array}
\tag{1.20}
$$

Any expression that begins with a bra and ends with a ket is closed and is a number.
The coefficients in equation (1.16) can be written

$$a_i = \langle i|v\rangle. \tag{1.21}$$

A set of basis vectors is called complete if every vector in the space can be written as a linear combination of them. A unit operator can be written

$$\mathbf{I} = \sum_i | i \rangle \langle i |, \tag{1.22}$$

if the set $\{|i\rangle\}$ is complete and orthonormal. The ket–bra order of the vectors indicates an operator. This relation is called the resolution of the identity. Writing $|v\rangle = \mathbf{I}|v\rangle$ and using equation (1.22) leads to equation (1.16).

An operator in the Dirac notation changes the magnitude and direction of a ket

$$| u \rangle = \mathbf{A} | v \rangle. \tag{1.23}$$

Writing this equation $\mathbf{I}|u\rangle = \mathbf{IAI}|v\rangle$ leads to equation (1.10), where

$$A_{ij} = \langle i | \mathbf{A} | j \rangle. \tag{1.24}$$

1.5 Rigged Hilbert space

To this point, the Dirac formalism has been used to reproduce the equations for the Hilbert space in section 1.2. It is possible to use it to go further and define a rigged Hilbert space that includes the ordinary Hilbert space but also includes the possibility of a space for which the dimension is uncountably infinite. [3] A specific example is to define a space in which the basis vectors $|x\rangle$ correspond to every point on line. These vectors are eigenvectors of the position operator

$$\mathbf{x} | x \rangle = x | x \rangle. \tag{1.25}$$

In a complete space spanned by these basis vectors, any vector can be written

$$|f\rangle = \int_{-\infty}^{\infty} f(x) | x \rangle dx, \tag{1.26}$$

and the function $f(x)$ plays the role of coefficients. It may be written $f(x) = \langle x|f \rangle$. It follows that the resolution of identity in this space is

$$\mathbf{I} = \int_{-\infty}^{\infty} | x \rangle \langle x | \, dx. \tag{1.27}$$

These equations are only consistent if

$$\langle x'|f \rangle = f(x') = \int_{-\infty}^{\infty} f(x) \langle x'|x \rangle dx, \tag{1.28}$$

and $\langle x'|x \rangle$ is the Dirac delta function

$$\langle x'|x \rangle = \delta(x' - x). \tag{1.29}$$

This formalism was known to physicists at the time of Dirac, but it was not accepted because there was no function that could have the properties attributed to the delta function. Because of this, a proper theory of rigged Hilbert spaces had to wait for a mathematically rigorous treatment of the equations in this paragraph.

The French mathematician Laurent Schwarz was the first to take the delta function seriously. In his book *Théorie des distributions* [4] he noted that, for example, the function

$$G(d, x, x_0) = \left(\frac{1}{\pi d^2}\right)^{\frac{1}{2}} e^{-\frac{(x-x_0)^2}{d^2}} \tag{1.30}$$

has no limit in the usual sense when d goes to zero. However, the limit of the integral of the product of this function with a sufficiently smooth function does exist

$$\lim_{d \to 0} \int_{-\infty}^{\infty} F(x)G(d, x, x_0)dx = F(x_0). \tag{1.31}$$

It can thus be said that the limit of $G(d, x, x_0)$ is a generalized function that has all the features of the delta function $\delta(x - x_0)$. It turns out that there are several functions that converge to the generalized function $\delta(x - x_0)$ in this same sense. For example, the rectangular function

$$G(d, x, x_0) = \frac{1}{2d} \text{ if } x_0 - d \leqslant x \leqslant x_0 + d \text{ and } = 0 \text{ otherwise,} \tag{1.32}$$

or

$$G(d, x, x_0) = \frac{1}{\pi(x - x_0)} \sin \frac{x - x_0}{d}. \tag{1.33}$$

It must be kept in mind that a generalized function has no meaning when it is removed from under an integral. With this caveat in mind, derivations like the one that led to equation (1.29) are legitimate.

Using the resolution of the identity in equation (1.27), an operator equation becomes

$$u(x) = \int_{-\infty}^{\infty} A(x, x')v(x')dx', \tag{1.34}$$

where

$$A(x, x') = \langle x \mid \mathbf{A} \mid x' \rangle. \tag{1.35}$$

Many of the operators in quantum mechanics are functions of the position operator $\mathbf{x}$ and the momentum operator $\mathbf{p}$. For $\mathbf{A} = \mathbf{x}$, equation (1.34) becomes

$$u(x) = xv(x). \tag{1.36}$$

The assumption that underpins Heisenberg's formulation of quantum mechanics, couched in the DvN language, is that the position and momentum of a particle cease to be numbers and become abstract Hermitean operators. Furthermore, the commutation relation between the position and momentum operators is assumed to be

$$\mathbf{xp-px} = i\hbar\mathbf{I}. \tag{1.37}$$

Premultiplying this equation with $\langle x'|$ and postmultiplying it with $|x\rangle$ leads to

$$(x' - x)\langle x'| p | x\rangle = i\hbar\delta(x' - x), \tag{1.38}$$

and hence, with the help of equation (1.27),

$$u(x') = \int_{-\infty}^{\infty} \langle x'| p | x\rangle v(x)dx = i\hbar \int_{-\infty}^{\infty} \frac{v(x)}{x' - x}\delta(x' - x)dx. \tag{1.39}$$

The appearance of the delta function implies that $v(x)$ must be known in the vicinity of x'. Inserting the Taylor's expansion of $v(x)$ into the integral leads to

$$u(x') = i\hbar v(x') \int_{-\infty}^{\infty} \frac{\delta(x' - x)}{x' - x}dx + i\hbar\frac{dv(x')}{dx'} \int_{-\infty}^{\infty} \frac{x - x'}{x' - x}\delta(x' - x)dx. \tag{1.40}$$

The first term is zero because the delta function is an even function of $(x - x')$ and the denominator is an odd function. Thus, premultiplying $|u\rangle = \mathbf{p}|v\rangle$ with $\langle x'|$ leads to

$$u(x') = -i\hbar\frac{dv(x')}{dx'}. \tag{1.41}$$

The abstract algebra version of the momentum eigenvalue equation is

$$\mathbf{p} | p\rangle = p | p\rangle. \tag{1.42}$$

Using the resolution of the identity and equation (1.41), this equation becomes

$$\frac{d\phi_p(x)}{dx} = \frac{ip}{\hbar}\phi_p(x), \tag{1.43}$$

which has the solution

$$\phi_p(x) = \langle x|p\rangle = ce^{\frac{ipx}{\hbar}}. \tag{1.44}$$

The eigenvectors $|p\rangle$ form a complete set of basis functions in the rigged Hilbert space, just as the eigenvectors of the position operators $|x\rangle$ do. Therefore, the resolution of the identity using them is

$$\mathbf{I} = \int_{-\infty}^{\infty} | p\rangle\langle p | dp. \tag{1.45}$$

When abstract algebra is used as the language of quantum mechanics, the physics information of a problem is described by the abstract vector $|\psi\rangle$. Using the resolutions of the identity in equations (1.27) and (1.45), this vector can be written in two equivalent ways

$$| \psi\rangle = \int_{-\infty}^{\infty} \psi(x) | x\rangle dx = \int_{-\infty}^{\infty} \psi(p) | p\rangle dp. \tag{1.46}$$

These coefficients must be related to each other. Using equations (1.45) and (1.44) the relationship is

$$\psi(x) = \langle x|\psi \rangle = c \int_{-\infty}^{\infty} e^{\frac{ipx}{\hbar}} \psi(p)dp. \tag{1.47}$$

Defining a parameter k by $p = \hbar k$ and investigating the coefficient $\langle p|\psi \rangle$ in the same manner shows that the relations can be written as Fourier transforms

$$\psi(x) = \frac{1}{\sqrt{2\pi}} \int_{-\infty}^{\infty} e^{ikx} \psi(k)dk$$
$$\psi(k) = \frac{1}{\sqrt{2\pi}} \int_{-\infty}^{\infty} e^{-ikx} \psi(x)dx \tag{1.48}$$

This connection has the advantage that all of the studies on the convergence of Fourier transforms are immediately available to this quantum mechanical problem. It also gives a value for the constant c.

The same physics can be described by the wave function $\psi(x)$ or $\psi(p)$. Working in the 'position representation' means that the function $\psi(x)$ is used, and $\psi(p)$ is used in the 'momentum representation'.

The use of the word rigged may be strange to Americans because it has gotten a bad connotation from its use in terms like a rigged election or a rigged gambling game. The Russian originators of the theory [5] had a more old fashioned use of the term in mind. A full rigged ship like the one in figure 1.1 is one that has the maximum set of sails and equipment on board. In that sense a rigged Hilbert space is one that includes all of the formalism that is needed in any possible quantum mechanical context.

Figure 1.1. A full rigged ship. Credit: Nona Lohr (CC0 Public Domain).

1.6 Observables and Hermitean operators

One of the axioms of DvN is that observables are represented by operators. The only possible result of a measurement of the observable is an eigenvalue a_i of the operator

$$\mathbf{A}|a_i\rangle = a_i|a_i\rangle, \tag{1.49}$$

where $|a_i\rangle$ is the corresponding eigenvector. The conjugate of the vector $|u\rangle = \mathbf{A}|v\rangle$ is

$$\langle u| = \langle v|\,\mathbf{A}^\dagger, \tag{1.50}$$

and $\mathbf{A}^\dagger$ is called the Hermitean conjugate of the operator $\mathbf{A}$. The conjugate of equation (1.49) is

$$\langle a_i|a_i^* = \langle a_i|\mathbf{A}^\dagger. \tag{1.51}$$

Premultiplying equation (1.49) with $\langle a_i|$ and postmultiplying equation (1.51) with $|a_i\rangle$ and subtracting the results leads to

$$a_i - a_i^* = \langle a_i|\mathbf{A}|a_i\rangle - \langle a_i|\mathbf{A}^\dagger|a_i\rangle. \tag{1.52}$$

Obviously if the operator $\mathbf{A}$ is Hermitean

$$\mathbf{A} = \mathbf{A}^\dagger, \tag{1.53}$$

the eigenvalues are real. In quantum mechanics, observations must give real numbers, therefore the operators that correspond to observables must be Hermitean.

By premultiplying equation (1.49) with $\langle a_j|$ and postmultiplying a version of equation (1.51) with $i = j$ with $|a_i\rangle$ and subtracting the two resulting equations leads to

$$\langle a_j|a_i\rangle(a_i - a_j) = 0, \tag{1.54}$$

for the case that $\mathbf{A}$ is Hermitean. Thus, the eigenvectors of a Hermitean operator corresponding to different eigenvalues are orthogonal to each other. Since it is easy enough to normalize a vector, the collection of eigenvectors of any Hermitean operator is an orthonormal set.

For measurements on the observable $\mathbf{A}$ it is useful to expand any state function in terms of the eigenvectors of that observable

$$|\psi\rangle = \sum_i c_i|a_i\rangle. \tag{1.55}$$

It was demonstrated above that the eigenvectors are an orthonormal set, but are they a complete set? The physicist's argument for this is that the only Hilbert space of interest is one that is spanned by them. Another way to state that the $|a_i\rangle$ are a complete orthonormal set is that the resolution of the identity is

$$\mathbf{I} = \sum_i |a_i\rangle\langle a_i|. \tag{1.56}$$

By operating **A** on this the Hermitean operator can be written

$$\mathbf{A} = \sum_i a_i |a_i\rangle\langle a_i|. \tag{1.57}$$

The statistical interpretation of the state vector $|\psi\rangle$ states that the absolute square $|c_i|^2$ of the coefficient

$$c_i = \langle a_i|\psi\rangle, \tag{1.58}$$

is the probability that a measurement of a state described by that vector will yield the eigenvalue a_i. This interpretation of the expansion coefficients in equation (1.55) is called the Born rule because it was first formulated in a paper by Max Born [6]. He was given a Nobel prize for his statistical interpretation of quantum mechanical wave functions.

The operator

$$\mathbf{P}(a_i) = |a_i\rangle\langle a_i|, \tag{1.59}$$

is called a projection operator because when it is operated on any state vector it projects it into the eigenstate $|a_i\rangle$. The expectation of $\mathbf{P}(a_i)$ in the state $|\psi\rangle$ is the probability that a measurement of **A** in that state will give the value a_i

$$\langle\psi|\,\mathbf{P}(a_i)|\,\psi\rangle = |c_i|^2 = p(a_i). \tag{1.60}$$

The above discussion was carried out using an ordinary Hilbert space. As pointed out in the preceding section, for measurements on an observable that can give an uncountable infinity of answers such as the position of a particle on a line, it is necessary to use a rigged Hilbert space. The completeness of the basis made up of the set of eigenvectors $|x\rangle$ is again trivial because the space of interest can be spanned by them

$$|\,\psi\rangle = \int_{-\infty}^{\infty} \psi(x)\,|\,x\rangle dx. \tag{1.61}$$

The resolution of the identity is in equation (1.27), and the operator can be written

$$\mathbf{x} = \int_{-\infty}^{\infty} x\,|\,x\rangle\langle x\,|\,dx. \tag{1.62}$$

The projection operator is

$$\mathbf{P}(x) = |\,x\rangle\langle x\,|, \tag{1.63}$$

and the probability that the particle will be found between x and $x + dx$ is

$$\langle\psi|\,\mathbf{P}(x)\,|\,\psi\rangle dx = \langle\psi\,||\,x\rangle\langle x\,||\,\psi\rangle dx = |\psi(x)|^2 dx = p(x)dx. \tag{1.64}$$

This is the version of the Born rule that applies to a continuous spectrum of eigenvalues. Prior to Born's assertion, it was thought that the particle was spread out in space with a density given by $|\psi(x)|^2$.

The energy plays a special role in quantum mechanics as it always has in classical mechanics. For that reason, the space for a particular quantum problem is usually defined by choosing the eigenvectors of a Hamiltonian to be the basis functions. The eigenvectors of the one-dimensional particle in a box Hamiltonian in equation (1.12) span an ordinary Hilbert space. The completeness of the basis set rests on the mathematical proofs that any smooth function that is zero outside the range $0 \leqslant x \leqslant L$ can be expanded in terms of the functions in equation (1.13). This is just Fourier series theory. The eigenvectors of the free-electron Hamiltonian in equation (1.14) are the same as the eigenvectors of the momentum operator in equation (1.44). As seen in connection with equation (1.48), the dimension of the vector space that they span is an uncountable infinity, which means that the space is a rigged Hilbert space. The proof of completeness is the one used in the theory of Fourier series.

1.7 The uncertainty relation

A lemma that is useful in proving the uncertainty relation is known as the Schwarz inequality. The norm of the vectors in any kind of Hilbert space must be positive. This fact can be used to prove a useful relation. Consider any two vectors, $|\alpha\rangle$ and $|\beta\rangle$. Define another vector $|v\rangle = |\alpha\rangle + \lambda|\beta\rangle$. Clearly,

$$\langle v|v\rangle \geqslant 0 = ((\langle\alpha| + \lambda^*\langle\beta|)(|\alpha\rangle + \lambda|\beta\rangle)), \tag{1.65}$$

and

$$\langle\alpha|\alpha\rangle + \lambda\lambda^*\langle\beta|\beta\rangle + \lambda\langle\alpha|\beta\rangle + \lambda^*\langle\beta|\alpha\rangle \geqslant 0. \tag{1.66}$$

Setting $\lambda = -\langle\beta|\alpha\rangle/\langle\beta|\beta\rangle$,

$$|\langle\alpha|\beta\rangle|^2 \leqslant \langle\alpha|\alpha\rangle\langle\beta|\beta\rangle. \tag{1.67}$$

This is called the Schwarz inequality and it makes it possible to define an angle between two vectors in a Hilbert space of any dimension

$$\frac{|\langle\alpha|\beta\rangle|}{\sqrt{\langle\alpha|\alpha\rangle}\sqrt{\langle\beta|\beta\rangle}} = \cos\theta \leqslant 1. \tag{1.68}$$

One of the reasons that the DvN algebra of operators on vector spaces is more useful for quantum mechanics than an ordinary algebra with numbers is that it is able to describe the fact that the measurement process on certain observables can be mutually exclusive. As the error in the measurement of one observable is reduced, the error in the measurement of its conjugate observable increases.

Let us consider two Hermitean operators that do not commute

$$\mathbf{AB} - \mathbf{BA} = [\mathbf{A}, \mathbf{B}] = ic\mathbf{I}. \tag{1.69}$$

First, define $|\alpha\rangle = (A - \langle A\rangle)|\psi\rangle = \Delta A|\psi\rangle$, where $\langle A\rangle = \langle\psi|A|\psi\rangle$. Similarly, define $|\beta\rangle = (B - \langle B\rangle)|\psi\rangle = \Delta B|\psi\rangle$, where $\langle B\rangle = \langle\psi|B|\psi\rangle$. Because of the Hermiticity of $\mathbf{A}$ and $\mathbf{B}$

$$\langle \alpha | \alpha \rangle = \langle \psi | (A - \langle A \rangle)^2 | \psi \rangle = \langle (A - \langle A \rangle)^2 \rangle$$
$$\langle \beta | \beta \rangle = \langle \psi | (B - \langle B \rangle)^2 | \psi \rangle = \langle (B - \langle B \rangle)^2 \rangle$$

(1.70)

Using the Schwarz inequality in equation (1.67) leads to

$$\langle (\Delta A)^2 \rangle \langle (\Delta B)^2 \rangle \geqslant |\langle \psi | \Delta A \Delta B | \psi \rangle|^2$$
$$= (\mathrm{Re} \langle \psi | \Delta A \Delta B | \psi \rangle)^2 + (\mathrm{Im} \langle \psi | \Delta A \Delta B | \psi \rangle)^2.$$

(1.71)

The complex conjugate of the expectation value of $\Delta A \Delta B$ is

$$\langle \Delta A \Delta B \rangle^* = \langle \psi | \Delta A \Delta B | \psi \rangle^* = \langle \psi | \Delta B \Delta A | \psi \rangle.$$

(1.72)

It follows that the expectation value of the commutator of ΔA with ΔB is

$$\langle [\Delta A, \Delta B] \rangle = \langle \Delta A \Delta B \rangle - \langle \Delta A \Delta B \rangle^* = 2i\mathrm{Im}\langle \Delta A \Delta B \rangle.$$

(1.73)

The anticommutator of two operators is like the commutator but the products are added

$$\mathbf{AB + BA = \{A, B\}}.$$

(1.74)

The expectation value of the anticommutator of ΔA with ΔB is

$$\langle \{\Delta A, \Delta B\} \rangle = \langle \Delta A \Delta B \rangle + \langle \Delta A \Delta B \rangle^* = 2 \, \mathrm{Re} \, \langle \Delta A \Delta B \rangle.$$

(1.75)

The two terms on the right side of equation (1.71) are both positive, so, as far as mathematics is concerned, either of them can be ignored and the inequality would only be stronger. In quantum mechanics the commutator of two operators plays a more central role than the anticommutator, so

$$\langle (\Delta A)^2 \rangle \langle (\Delta B)^2 \rangle \geqslant (\mathrm{Im}\langle \psi | \Delta A \Delta B | \psi \rangle)^2,$$

(1.76)

is chosen. It can easily be seen that adding a constant to an operator has no effect on the commutator. Therefore, using equation (1.73), the above becomes

$$\langle (\Delta A)^2 \rangle \langle (\Delta B)^2 \rangle \geqslant \frac{1}{4} |\langle [A, B] \rangle|^2.$$

(1.77)

Taking the square root of the above equation leads to

$$\Delta A \Delta B \geqslant \frac{1}{2}|\langle [A, B] \rangle|,$$

(1.78)

where $\Delta A = \sqrt{\langle (\Delta A)^2 \rangle}$ and $\Delta B = \sqrt{\langle (\Delta B)^2 \rangle}$. In words, this equation says that the root mean square (rms) error in the measurement of $\mathbf{A}$ times the rms error in the measurement of $\mathbf{B}$ is greater than or equal to the commutator of the two observables divided by two. Returning to the first paragraph of this section, this shows that $\mathbf{A}$ and $\mathbf{B}$ are conjugate variables if their commutator is not zero. The mathematical starting point for this derivation is the Schwarz inequality, which hold for the rigged Hilbert space.

As pointed out above, Heisenberg based his formulation of quantum mechanics on the commutation relation

$$[\mathbf{x}, \mathbf{p}] = i\hbar. \tag{1.79}$$

From equation (1.78)

$$\Delta x \Delta p \geqslant \frac{\hbar}{2}, \tag{1.80}$$

is obtained. This is the version of the uncertainty relation discussed by Heisenberg, although the derivation used here could not be carried out until the DvN theory was developed. He explained it by imagining an experiment in which a microscope is used to 'find' a particle. In order to localize the particle and hence reduce the rms error in the position, it would be necessary to reduce the wavelength of the light used by the microscope. DeBroglie's relation states that the momentum of a photon is equal to Planck's constant divided by the wavelength, so a reduction of the error in the position is accompanied by an increase in the momentum that the particle receives from the measuring device. It follows that the error in the momentum would increase. This physical argument provides an understanding of the effect, but it does not lead to the precise relation in equation (1.80).

1.8 Commuting observables

Let us consider two operators that do commute

$$\mathbf{AB} - \mathbf{BA} = [\mathbf{A}, \mathbf{B}] = 0. \tag{1.81}$$

The operator $\mathbf{A}$ has eigenvalues a_i and eigenvectors $|a_i\rangle$, and for $\mathbf{B}$ they are b_i and $|b_i\rangle$. Thus,

$$\mathbf{AB}|a_i\rangle = \mathbf{BA}|a_i\rangle = a_i\mathbf{B}|a_i\rangle, \tag{1.82}$$

so $\mathbf{B}|a_i\rangle$ is also an eigenvector of $\mathbf{A}$ with eigenvalue a_i. A similar argument leads to the result that $\mathbf{A}|b_i\rangle$ is an eigenvector of $\mathbf{B}$ with the eigenvalue b_i. Ignoring the trivial case that $\mathbf{A}$ or $\mathbf{B}$ is a constant times the unit matrix, there must be a set of vectors $|a_i, b_j\rangle$ that are simultaneously eigenvectors $\mathbf{A}$ and $\mathbf{B}$.

This result can be stated in other ways. It can be said that there is a basis in which the commuting operators are simultaneously diagonalized. It can also be discussed in the framework of degeneracies. An eigenstate is said to be degenerate when there is more that one eigenvector for a given eigenvalue. This is a situation that will arise numerous times in the following discussions.

1.9 Unitary operators

Undergraduate physicists learn about orthogonal transformations on vectors in a three-dimensional Euclidean space. These transformations can be looked upon as changing the axes for the Cartesian coordinate system by a simple rotation. Under such a transformation, the distance between any two points, $\vec{r} = \vec{r_1} - \vec{r_2}$, is unchanged. Also, the angle between two vectors, $\cos\theta = \vec{r_1} \cdot \vec{r_2}/|\vec{r_1}||\vec{r_2}|$. The analog of this transformation in Hilbert space is called a unitary transformation.

Consider an operator $\mathbf{U}$ with the property that

$$\mathbf{U}^\dagger \mathbf{U} = \mathbf{I}, \tag{1.83}$$

where $\mathbf{U}^\dagger$ is the Hermitean conjugate operator an $\mathbf{I}$ is the unit operator. In the DvN notation, all of the vectors in a Hilbert space can be replaced by new vectors

$$| \alpha' \rangle = \mathbf{U} | \alpha \rangle \tag{1.84}$$

The length of the vectors are unchanged

$$\langle \alpha' | \alpha' \rangle = \langle \alpha | \mathbf{U}^\dagger \mathbf{U} | \alpha \rangle = \langle \alpha | \alpha \rangle, \tag{1.85}$$

which is equivalent to the assertion that the distance between points is unchanged. The angle between vectors in Hilbert space was defined in equation (1.68). The invariance of this quantity under a unitary transformation is seen by

$$\frac{|\langle \alpha' | \beta' \rangle|}{\sqrt{\langle \alpha' | \alpha' \rangle} \sqrt{\langle \beta' | \beta' \rangle}} = \frac{\langle \alpha | \mathbf{U}^\dagger \mathbf{U} | \beta \rangle}{\sqrt{\langle \alpha | \mathbf{U}^\dagger \mathbf{U} | \alpha \rangle} \sqrt{\langle \beta | \mathbf{U}^\dagger \mathbf{U} | \beta \rangle}} = \frac{|\langle \alpha | \beta \rangle|}{\sqrt{\langle \alpha | \alpha \rangle} \sqrt{\langle \beta | \beta \rangle}}. \tag{1.86}$$

1.10 The Gaussian wave packet

There exists a state vector $|\psi\rangle$ with the property that the uncertainty relation for a particle in that state is a minimum. That is, the inequality in equation (1.78) becomes an equality. Two steps are required to achieve this. First, the Schwartz inequality is an equality if $|\beta\rangle$ is parallel to $|\alpha\rangle$. That means $|\beta\rangle = c|\alpha\rangle$, where c is a possibly complex number. Second, the anticommutator in equation (1.75) must be zero.
    If

$$\begin{aligned} | \alpha \rangle &= (x - x_0) | \psi \rangle \\ | \beta \rangle &= (p - p_0) | \psi \rangle \end{aligned}, \tag{1.87}$$

the first condition leads to

$$\frac{\hbar}{i} \frac{d\psi}{dx} = \left[p_0 + c(x - x_0) \right] \psi. \tag{1.88}$$

The solution of this equation is

$$\psi = e^{\frac{i}{\hbar} \left[p_0 x + \frac{c(x - x_0)^2}{2} \right]}. \tag{1.89}$$

The second condition is

$$\langle \alpha | \beta \rangle + \langle \beta | \alpha \rangle = 0 = 2 \operatorname{Re} c \langle \alpha | \alpha \rangle, \tag{1.90}$$

which means that c must be purely imaginary. Writing it in the form $c = i\hbar/\sigma^2$ leads to the equation

$$\psi(x) = e^{\frac{ip_0 x}{\hbar} - \frac{(x-x_0)^2}{2\sigma^2}}. \tag{1.91}$$

This wave function describes a Gaussian wave packet with a peak at x_0 and a variance of σ. Time dependence can be added to this function, and it will be seen to move to the right with velocity p_0/m.

1.11 Two-dimensional Hilbert space

The smallest non-trivial Hilbert space is the two-dimensional one with two basis functions. It is easier to explain quantum mechanics in such a space because the arguments are not obscured by irrelevant mathematical details. For this reason, it is frequently used as a toy model in discussions of the measurement.

The two-dimensional Hilbert space can be used to describe an idealized version of the behavior of the spin of an electron. Spin is an intrinsic attribute of a particle like mass or charge. Particles with spin one half are predicted by Dirac's relativistic wave equation. Fundamental particles with spin one half are called fermions, and the most common example of a fermion is the electron. Measurements of the spin of an electron in ordinary three-dimensional Euclidean space are represented by three operators S_x, S_y, and S_z.

In order to measure a spin it is necessary to interact with it in some way. According to Dirac's theory, spins interact with a magnetic field as if they were a magnetic dipole. The Hamiltonian for a spin in a magnetic field $\vec{B}$ is

$$H = -\frac{e}{mc}\vec{S} \cdot \vec{B}, \tag{1.92}$$

where e is the electron charge, m is the mass of the electron, and c is the velocity of light. The operator $\vec{S}$ can be written

$$\vec{S} = S_x\hat{x} + S_y\hat{y} + S_z\hat{z} \tag{1.93}$$

where $\hat{x}$, $\hat{y}$, and $\hat{z}$ are unit vectors in ordinary Euclidean space. The magnetic field is a vector function $\vec{B} = B_x\hat{x} + B_y\hat{y} + B_z\hat{z}$. According to Ehrenfest's theorem, the derivative of the expectation value of the momentum operator with respect to time is equal to the expectation value of the minus the gradient of the potential operator

$$\frac{d\langle \mathbf{p}\rangle}{dt} = -\langle \nabla V(\mathbf{x})\rangle. \tag{1.94}$$

From this equation it follows that a beam of fermions moving in the x direction in an inhomogeneous magnetic field with $B_x = B_y = 0$ and $B_z = B_z(x)$ would be split into two beams, one with z component of spin up and one with it down. This was first tried by Otto Stern and Walter Gerlach in 1922, although the theoretical premise they used to explain the results of their experiment was incorrect.

Today, spins can be measured in many different ways, and the construction of devices based on the manipulation of spins is an area of experimental study known as spintronics. Thus, a thought experiment in which Fermions could be separated into different bins depending on their spin can be realized in a laboratory.

An axiom of quantum mechanics is that spin operators obey the commutation rules

$$[\mathbf{S}_x, \mathbf{S}_y] = i\hbar\mathbf{S}_z$$
$$[\mathbf{S}_y, \mathbf{S}_z] = i\hbar\mathbf{S}_x. \tag{1.95}$$
$$[\mathbf{S}_z, \mathbf{S}_x] = i\hbar\mathbf{S}_y$$

The anticommutation relations are

$$\{\mathbf{S}_\alpha, \mathbf{S}_\beta\} = 0 \text{ if } \alpha \neq \beta, \tag{1.96}$$

where α and β stands for $x, y,$ or z. Finally, the square of any of the spin operators is proportional to the identity operator

$$\mathbf{S}_\alpha^2 = 1/4\hbar^2\mathbf{I}. \tag{1.97}$$

From the results of the preceding section, it follows that the components of the spin cannot be accurately measured at the same time. The spin operator $\vec{\mathbf{S}}$ can be written in the form of a vector in three-dimensional Euclidean space with components that are the spin operators

$$\vec{\mathbf{S}} = \mathbf{S}_x\hat{\mathbf{x}} + \mathbf{S}_y\hat{\mathbf{y}} + \mathbf{S}_z\hat{\mathbf{z}}. \tag{1.98}$$

From equation (1.97), it is obvious that any one of the spin operators commutes with the operator

$$\vec{\mathbf{S}} \cdot \vec{\mathbf{S}} = \mathbf{S}^2 = \mathbf{S}_x^2 + \mathbf{S}_y^2 + \mathbf{S}_z^2, \tag{1.99}$$

or

$$[\mathbf{S}^2, \mathbf{S}_\alpha] = 0. \tag{1.100}$$

Thus, the measurement of a component of spin does not change the magnitude of the spin. From equation (1.97) it follows that

$$\mathbf{S}^2 \,|\pm\rangle = \frac{3}{4}\hbar^2 \,|\pm\rangle. \tag{1.101}$$

Let us choose as the basis of a two-dimensional Hilbert space the eigenvectors of $\mathbf{S}_z$, $|+\rangle$ and $|-\rangle$ with

$$\mathbf{S}_z \,|\pm\rangle = \pm\frac{1}{2}\hbar \,|\pm\rangle. \tag{1.102}$$

The exact eigenvalues are not very important, but the $\pm 1/2\hbar$ are correct for the spin of an electron and are consistent with equation (1.97). From the theorem proved with equation (1.52) these vectors are orthogonal, and can be chosen to be normalized. Even though the eigenvectors of $\mathbf{S}_z$ have been chosen as a basis, there is nothing special about the z direction in Euclidean space. The eigenvector equations for the other two operators must be the same as the above

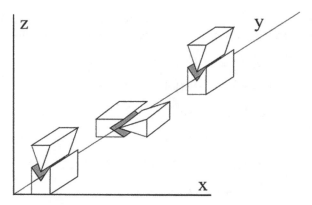

Figure 1.2. A hypothetical set of Stern–Gerlach experiments. The particle traveling in direction y is measured by an experiment that measures S_z then S_x and then S_z again.

$$\mathbf{S}_x \mid x, \pm\rangle = \pm 1/2\hbar \mid x, \pm\rangle$$
$$\mathbf{S}_y \mid x, \pm\rangle = \pm 1/2\hbar \mid y, \pm\rangle \tag{1.103}$$

It will not be a surprise that if a measurement of $\mathbf{S}_z$ gives the result that the spin is in state $|+\rangle$, a subsequent measurement of $\mathbf{S}_z$ will find the spin in the same state. The sequence of measurements shown in figure 1.2 is more complicated. The first measurement is said to prepare the state of the spin in the state $|+\rangle$. After the spin is prepared in the state $|+\rangle$, a measurement of $\mathbf{S}_x$ will find that the spin is in $|x, +\rangle$ or $|x, -\rangle$, Because of the symmetry of Euclidean space, the probability of either result will be 50%. If it happens to be $|x, +\rangle$, the spin at this point is prepared in that state. Now, a subsequent measurement of $\mathbf{S}_z$ will find the spin in $|+\rangle$ or $|-\rangle$ with equal probability. The neophyte to quantum mechanics will find it surprising that the order in which measurements are made can wipe out all memory of the original state of the spin.

It is possible to use the same experiments to find the eigenfunctions of S_x. The probability that a spin prepared in $|+\rangle$ is subsequently found in $|x, +\rangle$ by a measurement of $\mathbf{S}_x$ is 50%, so

$$|\langle +|x, +\rangle|^2 = \frac{1}{2}. \tag{1.104}$$

Given that it must be possible to write

$$\mid x, +\rangle = a_x^+(+) \mid +\rangle + a_x^+(-) \mid -\rangle$$
$$\mid x, -\rangle = a_x^-(+) \mid +\rangle + a_x^-(-) \mid -\rangle \tag{1.105}$$

the problem is to evaluate the coefficients. The first condition that can be used to determine this coefficients is to prepare a state in which it is known that the spin is in $|x, +\rangle$ or $|x, -\rangle$. This is achieved by measuring the observable $\mathbf{S}_x$. If the spin is prepared in $|x, +\rangle$, what is the probability that a subsequent measurement of $\mathbf{S}_z$ will find the electron in $|+\rangle$ or $|+\rangle$. The symmetry of Euclidean space requires that the probabilities would be 50%, so the magnitudes of the coefficients a_x^+ and a_x^- must be

$1/\sqrt{2}$. The same argument can be applied to a spin that is know to be in the state $|x, -\rangle$. Therefore, ignoring a unimportant phase shift, these vectors can be written

$$
\begin{aligned}
|x, +\rangle &= 1/\sqrt{2}(|+\rangle + e^{i\delta_+}|-\rangle) \\
|x, -\rangle &= 1/\sqrt{2}(|+\rangle - e^{i\delta_-}|-\rangle).
\end{aligned}
\tag{1.106}
$$

These vectors must be orthogonal, so

$$
\langle x, -|x, +\rangle = 1/2(1 - e^{i(\delta_+ - \delta_-)}),
\tag{1.107}
$$

must be zero. The easiest way to achieve this is to set $\delta_+ = \delta_- = 0$, which leads to

$$
\begin{aligned}
|x, +\rangle &= 1/\sqrt{2}(|+\rangle + |-\rangle) \\
|x, -\rangle &= 1/\sqrt{2}(|+\rangle - |-\rangle)
\end{aligned}.
\tag{1.108}
$$

Using the same arguments, $|y, +\rangle$ and $|y, -\rangle$ are written

$$
\begin{aligned}
|y, +\rangle &= 1/\sqrt{2}(|+\rangle + e^{i\delta_+}|-\rangle) \\
|y, -\rangle &= 1/\sqrt{2}(|+\rangle - e^{i\delta_-}|-\rangle)
\end{aligned}.
\tag{1.109}
$$

From equation (1.107), a way to insure the orthogonality of $|y, +\rangle$ and $|y, -\rangle$ is to choose $\delta_+ = \delta_- = \delta$. The argument that led to equation (1.106) leads to the conclusion that, if the spin of a particle is prepared in the state $|x, +\rangle$, the probability that a subsequent measurement of $\mathbf{S}_y$ will find it in the state $|y, +\rangle$ or $|y, -\rangle$ must be 50%. Since

$$
\begin{aligned}
|\langle y, +|x, +\rangle|^2 &= 1/2(1 + \cos \delta) \\
|\langle y, -|x, +\rangle|^2 &= 1/2(1 - \cos \delta)
\end{aligned},
\tag{1.110}
$$

this can only achieved by setting $\delta = \pi/2$. The conclusion is then

$$
\begin{aligned}
|y, +\rangle &= 1/\sqrt{2}(|+\rangle + i|-\rangle) \\
|y, -\rangle &= 1/\sqrt{2}(|+\rangle - i|-\rangle)
\end{aligned},
\tag{1.111}
$$

and a combination of mathematical and physical arguments have made it possible to find the eigenvalues for all three of the spin operators.

Using equation (1.57), the spin operators can be written in terms of projection operators

$$
\begin{aligned}
\mathbf{S}_x &= \frac{\hbar}{2}(|+\rangle\langle-|+|-\rangle\langle+|) \\
\mathbf{S}_y &= \frac{i\hbar}{2}(-|+\rangle\langle-|+|-\rangle\langle+|). \\
\mathbf{S}_x &= \frac{\hbar}{2}(|+\rangle\langle+|-|-\rangle\langle-|)
\end{aligned}
\tag{1.112}
$$

A little algebra shows that these operators satisfy the commutation rules in equations (1.95) through (1.100).

The two-dimensional Hilbert space is an excellent source of toy models of the measurement process in quantum mechanics. For example, if a state $|+\rangle$ is prepared by measuring $\mathbf{S}_z$, and then $\mathbf{S}_z$ is measured again, the probability that the outcome of the measurement will be $|+\rangle$ is 100% and the probability it will be $|-\rangle$ is zero. This is not surprising, but people who have not thought a lot about quantum mechanics may be surprised to find that an intervening measurement of, for example, $\mathbf{S}_x$ followed by the measurement of $\mathbf{S}_z$ will completely change these probabilities. No matter whether the measurement of $\mathbf{S}_x$ gives $|x, +\rangle$ or $|x, -\rangle$, the probability of ending in the states $|+\rangle$ or $|-\rangle$ are the same and are 25%. This can be shown easily using the formulae derived in this section and the definition of probabilities in equation (1.60).

General expressions for measuring processes can be written for an n-dimensional Hilbert space. Let us assume the system is prepared in the state $|a\rangle$. The probability that a later measurement will find it in $|c\rangle$ is

$$|\langle a|c\rangle|^2 = \langle c|a\rangle\langle a|c\rangle = P. \tag{1.113}$$

The resolution of the identity was given it equation (1.22). Inserting $\mathbf{I}$ into the preceding equation leads to

$$\sum_{i=1}^{n}\sum_{j=1}^{n}\langle c|i\rangle\langle i|a\rangle\langle a|j\rangle\langle j|c\rangle = P, \tag{1.114}$$

which is a complicated way to rewrite equation (1.113). It can be broken into two parts, the first of which is

$$\sum_{i=1}^{n}\langle c|i\rangle\langle i|a\rangle\langle a|i\rangle\langle i|c\rangle = p. \tag{1.115}$$

This equation can be interpreted as the result of a number of experiments in which intervening measurements lead to results $|i\rangle$. It is the sum over all possible outcomes. The next part is the terms in equation (1.114) not in the preceding equation

$$\sum_{j=2}^{n-1}\sum_{i=1}^{j-1}[\langle c|i\rangle\langle i|a\rangle\langle a|j\rangle\langle j|c\rangle + c.\,c.] = r, \tag{1.116}$$

where $c.c.$ means the complex conjugate. The remainder r is a real number that is not, in general, zero. The fact that $p \neq P$ highlights a very important fact about measurements in quantum mechanics. It is hard to think of any semi-classical picture that would lead to the same result.

An example of the application of this equation in the two-dimensional Hilbert space is to take $|a\rangle = |+\rangle$ and $|c\rangle = |-\rangle$. The probability of a state prepared in $|+\rangle$ will be found by a later measurement of $\mathbf{S}_z$ to be in $|-\rangle$ is zero

$$|\langle +|-\rangle|^2 = \langle +|-\rangle\langle -|+\rangle = P = 0. \tag{1.117}$$

Take the states used in the resolution of the identity to be

$$| 1 \rangle = | x, + \rangle, | 2 \rangle = | x, - \rangle. \qquad (1.118)$$

Since

$$\langle 1|+ \rangle = \frac{1}{\sqrt{2}} \langle 2|+ \rangle = \frac{1}{\sqrt{2}} \langle 1|- \rangle = \frac{1}{\sqrt{2}} \langle 2|- \rangle = -\frac{1}{\sqrt{2}}, \qquad (1.119)$$

it is found that $p = \frac{1}{2}$. The remainder is $r = -\frac{1}{2}$, so the sum is $p + r = 0$ as it should be. Again, there is no semi-classical explanation for these results.

A unit vector in Euclidean space that makes an angle ϑ with the z axis and an angle φ with the x–z plane can be written

$$\hat{e} = \sin \vartheta \cos \varphi \hat{x} + \sin \vartheta \sin \varphi \hat{y} + \cos \vartheta \hat{z}. \qquad (1.120)$$

Writing the spin operators as in equation (1.93), an expression for the component of the spin operator in an arbitrary direction can be written

$$\vec{S} \cdot \hat{e} = \frac{\hbar}{2}[\sin \vartheta (e^{-i\varphi} | + \rangle\langle -|+e^{i\varphi} | - \rangle\langle +|) + \cos \vartheta (| + \rangle\langle +|-|-\rangle\langle -|)]. \qquad (1.121)$$

The eigenvectors of this operator are

$$\begin{aligned} | \hat{e}, + \rangle &= \cos \frac{\vartheta}{2} | + \rangle + e^{i\varphi} \sin \frac{\vartheta}{2} | - \rangle \\ | \hat{e}, - \rangle &= \sin \frac{\vartheta}{2} | + \rangle - e^{i\varphi} \cos \frac{\vartheta}{2} | - \rangle \end{aligned}, \qquad (1.122)$$

which are an orthonormal set. Obviously, the results for the x, y, or z directions are obtained by evaluating the above for $\vartheta = \pi/2$ and $\varphi = 0$, $\vartheta = \pi/2$ and $\varphi = \pi/2$, or $\vartheta = 0$. Equation (1.122) will be very useful at a later point.

The preceding discussion can be reproduced using matrices rather than abstract operators and vectors. The spin operators are replaced with matrices as follows

$$S_\alpha \leftrightarrow \hbar/2\sigma_\alpha, \qquad (1.123)$$

where the σ_α are the Pauli matrices

$$\sigma_x = \begin{pmatrix} 0 & 1 \\ 1 & 0 \end{pmatrix}, \quad \sigma_y = \begin{pmatrix} 0 & -i \\ i & 0 \end{pmatrix}, \quad \sigma_z = \begin{pmatrix} 1 & 0 \\ 0 & -1 \end{pmatrix}. \qquad (1.124)$$

The commutation and anticommutation rules for these matrices are the same as equations (1.95) through (1.101), except for the constant $\hbar/2$. The vectors are replaced with

$$| + \rangle = \begin{pmatrix} 1 \\ 0 \end{pmatrix}, \quad | - \rangle = \begin{pmatrix} 0 \\ 1 \end{pmatrix}. \qquad (1.125)$$

The σ_α are obviously Hermitean. It can be shown that any 2×2 Hermitean matrix can be written in terms of the Pauli matrices and the 2×2 unit matrix

$$A = a_0 I + a_x \sigma_x + a_y \sigma_y + a_z \sigma_z, \tag{1.126}$$

with real coefficients. It follows that any 2×2 unitary matrix can be written by exponentiating the Hermitean matrix

$$U = e^{iA}. \tag{1.127}$$

This is proved by construction

$$U^\dagger U = (e^{iA})^\dagger e^{iA} = e^{-iA^\dagger} e^{iA} = e^{-i(A^\dagger - A)} = 1. \tag{1.128}$$

The unitary matrices can be used to describe rotations in three dimensions, but this is a complicated story that will be put off until chapter 4.

1.12 Pairs of spins

There are cases in quantum mechanics in which the spins of two particles are coupled. The most obvious of these is the alignment of the spins in the ground state of the helium atom. The spins on the two electrons that bind the hydrogen molecule are similarly aligned. The theory of the covalent bond that was first worked out for this molecule is fundamental for the theoretical understanding of the field of organic chemistry. It will be seen that this bond cannot be explained without a knowledge of the eigenfunctions of two coupled spins.

In the preceding section it was seen that the spin operator can be expressed as a vector in three-dimensional Euclidean space. The spin operator for two spins can likewise be written as the sum of the two spins

$$\vec{S} = \vec{S}_1 + \vec{S}_2, \tag{1.129}$$

so

$$S_z = S_{z1} + S_{z2}, \tag{1.130}$$

and

$$S^2 = S_1^2 + S_2^2 + 2S_{x1}S_{x2} + 2S_{y1}S_{y2} + 2S_{z1}S_{z2}. \tag{1.131}$$

It will be seen that it is convenient to use the raising and lowering operators

$$S_{\pm i} = S_{xi} \pm iS_{yi}, \tag{1.132}$$

where

$$\begin{array}{ll} S_{+i} \mid + \rangle = 0 \mid + \rangle & S_{+i} \mid - \rangle = \hbar \mid + \rangle \\ S_{-i} \mid + \rangle = \hbar \mid - \rangle & S_{-i} \mid - \rangle = 0 \mid - \rangle \end{array}. \tag{1.133}$$

With this notation, equation (1.131) can be rewritten

$$S^2 = S_1^2 + S_2^2 + 2S_{z2}S_{z1} + S_{+1}S_{-2} + S_{-1}S_{+2}. \tag{1.134}$$

The wave functions for these operators are linear combinations of the couplets $|\alpha\rangle|\beta\rangle$ where the convention is that an operator with a subscript 1 operates on the ket in the first position and the one with a subscript 2 operates on the second. Testing the states $|+\rangle|+\rangle$ and $|-\rangle|-\rangle$

$$
\begin{aligned}
\mathbf{S}_z |+\rangle|+\rangle &= \hbar|+\rangle|+\rangle \quad \mathbf{S}^2|+\rangle|+\rangle = 2\hbar^2|+\rangle|+\rangle \\
\mathbf{S}_z |-\rangle|-\rangle &= -\hbar|-\rangle|-\rangle \quad \mathbf{S}^2|-\rangle|-\rangle = 2\hbar^2|-\rangle|-\rangle
\end{aligned}
\tag{1.135}
$$

shows that they are eigenstates of $\mathbf{S}_z$ and $\mathbf{S}^2$. Using the notation

$$
S_z|j, m\rangle = m\hbar|j, m\rangle \quad S^2|j, m\rangle = j(j+1)\hbar^2|j, m\rangle,
\tag{1.136}
$$

these states are identified as

$$
|1, 1\rangle = |+\rangle|+\rangle \quad |1, -1\rangle = |-\rangle|-\rangle.
\tag{1.137}
$$

Testing the states $|+\rangle|-\rangle$ and $|-\rangle|+\rangle$ leads to

$$
\begin{aligned}
\mathbf{S}_z|+\rangle|-\rangle &= 0|+\rangle|-\rangle \quad \mathbf{S}^2|+\rangle|-\rangle = \hbar^2|+\rangle|-\rangle+\hbar^2|-\rangle|+\rangle \\
\mathbf{S}_z|-\rangle|+\rangle &= 0|-\rangle|+\rangle \quad \mathbf{S}^2|-\rangle|+\rangle = \hbar^2|-\rangle|+\rangle+\hbar^2|+\rangle|-\rangle
\end{aligned}
\tag{1.138}
$$

These states are obviously not eigenstates of $\mathbf{S}^2$, but by adding and subtracting them normalized eigenstates can be constructed

$$
|0, 0\rangle = 1/\sqrt{2}(|+\rangle|-\rangle-|-\rangle|+\rangle),
\tag{1.139}
$$

and

$$
|1, 0\rangle = 1/\sqrt{2}(|+\rangle|-\rangle+|-\rangle|+\rangle).
\tag{1.140}
$$

With this process all of the eigenstates of $\mathbf{S}_z$ and $\mathbf{S}^2$ for two electrons have been identified. In the following section, it will be shown how these eigenstates became the basis for one of the most famous debates concerning the measurement process in quantum mechanics.

1.13 Einstein, Podolsky, and Rosen

During the 'annus mirabilis' of 1905, Albert Einstein published four papers that revolutionized the field of physics. One of these papers was an explanation of the photoelectric effect in which shining ultraviolet light on the surface of a metal causes electrons to be ejected. In retrospect, the assumptions that Einstein was forced to make in order to explain the experimental observations were some of the first steps in the development of quantum mechanics. In spite of the fact that Einstein could be considered to be a pioneer of the field, he never liked quantum mechanics and tried to find counter-arguments that would show the flaws in its foundations.

In 1935, Einstein enlisted the aid of two of his colleagues at the Institute for Advanced Studies in Princeton, New Jersey, Boris Podolsky and Nathan Rosen, to write a paper entitled *Can Quantum-Mechanical Description of Physical Reality be Considered Complete?* From the title of the paper, it is clear what Einstein thought was the answer to the rhetorical question.

The authors posed the following thought experiment. Suppose that two electrons are bound to a positive charge at the origin. Think a helium atom or diatomic hydrogen molecule. Quantum mechanical calculations show that the electrons will be in a state that is the product of a spatial function $\phi(r_1, r_2)$ and a spin function $\chi(s_1, s_2)$

$$\psi(r_1, r_2, s_1, s_2) = \phi(r_1, r_2)\chi(s_1, s_2). \qquad (1.141)$$

All of the calculations conclude that the electrons are in the antisymmetrical state $|0, 0\rangle$ in equation (1.139). Now imagine that the positive charge disappears in some way, and the coulomb repulsion between the electrons will cause the position function to spread out without changing the spin function.

Suppose that two observers Alice and Bob are at positions A and B a large distance apart and they both have a device to measure spins. If the outcome of the measurement by Bob is that the spin is in the spin up state $|+\rangle$, the measurement collapses the spin function so that a measurement at position A must give the spin down state $|-\rangle$. There is a 50% chance of finding the spin up or down at B, but, once the measurement has been made, the measurement at A can only give the opposite spin.

Einstein believed that the quantum mechanical analysis in the preceding paragraph had to be flawed. This came to be known as the EPR paradox. It seemed that Bob's measurement sent a signal to Alice's apparatus restricting her measurement, and that this signal traveled faster than the speed of light. Such a result contradicts another of Einstein's discoveries published in 1905, the special theory of relativity.

Einstein was confident that the EPR paradox would never be resolved by experiment, but it has been. Experimentalists today have the techniques to make all of the requisite measurements. The predictions of quantum mechanics have been completely validated.

What about Einstein's worry concerning a signal traveling faster than the speed of light? A more complete analysis demonstrates that it never happens. Suppose the observers at A and B spend an afternoon measuring spin directions and writing their results on a paper. Since they know accurately their distances from the origin, they can do a calculation to find which of the spins are from the same wave function. However, neither of them see anything strange about their data. Their measurements give a random distribution of up and down spins, like a coin flip. It is only when they compare their lists side-by-side that they see the correlations caused by fact that they are part of the same wave function. However, the comparison of their lists requires that they be communicated to each other in some manner, such as a radio. That communication cannot travel faster than the speed of light.

Problems

PI.1 Write a half-page analysis of the usefulness of a graduate level course on quantum mechanics for obtaining a job on graduation.

PI.2 The transpose of a matrix is one for which the indices of the elements are interchanged, $\tilde{A}_{ij} = A_{ji}$. Prove that the transpose of the product of two matrices is the product of their transposes, but in opposite order $\widetilde{AB} = \tilde{B}\tilde{A}$.

PI.3 There are an infinite number of eigenvalues for the one-dimensional particle in a box and the one-dimensional harmonic oscillator. For the hydrogen atom, there are an infinite number of eigenvalues with $E < 0$ and an infinity of eigenvalues for $E > 0$. What are the aleph numbers for these four infinities?

PI.4 Prove that any operator A can be written as a sum of terms in which each term is a ket followed by a number followed by a bra.

PI.5 Expand $F(x)$ in a Taylor's series near x_0. Do the integrals to show that the functions in equations (1.30), (1.32), and (1.33) approach $\delta(x - x_0)$ in the sense of a distribution.

PI.6 Show that an operator that is not Hermitean can have eigenvalues and eigenvectors. Are the eigenvalues real?

PI.7 Describe a hypothetical thought experiment in which a microscope is used to locate a small particle. Show that it leads to an uncertainty principle like the one derived above.

PI.8 Show that the eigenvalues u_i of a unitary operator $\mathbf{U}$ are such that $|u_i|^2 = 1$.

PI.9 Convert the wave function in equation (1.91) to the momentum representation.

PI.10 Test the wave functions in equations (1.139) and (1.140) and demonstrate algebraically that they are eigenstates of $\mathbf{S}_z$ and $\mathbf{S}^2$.

PI.11 The EPR paper describes the principle of locality as asserting that physical processes occurring at one place should have no immediate effect on the elements of reality at another location, and that an affirmative result for the experiment would violate this principle. What is wrong with this argument?

References

[1] Dirac P A M 1930 *The Principles of Quantum Mechanics* (Oxford: Oxford University Press)
[2] von Neumann J 1952 *Mathematische Grundlagen der Quantenmechanik* (Berlin: Springer) English translation by Beyer R T 1955 (Princeton, NJ: Princeton University)
[3] Maurin K 1968 *Generalized Eigenfunction Expansions and Unitary Representations of Toppological Groups* (Warsaw: Polish Scientific Publishers)
[4] Schwartz L 1950/1951 *Théorie des Distributions* (Paris: Hermann), 2 vols
[5] Gelfand I M and Vilenkin N Y 1964 *Generalized Functions* vol IV (New York: Academic)
[6] Born M 1926 *Z. Phys.* **37** 863

IOP Publishing

Modern Quantum Mechanics and Quantum Information

J S Faulkner

Chapter 2

Non-relativistic quantum mechanics

2.1 Heisenberg's matrix mechanics

Planck assumed in 1900 that the electromagnetic waves bouncing inside a cavity heated to some temperature T behave like particles of energy $E = h\nu$, where ν is the frequency of the waves and h is the famous Planck constant. Planck stated that he was unhappy with this assumption and only made it out of despair. It was the only way that he could obtain a formula that could adequately describe the spectrum of the black-body radiation streaming out of a hole in the side of the cavity.

In 1905, Einstein was forced to assume that the light shone of the surface of a metal had to be treated like a collection of photons of energy $E = h\nu$. This was the only way he could explain the energy distribution of the photoelectrons that are knocked off the surface. In 1907, Einstein explained the specific heat of solids at low temperatures by assuming that the vibrational energies of the atoms are quantized with $E = nh\nu$, where n is an integer and ν is the vibrational frequency.

In 1915, Bohr proposed a model of the atom that explains the line spectrum of the light emitted when an excited atom loses energy. He modified Newton's law of planetary motion by assuming the electrons circling the nucleus could only be in orbits for which the angular momentum was $l = nh/2\pi$. The frequency of the photon emitted when the electron jumped from an orbit with energy E_i to one with energy E_f is $\nu = (E_i - E_f)/h$.

In 1924, de Broglie pointed out that the relationship between the wavelength and the momentum $\lambda = h/p$ is true for particles like electrons as well as for photons. This wavelength would lead to interference fringes in a two-slit diffraction experiment.

Although the theories for black-body radiation, the photoelectric effect, low-temperature specific heat, atomic spectra, and electron diffraction presaged a new fundamental theory, it was not until 1925 that Heisenberg put forward a theory called quantum mechanics that could explain all of these disparate phenomena and more [1]. Heisenberg had the basic idea, but his mathematical skills did not allow him to express his vision very well. He was aided by his colleagues M Born and

doi:10.1088/978-0-7503-2167-9ch2

P Jordan who suggested that matrices provide an appropriate mathematical language for Heisenberg's theory. The matrix theory of Heisenberg, Born, and Jordan was revolutionary, but Heisenberg's theory can be described even better using the terminology of Dirac and von Neuman (DvN) described in the previous chapter. In that terminology, Heisenberg's ground breaking observation is that Hamilton's form of classical mechanics is retained but the position and momentum of the particle lose their classical meaning and become abstract operators. In addition, they are conjugate operators, which means they do not commute. Rather they satisfy the relation

$$[\mathbf{x}, \mathbf{p}] = i\hbar, \tag{2.1}$$

as was anticipated in chapter 1.

In the following exposition, $h\nu$ is normally written $h\nu = \hbar\omega$, where $\hbar = h/2\pi$ and $\omega = 2\pi\nu$.

2.2 The one-dimensional harmonic oscillator

Finding the quantum mechanical behavior of the one-dimensional harmonic oscillator is a model problem that is particularly well suited for Heisenberg's quantum mechanics. The abstract energy operator is the classical Hamiltonian function with the position and momentum replaced with abstract operators

$$\mathbf{H} = \frac{\mathbf{p}^2}{2m} + \frac{k\mathbf{x}^2}{2}. \tag{2.2}$$

The problem is to find the eigenvalues and eigenvectors of this operator

$$\mathbf{H}|\psi_E\rangle = E|\psi_E\rangle. \tag{2.3}$$

Later, it will be shown how this problem relates to diagonalizing a matrix.

It has been found that the easiest way to solve this problem is to define an ancillary operator $\mathbf{a}$ and its Hermitean conjugate

$$\mathbf{a} = \sqrt{\frac{m\omega}{2\hbar}}\left(\mathbf{x} + \frac{i\mathbf{p}}{m\omega}\right)$$
$$\mathbf{a}^\dagger = \sqrt{\frac{m\omega}{2\hbar}}\left(\mathbf{x} - \frac{i\mathbf{p}}{m\omega}\right), \tag{2.4}$$

where $\mathbf{x}$ and $\mathbf{p}$ are Hermitean. Multiplying these operators and using equation (2.1) leads to

$$\mathbf{a}^\dagger\mathbf{a} = \frac{\mathbf{H}}{\hbar\omega} - \frac{1}{2}, \tag{2.5}$$

It is useful to consider the product introduced in the equation (2.5) as an operator in its own right

$$\mathbf{N} = \mathbf{a}^\dagger\mathbf{a}, \tag{2.6}$$

which is obviously Hermitean and thus has real eigenvalues n. With the help of equation (2.1), it can be proved that

$$[\mathbf{N}, \mathbf{a}^\dagger] = \mathbf{a}^\dagger, \ [\mathbf{N}, \mathbf{a}] = -\mathbf{a}$$

$$\mathbf{a}^\dagger\mathbf{a} - \mathbf{a}\mathbf{a}^\dagger = \frac{2i}{2\hbar}i\hbar = -1 \ . \tag{2.7}$$

$$\mathbf{a}\mathbf{a}^\dagger = \mathbf{a}^\dagger\mathbf{a} + 1$$

Now consider the eigenvector of $\mathbf{N}$,

$$\mathbf{N} \mid n\rangle = n \mid n\rangle. \tag{2.8}$$

With the help of equation (2.7) it can be shown that $\mathbf{a}^\dagger|n\rangle$ is also an eigenvector of $\mathbf{N}$, and it is proportional to the eigenvector that corresponds to the eigenvalue $n + 1$,

$$\mathbf{N}\mathbf{a}^\dagger \mid n\rangle = (n + 1)\mathbf{a}^\dagger \mid n\rangle. \tag{2.9}$$

A similar argument leads to

$$\mathbf{N}\mathbf{a} \mid n\rangle = (n - 1)\mathbf{a} \mid n\rangle. \tag{2.10}$$

Let us now assume that the operator $\mathbf{N}$ is bounded from below. That means that there is a lowest eigenvalue of $\mathbf{N}$, and the corresponding eigenvector will be called $|\text{lowest}\rangle$. Operating on this vector with $\mathbf{a}$ must give the null vector $\mathbf{a}|\text{lowest}\rangle = |\text{null}\rangle$, because otherwise $\mathbf{N}$ would not be bounded. It follows that the expectation of $\mathbf{N}$ in the lowest state is zero, so it is written as $|\text{lowest}\rangle = |0\rangle$. From equation (2.9), it is clear that the vector $\mathbf{a}^\dagger|0\rangle$ is proportional to the eigenvector $|1\rangle$ corresponding to the eigenvalue 1 and $(\mathbf{a}^\dagger)^n|0\rangle$ is proportional to $|n\rangle$. At this point $\mathbf{N}$ can be called the number operator because it exists in a Hilbert space spanned by a countable infinity of eigenvectors with eigenvalues that are the positive integers.

The vector $|0\rangle$ is normalized $\langle0|0\rangle = 1$. The commutation relation for the $\mathbf{a}$ operators are

$$[\mathbf{a}^\dagger, \mathbf{a}] = -1. \tag{2.11}$$

From this it can be shown that

$$\langle n \mid \mathbf{a}\mathbf{a}^\dagger \mid n\rangle = n + 1. \tag{2.12}$$

It follows that the vectors generated by operating using $\mathbf{a}^\dagger$ can all be normalized if

$$\mathbf{a}^\dagger \mid n\rangle = \sqrt{n + 1} \mid n + 1\rangle, \tag{2.13}$$

and

$$\mid n\rangle = \frac{(\mathbf{a}^\dagger)^n}{\sqrt{n!}} \mid 0\rangle. \tag{2.14}$$

Having worked out the theory for the number operator $\mathbf{N}$, it is time to address the problem of interest, namely to find the eigenvalues and eigenvectors of the energy operator $\mathbf{H}$. Equation (2.5) can be rewritten

$$\mathbf{H} = \hbar\omega(\mathbf{N} + 1/2). \tag{2.15}$$

The eigenvectors $|n\rangle$ are also eigenvectors of the energy

$$\mathbf{H} \mid n\rangle = E_n \mid n\rangle, \tag{2.16}$$

corresponding to the eigenvalues

$$E_n = \hbar\omega(n + 1/2). \tag{2.17}$$

Physicists are so accustomed to this expression for the quantized energies of the harmonic oscillator that it may seem obvious, but the fact that the lowest energy state of the system is not zero has many ramifications. The fields of statistical mechanics and thermodynamics are profoundly changed because it had been thought that the atoms in a solid ceased their vibrations when the temperature reached absolute zero. Zero point vibrations in molecules have to be taken into account in chemical reactions. Helium cannot freeze into a solid due to temperature alone because of zero point vibrations.

A calculation related to the uncertainty principle can be carried out exactly using the results in this section. Inverting equation (2.4) leads to

$$\mathbf{x} = \sqrt{\frac{\hbar}{2m\omega}}(\mathbf{a}^\dagger + \mathbf{a})$$

$$\mathbf{p} = i\sqrt{\frac{m\hbar\omega}{2}}(\mathbf{a}^\dagger - \mathbf{a}) \tag{2.18}$$

Since $\langle n'|n\rangle$ is zero when $n' \neq n$, it is obvious that the expectation values or $\mathbf{x}$ and $\mathbf{p}$ in state $|n\rangle$ are zero. The expectations of $\mathbf{x}^2$ and $\mathbf{p}^2$ are

$$\langle n \mid \mathbf{x}^2 \mid n\rangle = \frac{\hbar}{m\omega}(n + 1/2)$$

$$\langle n \mid \mathbf{p}^2 \mid n\rangle = m\omega\hbar(n + 1/2) \tag{2.19}$$

Since $\Delta x_n = \sqrt{\langle n|\mathbf{x}^2|n\rangle}$ and $\Delta p_n = \sqrt{\langle n|\mathbf{p}^2|n\rangle}$, the uncertainty relation is

$$\Delta x_n \Delta p_n = (n + 1/2)\hbar = E_n/\omega. \tag{2.20}$$

Comparing this equation with the general uncertainty relation derived in chapter 1 shows that the ground state of the harmonic oscillator has the minimum possible uncertainty. The greater than or equal to relation is an equality in this state. The zero point energy of the oscillator is required by the uncertainty relation. In the classical ground state, the particle is at $x = 0$ with momentum $p = 0$. This state would clearly not be consistent with the uncertainty relation.

2.3 Schrödinger's wave mechanics

Only a few months after Heisenberg put forward his matrix mechanics, Schrödinger announced his wave mechanics [2]. He arrived at his theory by focusing on de

Broglie's hypothesis that particles must have a wave nature. The Schrödinger theory, like Heisenbergs, is a tool that can be used to solve any problem in quantum mechanics, which distinguishes it from all previous applications of the quantum idea.

The best way to describe Schrödinger's concept today is to use the DvN structure of abstract operators and vector spaces. It was pointed out in the previous chapter that, if an abstract vector $|u\rangle$ is related to a vector $|v\rangle$ by

$$| u\rangle = \mathbf{A} \mid v\rangle, \tag{2.21}$$

then the function $u(x) = \langle x|u\rangle$ is related to $v(x) = \langle x|v\rangle$ by

$$u(x) = \int_{-\infty}^{\infty} A(x, x')v(x')dx', \tag{2.22}$$

where

$$A(x, x') = \langle x \mid \mathbf{A} \mid x'\rangle. \tag{2.23}$$

The resolution of the identity, $\mathbf{I} = \int_{-\infty}^{\infty} |x\rangle\langle x|dx$ was used. Many of the energy operators in quantum mechanics are functions of the position operator $\mathbf{x}$ and the momentum operator $\mathbf{p}$.

For $\mathbf{A} = \mathbf{x}$,

$$\langle x \mid \mathbf{x} \mid x'\rangle = x\delta(x - x') \tag{2.24}$$

and equation (2.22) becomes

$$u(x) = xv(x). \tag{2.25}$$

For $\mathbf{A} = \mathbf{p}$, the commutation rule in equation (2.1) is used to show that

$$\langle x \mid \mathbf{p} \mid v\rangle = \int_{-\infty}^{\infty} \langle x \mid \mathbf{p} \mid x'\rangle\langle x'|v\rangle dx' = -i\hbar\frac{dv(x)}{dx}, \tag{2.26}$$

and

$$u(x) = -i\hbar\frac{dv(x)}{dx}. \tag{2.27}$$

The one-dimensional eigenvalue equation for the energy operator in the abstract operator and vector form is

$$\mathbf{H} \mid \psi\rangle = E \mid \psi\rangle, \tag{2.28}$$

where the energy operator is the classical Hamiltonian with position and momentum converted into abstract operators

$$\mathbf{H} = \frac{1}{2m}\mathbf{p}^2 + V(\mathbf{x}). \tag{2.29}$$

Using equation (2.22) with $\mathbf{A} = \mathbf{H}$ and also invoking equations (2.25) and (2.27) leads to

$$-\frac{\hbar^2}{2m}\frac{d^2\psi(x)}{dx^2} + V(x)\psi(x) = E\psi(x). \tag{2.30}$$

This is the one-dimensional form of the time-independent equation that underpins Schrödinger's wave mechanics.

2.4 The one-dimensional harmonic oscillator (again)

The best way to compare the Heisenberg and Schrödinger mechanics is to apply both of them to the one-dimensional harmonic oscillator. From equation (2.2), the Schrödinger equation for this problem is

$$-\frac{\hbar^2}{2m}\frac{d^2\psi(x)}{dx^2} + \frac{m\omega^2 x^2}{2}\psi(x) = E\psi(x), \tag{2.31}$$

where an angular frequency $\omega = 2\pi\nu = \sqrt{k/m}$ is defined. Defining the variables

$$\xi = \sqrt{\frac{m\omega}{\hbar}}\,x, \tag{2.32}$$

and

$$\varepsilon = \frac{2}{\hbar\omega}E, \tag{2.33}$$

allows equation (2.31) to be written in a dimensionless form

$$-\frac{d^2\psi(\xi)}{d\xi^2} + \xi^2\psi(\xi) = \varepsilon\psi(\xi). \tag{2.34}$$

Although the physical meaning of the wave function ψ was not immediately obvious, the physicists of 1925 were happy to see equation (2.34) because it is a simple example of a Sturm–Liouville system. Such systems had been widely studied during the previous century. Also, the most famous book on theoretical physics, Lord Rayleigh's *Theory of Sound* published in 1877, contains many applications of wave equations.

The first step in the solution of a Sturm–Liouville equation is to eliminate any singular points at which the coefficient of ψ becomes infinite. The singular points in equation (2.34) are at $\xi = \pm\infty$. They can be eliminated by introducing a new function by

$$\psi(\xi) = e^{-\xi^2/2}h(\xi). \tag{2.35}$$

Inserting this into the above equation leads to

$$\frac{d^2h(\xi)}{d\xi^2} - 2\xi\frac{dh(\xi)}{d\xi} + (\varepsilon - 1)h(\xi) = 0, \tag{2.36}$$

which has no singular points. The advantage of an equation with no singular points is that it can be shown that the solution can be expressed as a power series, and in general the power series will terminate to give a polynomial. When

$$h(\xi) = \sum_{j=0}^{\infty} a_j \xi^j, \tag{2.37}$$

is inserted into equation (2.36), the coefficients are seen to satisfy a recursion relation

$$a_{j+2} = \frac{(2j + 1 - \varepsilon)}{(j + 1)(j + 2)} a_j. \tag{2.38}$$

If the power series does not terminate at some finite power of ξ, $h(\xi)$ will become infinite at $\xi = \pm\infty$ and $\psi(\xi)$ will not be normalizable. The power series will terminate at some ξ^n for energies

$$E_n = \hbar\omega(n + 1/2), \tag{2.39}$$

which gives exactly the same energy eigenvalues as equation (2.17). The solution of equation (2.36) corresponding to this eigenvalue is obtained from the equation

$$\frac{d^2 H_n(\xi)}{d\xi^2} - 2\xi \frac{dH_n(\xi)}{d\xi} + 2n H_n(\xi) = 0, \tag{2.40}$$

which was solved in 1864 by Hermite. The solutions are called Hermite polynomials, and their properties are well-known.

It follows from equation (2.35) that

$$\psi_n(\xi) = c_n e^{-\xi^2/2} H_n(\xi). \tag{2.41}$$

The integral over all ξ of products of these polynomials is known to be

$$\int_{-\infty}^{\infty} e^{-\xi^2} H_m(\xi) H_n(\xi) d\xi = \sqrt{\pi} 2^n n! \delta_{mn}, \tag{2.42}$$

so the wave functions $\psi_n(x)$ form a complete orthonormal set when

$$\psi_n(x) = \left(\frac{m\omega}{\pi\hbar}\right)^{1/4} (2^n n!)^{-1/2} H_n\left(\sqrt{\frac{m\omega}{\hbar}} x\right) e^{-m\omega x^2/2\hbar}. \tag{2.43}$$

The expectation value of **x** in the state n is

$$\langle x \rangle_n = \int_{-\infty}^{\infty} \psi_n^*(x) x \psi_n(x) dx = \sqrt{\frac{\hbar}{m\omega}} c_n^2 \int_{-\infty}^{\infty} e^{-\xi^2} H_n(\xi) \xi H_n(\xi) d\xi, \tag{2.44}$$

and for $\mathbf{x}^2$ it is

$$\langle x^2 \rangle_n = \int_{-\infty}^{\infty} \psi_n^*(x) x^2 \psi_n(x) dx = \frac{\hbar}{m\omega} c_n^2 \int_{-\infty}^{\infty} e^{-\xi^2} H_n(\xi) \xi^2 H_n(\xi) d\xi. \tag{2.45}$$

From theoretical studies of Hermite polynomials

$$\xi H_n = (1/2 H_{n+1} + n H_{n-1}),$$ (2.46)

and

$$\xi^2 H_n = (n + 1/2) H_n + 1/4 H_{n+2} + n(n - 1) H_{n-2}.$$ (2.47)

Inserting these relations in equations (2.44) and (2.45) shows that $\langle x \rangle_n = 0$ and

$$\langle x^2 \rangle_n = \frac{\hbar}{m\omega}(n + 1/2).$$ (2.48)

The momentum expectation is

$$
\begin{aligned}
\langle p \rangle_n &= -i\hbar \int_{-\infty}^{\infty} \psi_n^*(x) \frac{d\psi_n(x)}{dx} dx \\
&= -i\sqrt{m\omega\hbar}\, c_n^2 \int_{-\infty}^{\infty} e^{-\xi^2/2} H_n(\xi) \frac{d}{d\xi}[e^{-\xi^2/2} H_n(\xi)] d\xi
\end{aligned}
$$ (2.49)

and the expectation of the square of the momentum is

$$
\begin{aligned}
\langle p^2 \rangle_n &= -\hbar^2 \int_{-\infty}^{\infty} \psi_n^*(x) \frac{d^2\psi_n(x)}{dx^2} dx \\
&= -m\omega\hbar c_n^2 \int_{-\infty}^{\infty} e^{-\xi^2/2} H_n(\xi) \frac{d^2}{d\xi^2}[e^{-\xi^2/2} H_n(\xi)] d\xi
\end{aligned}
$$ (2.50)

It is known that

$$\frac{d}{d\xi} H_n = 2n H_{n-1},$$ (2.51)

and from this

$$\frac{d}{d\xi}[e^{-\xi^2/2} H_n] = e^{-\xi^2/2}(-1/2 H_{n+1} + n H_{n-1}),$$ (2.52)

and

$$\frac{d^2}{d\xi^2}[e^{-\xi^2/2} H_n] = e^{-\xi^2/2}[-(n + 1/2) H_n + 1/4 H_{n+2} + n(n - 1) H_{n-2}].$$ (2.53)

Inserting these relations into equations (2.49) and (2.50) shows that $\langle p \rangle_n = 0$ and

$$\langle p^2 \rangle_n = m\omega\hbar(n + 1/2).$$ (2.54)

These calculations lead to the same results as the ones in equation (2.19).

2.5 Comparison of Heisenberg and Schrödinger theories

There is every reason to believe that the matrix mechanics of Heisenberg and the wave mechanics of Schrödinger will give the same results when applied to any problem. Schrödinger proved their equivalence. However, wave mechanics is much more popular among physicists. One reason for this is that the most important problem when the two theories were put forward was to produce a complete theory of the hydrogen atom. Although Pauli found the spectrum of hydrogen with the Heisenberg method, [3] the Schrödinger approach is much better suited for treating this problem.

Another reason for the popularity of the Schrödinger wave mechanics is that it leads to a wave function. M Born published a statistical interpretation of quantum mechanics [4] shortly after Schrödinger proposed wave mechanics, and was awarded a Nobel prize for his work in 1954. He interpreted the absolute square of the Schrödinger wave function as a probability $P(x) = |\psi(x)|^2$. The probability for finding the particle between x and $x + dx$ is $P(x)dx$. The Fourier transform of $\psi(x)$, $\psi(p)$, was discussed in chapter 1. In Born's theory, the probability for finding the momentum of the particle between p and $p + dp$ is $P(p)dp$, where $P(p) = |\psi(p)|^2$. In classical mechanics, the state of a particle is described by a point in a phase space where the position and momentum is specified. Many of the early pioneers in quantum theory, such as Planck, Einstein, and de Broglie, were never happy with the statistical nature of quantum mechanics. However, all efforts to suggest an alternative have proved futile.

2.6 Wave mechanics in three dimensions

The position operators in three dimensions will be called $\mathbf{x}$, $\mathbf{y}$, and $\mathbf{z}$, and the corresponding momentum operators are $\mathbf{p}_x$, $\mathbf{p}_y$, and $\mathbf{p}_z$. They satisfy commutation relations that are obvious extensions of the one in equation (2.1),

$$[\mathbf{x}, \mathbf{p}_x] = i\hbar, \ [\mathbf{y}, \mathbf{p}_y] = i\hbar, \ [\mathbf{z}, \mathbf{p}_z] = i\hbar. \tag{2.55}$$

Of course, all other pairs of operators commute. The ordinary unit vectors for Euclidean space can be used with these operators, so that a position operator is

$$\vec{r} = \mathbf{x}\hat{x} + \mathbf{y}\hat{y} + \mathbf{z}\hat{z} \tag{2.56}$$

and a momentum operator is

$$\vec{p} = \mathbf{p}_x\hat{x} + \mathbf{p}_y\hat{y} + \mathbf{p}_z\hat{z}. \tag{2.57}$$

Straightforward extensions of the preceding arguments lead to the three-dimensional Schrödinger equation

$$\left[-\frac{\hbar^2}{2m}\nabla^2 + V(\vec{r}) \right]\psi(E, \vec{r}) = E\psi(E, \vec{r}), \tag{2.58}$$

where $\vec{r}$ is the position vector and the Laplacian operator is

$$\nabla^2 = \frac{\partial^2}{\partial x^2} + \frac{\partial^2}{\partial y^2} + \frac{\partial^2}{\partial z^2}. \tag{2.59}$$

2.7 Angular momentum

It is common in dealing with motion in three dimensions in classical mechanics that the angular momentum is conserved. The same is true in quantum mechanics. Using the notation developed above, the angular momentum operator can be written in the standard form,

$$\vec{\mathbf{L}} = \vec{\mathbf{r}} \times \vec{\mathbf{p}}, \tag{2.60}$$

which is equivalent to

$$\mathbf{L}_x = \mathbf{y}\mathbf{p}_z - \mathbf{z}\mathbf{p}_y, \quad \mathbf{L}_y = \mathbf{z}\mathbf{p}_x - \mathbf{x}\mathbf{p}_z, \quad \mathbf{L}_z = \mathbf{x}\mathbf{p}_y - \mathbf{y}\mathbf{p}_x. \tag{2.61}$$

Using the commutation rules for the position and momentum operators in equation (2.55), it can be seen that

$$[\mathbf{L}_x, \mathbf{L}_y] = i\hbar \mathbf{L}_z, [\mathbf{L}_y, \mathbf{L}_z] = i\hbar \mathbf{L}_x, [\mathbf{L}_z, \mathbf{L}_x] = i\hbar \mathbf{L}_y. \tag{2.62}$$

From these commutation rules, it follows that the square of the total angular momentum commutes with any of the individual components

$$[\mathbf{L}^2, \mathbf{L}_z] = [\mathbf{L}^2, \mathbf{L}_x] = [\mathbf{L}^2, \mathbf{L}_y] = 0, \tag{2.63}$$

where $\mathbf{L}^2 = \mathbf{L}_x^2 + \mathbf{L}_y^2 + \mathbf{L}_z^2$.

As pointed out in chapter 1, from these commutation rules it follows that there are simultaneous eigenvectors of $\mathbf{L}^2$ and one component of $\vec{\mathbf{L}}$, which is chosen to be $\mathbf{L}_z$. That is

$$\mathbf{L}^2 \mid a, b \rangle = a \mid a, b \rangle, \tag{2.64}$$

and

$$\mathbf{L}_z \mid a, b \rangle = b \mid a, b \rangle. \tag{2.65}$$

Our first observation is that b has a limit, $b^2 \leqslant a$, because

$$\begin{aligned} \langle a, b \mid (\mathbf{L}^2 - \mathbf{L}_z^2) \mid a, b \rangle &= (a - b^2) \\ &= \langle a, b \mid \mathbf{L}_x^2 \mid a, b \rangle + \langle a, b \mid \mathbf{L}_y^2 \mid a, b \rangle \geqslant 0 \end{aligned} \tag{2.66}$$

since the components of angular momentum are Hermitean and thus the expectation is the sum of two non-negative terms. The bound on b can be expressed in the form

$$b_{min} \leqslant b \leqslant b_{max}, \, b_{min} \geqslant -\sqrt{a}, b_{max} \leqslant \sqrt{a}. \tag{2.67}$$

Next, two ancillary operators

$$\mathbf{L}_+ = \mathbf{L}_x + i\mathbf{L}_y, \tag{2.68}$$

and

$$\mathbf{L}_- = \mathbf{L}_x - i\mathbf{L}_y = \mathbf{L}_+^\dagger, \tag{2.69}$$

are defined. It can be shown that

$$[\mathbf{L}_z, \mathbf{L}_\pm] = \pm\hbar\mathbf{L}_\pm, \tag{2.70}$$

and

$$[\mathbf{L}^2, \mathbf{L}_\pm] = 0. \tag{2.71}$$

From this last equation, the following relation is obtained

$$\mathbf{L}^2\mathbf{L}_\pm \mid a, b\rangle = a\mathbf{L}_\pm \mid a, b\rangle, \tag{2.72}$$

so $\mathbf{L}_\pm|a, b\rangle$ is an eigenvector of $\mathbf{L}^2$.

From equation (2.70)

$$\mathbf{L}_z\mathbf{L}_+ \mid a, b\rangle = \mathbf{L}_+\mathbf{L}_z \mid a, b\rangle + \hbar\mathbf{L}_+ \mid a, b\rangle = (b + \hbar)\mathbf{L}_+ \mid a, b\rangle, \tag{2.73}$$

and from the same equation

$$\mathbf{L}_z\mathbf{L}_- \mid a, b\rangle = \mathbf{L}_-\mathbf{L}_z \mid a, b\rangle - \hbar\mathbf{L}_- \mid a, b\rangle = (b - \hbar)\mathbf{L}_- \mid a, b\rangle. \tag{2.74}$$

It follows that $\mathbf{L}_+|a, b\rangle$ is an eigenvector of $\mathbf{L}_z$ with an eigenvalue increased by $\hbar$, and $\mathbf{L}_-|a, b\rangle$ is an eigenvector of $\mathbf{L}_z$ with an eigenvalue decreased by $\hbar$.

The question is, what happens if the operator $\mathbf{L}_+$ acts on an eigenvector $|a, b_{max}\rangle$ corresponding to the maximum eigenvalue. The answer is that it generates a null vector that has the length zero, as discussed in chapter 1

$$\mathbf{L}_+|a, b_{max}\rangle = 0. \tag{2.75}$$

The same argument leads to

$$\langle a, b_{max}|\mathbf{L}_-\mathbf{L}_+|a, b_{max}\rangle = 0 \tag{2.76}$$

It is easy enough to show that

$$\mathbf{L}_-\mathbf{L}_+ = \mathbf{L}^2 - \mathbf{L}_z^2 - \hbar\mathbf{L}_z. \tag{2.77}$$

Combining this with equation (2.76) gives

$$a = b_{max}(b_{max} + \hbar). \tag{2.78}$$

From an analogous argument,

$$\mathbf{L}_-|a, b_{min}\rangle = 0, \tag{2.79}$$

and arguments like the ones above lead to an equation for the eigenvalue a

$$a = b_{min}(b_{min} - \hbar). \tag{2.80}$$

The only way that equations (2.78) and (2.80) can be satisfied simultaneously is for there to be an integer l such that

$$a = l(l + 1)\hbar^2, \tag{2.81}$$

and

$$b = -l\hbar, (-l + 1)\hbar, (-l + 2)\hbar, ...(l - 1)\hbar, l\hbar. \tag{2.82}$$

The Heisenberg style argument above has led to the result that, in quantum mechanics, the angular momentum can only take on a limited set of values. The square of the magnitude L^2 can only have the values

$$L^2 = l(l + 1)\hbar^2 \quad l = 0, 1, 2, \dots. \qquad (2.83)$$

and, for a given value of l, the z component of $\vec{L}$ can have the values

$$L_z = m\hbar \quad m = -l, -l + 1, \dots l. \qquad (2.84)$$

The operator equations become

$$\mathbf{L}^2 \mid l, m \rangle = l(l + 1)\hbar^2 \mid l, m \rangle, \qquad (2.85)$$

and

$$\mathbf{L}_z \mid l, m \rangle = m\hbar \mid l, m \rangle. \qquad (2.86)$$

In Cartesian coordinates, these operator equations can be written in the notation of wave mechanics like

$$\langle \vec{r} \mid \mathbf{L}_z \mid \psi \rangle = -i\hbar \left[x\frac{\partial}{\partial y} - y\frac{\partial}{\partial z} \right] \psi(x, y, z) = m\hbar\psi(x, y, z). \qquad (2.87)$$

It is more convenient to use spherical coordinates

$$x = r \sin \theta \cos \phi \quad y = r \sin \theta \sin \phi \quad z = r \cos \theta. \qquad (2.88)$$

With these coordinates

$$\langle \vec{r} \mid \mathbf{L}_z \mid \psi \rangle =$$
$$- i\hbar\frac{\partial \psi(r, \theta, \phi)}{\partial \phi} = m\hbar\psi(r, \theta, \phi), \qquad (2.89)$$

and

$$\langle \vec{r} \mid \mathbf{L}^2 \mid \psi \rangle =$$
$$- \hbar^2\left[\frac{1}{\sin^2 \theta}\frac{\partial^2}{\partial \phi^2} + \frac{1}{\sin \theta}\frac{\partial}{\partial \theta}\left(\sin \theta\frac{\partial}{\partial \theta} \right) \right] \psi(r, \theta, \phi) \qquad (2.90)$$
$$= l(l + 1)\hbar^2\psi(r, \theta, \phi).$$

The solutions of these equations have been known for many years. Since there is no r in them, they can be written

$$\psi_{l,m}(r, \theta, \phi) = R_l(r) Y_{lm}(\theta, \phi). \qquad (2.91)$$

The functions $Y_{lm}(\theta, \phi)$ are called spherical harmonics. They are typically written

$$Y_l^m(\theta, \phi) = (-1)^m \sqrt{\frac{(2l + 1)}{4\pi}\frac{(l - m)!}{l + m!}} \, P_{lm}(\cos \theta)e^{im\phi}, \qquad (2.92)$$

where the $P_{lm}(\cos\theta)$ are associated Legendre polynomials, which were originally found by A-M Legendre in 1782. The multiplicative constant is to insure that they are orthonormal,

$$\int_{\theta=0}^{\pi}\int_{\phi=0}^{2\pi} Y_{lm}^* Y_{l'm'} d\Omega = \delta_{ll'}\delta_{mm'}. \tag{2.93}$$

It is useful to know the spherical harmonics for the smallest values of l and they are listed in Table 2.1. The $Y_{l-m}(\theta, \phi)$ are not listed because they are the complex conjugate of the $Y_{lm}(\theta, \phi)$.

Table 2.1. Spherical harmonics for $l = 0, 1, 2$.

$$Y_{00}(\theta, \phi) = \frac{1}{2}\sqrt{\frac{1}{\pi}}$$

$$Y_{10}(\theta, \phi) = \frac{1}{2}\sqrt{\frac{3}{\pi}}\cos\theta, \quad Y_{11}(\theta, \phi) = \frac{-1}{2}\sqrt{\frac{3}{2\pi}}\sin\theta e^{i\phi}$$

$$Y_{20}(\theta, \phi) = \frac{1}{4}\sqrt{\frac{5}{\pi}}(3\cos^2\theta - 1), \quad Y_{21}(\theta, \phi) = \frac{-1}{2}\sqrt{\frac{15}{2\pi}}\sin\theta\cos\theta e^{i\phi},$$

$$Y_{22}(\theta, \phi) = \frac{1}{4}\sqrt{\frac{15}{2\pi}}\sin^2\theta e^{i2\phi}$$

It is easier to visualize these functions when they are written in Cartesian coordinates (table 2.2).

Table 2.2. Spherical harmonics for $l = 0, 1, 2$ in Cartesian coordinates.

$$Y_{00}(\theta, \phi) = \frac{1}{2}\sqrt{\frac{1}{\pi}}$$

$$Y_{10}(\theta, \phi) = \frac{1}{2}\sqrt{\frac{3}{\pi}}\frac{z}{r}, \quad Y_{11}(\theta, \phi) = \frac{-1}{2}\sqrt{\frac{3}{2\pi}}\frac{(x+iy)}{r}$$

$$Y_{20}(\theta, \phi) = \frac{1}{4}\sqrt{\frac{5}{\pi}}\frac{(2z^2 - x^2 - y^2)}{r^2}, \quad Y_{21}(\theta, \phi) = \frac{-1}{2}\sqrt{\frac{15}{2\pi}}\frac{(x+iy)z}{r^2}, \quad Y_{22}(\theta, \phi)$$

$$= \frac{1}{4}\sqrt{\frac{15}{2\pi}}\frac{(x+iy)^2}{r^2}$$

Contours on which these functions have a constant value can be plotted in three dimensions by a function such as Contour Plot 3D from Mathematica. Some of these are shown in table 2.3. The ones colored in purple are for a positive constant, and the green ones are for the negative of that constant. An exponential function is put in to speed up convergence in r. In applications, the radial function $R_l(r)$ always converges at least as rapidly as $\exp[-r]$. The contour for Y_{00} is in red because it is only positive.

Table 2.3. Contours on which the functions in table 2.2 are equal to some constant. For the purple contours the constant is positive, and for the green it is negative. The red contour is for a function that is only positive.

$Y_{00} = e^{-r}/r$		
$Y_{10} = e^{-r}z/r$	$\mathrm{Re}\,Y_{11} = e^{-r}x/r$	
$Y_{20} = e^{-r}(2z^2 - x^2 - y^2)/r^2$	$\mathrm{Re}\,Y_{21} = e^{-r}xz/r^2$	$\mathrm{Re}\,Y_{22} = e^{-r}(x^2 - z^2)/r^2$

2.8 Schrödinger equation for a spherically symmetric potential

The Laplacian operator in the Schrödinger equation in equation (2.58) can be written in spherical coordinates

$$\nabla^2 = \frac{1}{r^2}\frac{\partial}{\partial r}\left(r^2\frac{\partial}{\partial r}\right) + \frac{1}{r^2}\left[\frac{1}{\sin\theta}\frac{\partial}{\partial\theta}\left(\sin\theta\frac{\partial}{\partial\theta}\right) + \frac{1}{\sin^2\theta}\frac{\partial^2}{\partial\phi^2}\right]. \tag{2.94}$$

Doing this and writing $V(\mathbf{r}) = V(r)$ because the potential depends only on the magnitude r leads to

$$\left[-\frac{\hbar^2}{2m}\nabla^2 + V(r)\right]\psi(\mathbf{r}) = E\psi(\mathbf{r}). \tag{2.95}$$

Comparing equation (2.94) with equation (2.90) gives

$$\nabla^2 = \frac{1}{r^2}\frac{\partial}{\partial r}\left(r^2\frac{\partial}{\partial r}\right) - \frac{\mathbf{L}^2}{\hbar^2 r^2}. \tag{2.96}$$

It follows that the solution of Schrödinger's equation has the form

$$\psi(\mathbf{r}) = R_{nl}(r)\,Y_{lm}(\theta, \phi), \tag{2.97}$$

where the spherical harmonics are given in equation (2.92). The radial functions are solutions of

$$\left\{ -\frac{\hbar^2}{2m}\left[\frac{1}{r^2}\frac{\partial}{\partial r}\left(r^2\frac{\partial}{\partial r} \right) + \frac{l(l+1)}{r^2} \right] + V(r) \right\}R_{nl}(r) = ER_{nl}(r). \tag{2.98}$$

This last equation should not be a surprise. Applying Lagrange's identity to $\mathbf{L}^2$ using the definition in equation (2.60) leads to

$$\mathbf{L}^2 = \mathbf{r}^2\mathbf{p}^2 - (\mathbf{r}\cdot\mathbf{p})^2 = -\hbar^2 r^2\nabla^2 + \hbar^2 r^2(\hat{\mathbf{r}}\cdot\nabla)^2, \tag{2.99}$$

or

$$-\hbar^2\nabla^2 = -\hbar^2(\hat{\mathbf{r}}\cdot\nabla)^2 + \frac{\mathbf{L}^2}{r^2}. \tag{2.100}$$

This means that the three-dimensional differential equation (2.95) becomes a one-dimensional equation with an effective potential

$$V_{eff}(r) = V(r) + \frac{\mathbf{L}^2}{r^2}. \tag{2.101}$$

A similar situation occurs in the classical description of a particle moving in the field of a spherically symmetric potential.

2.9 Schrödinger equation for the hydrogen atom

It was known since the early eighteen hundreds that the spectrum of the light given off by heated hydrogen atoms was not distributed evenly over all of the wavelengths. Using a prism to diffract the light, it could be seen that the intensity was very high at certain discrete wavelengths and essentially zero at all others. As the structure of the atom became more clear in the early nineteen hundreds the line spectrum of hydrogen became even harder to understand. If, as seemed more and more likely, the hydrogen atom was made up of a heavy nucleus that is a proton with charge e circled by an electron with charge $-e$, classical mechanics gives the answer that an excited atom should lose energy by giving off light in a continuous range of wavelengths. In fact, the classical prediction is that the electron should lose all of its energy by radiation and spiral into the nucleus.

Bohr made some progress in explaining the line spectrum of hydrogen, but he achieved this by applying ad hoc restrictions to the classical description. The Schrödinger equation was obviously a massive step in the history of physics when it was put forward in 1926. It not only predicts the experimental results for hydrogen but also provides a theoretical scaffold for treating the entire range of microscopic phenomena

Starting from equation (2.98), the radial part of the Schrödinger equation for hydrogen is

$$\left\{ -\frac{\hbar^2}{2m}\left[\frac{1}{r^2}\frac{d}{dr}\left(r^2\frac{d}{dr} \right) + \frac{l(l+1)}{r^2} \right] - \frac{e^2}{r} \right\}R_{nl}(r) = ER_{nl}(r). \tag{2.102}$$

As has been described, the physicists of the day were well trained to deal with this type of Sturm–Liouville system. The first step is to define a different radial function

$$u_{nl}(r) = rR_{nl}(r) \tag{2.103}$$

which satisfies the equation

$$\left\{-\frac{\hbar^2}{2m}\left[\frac{d^2}{dr^2} + \frac{l(l+1)}{r^2}\right] - \frac{e^2}{r}\right\}u_{nl}(r) = Eu_{nl}(r). \tag{2.104}$$

It is convenient to rewrite this equation in dimensionless form. The goal is to find the negative energy bound states, so a positive number κ is defined by

$$\kappa = \frac{\sqrt{-2mE}}{\hbar}. \tag{2.105}$$

Using the dimensionless distance

$$\rho = \kappa r, \tag{2.106}$$

and the parameter

$$\xi = \frac{2me^2}{\kappa\hbar^2}, \tag{2.107}$$

Equation (2.102) becomes

$$\left[-\frac{d^2}{d\rho^2} + \frac{l(l+1)}{\rho^2} - \frac{\xi}{\rho}\right]u_{nl}(\rho) = -u_{nl}(\rho). \tag{2.108}$$

The boundary condition is that

$$\psi(\mathbf{r}) = \frac{u_{nl}(r)}{r}Y_{lm}(\theta, \phi), \tag{2.109}$$

must be normalizable.

Equation (2.108) has a singular point at $\rho = 0$, so it will be investigated near that point where it can be written

$$\left[-\rho^2\frac{d^2}{d\rho^2} + l(l+1)\right]u_{nl}(\rho) = 0. \tag{2.110}$$

This equation has a regular solution if $u_{nl}(\rho) \rightarrow a\rho^{l+1}$ as $\rho \rightarrow 0$. The solution has to be normalizable, so $u_{nl}(\rho) \rightarrow 0$ as $\rho \rightarrow \infty$. Equation (2.108) becomes

$$-\frac{d^2u_{nl}(\rho)}{d\rho^2} = -u_{nl}(\rho), \tag{2.111}$$

as $\rho \to \infty$ and the solution is acceptable only if $u_{nl}(\rho) \to be^{-\rho}$. It follows that $u_{nl}(\rho)$ will behave correctly near the singular points if

$$u_{nl}(\rho) = e^{-\rho}\rho^{l+1}v_{nl}(\rho), \qquad (2.112)$$

where $v_{nl}(\rho)$ is a polynomial of finite order.

The function $v_{nl}(\rho)$ satisfies

$$\frac{d^2v_{nl}(\rho)}{d\rho^2} - 2\left(1 - \frac{l+1}{\rho}\right)\frac{dv_{nl}(\rho)}{d\rho} + \left(\frac{\xi - 2l - 2}{\rho}\right)v_{nl}(\rho) = 0. \qquad (2.113)$$

This equation was studied many decades before the invention of quantum mechanics by Edmond Laguerre. It is known that the solutions are nonsingular if and only if the quantity ξ has specific values

$$\xi = 2n, \qquad (2.114)$$

where n is a positive integer such that

$$n \geqslant l + 1. \qquad (2.115)$$

The solutions are known as generalized Laguerre polynomials, and, for the above equation, they take the form

$$v_{nl}(\rho) = L_{n-l-1}^{2l+1}(2\rho). \qquad (2.116)$$

The following table shows the first few associated Laguerre polynomials (table 2.4).

Table 2.4. Table of Laguerre polynomials.

		Laguerre polynomials	
	$l = 0$	$l = 1$	$l = 2$
$n = 1$	$L_0^1(2\rho) = 1$		
$n = 2$	$L_1^1(2\rho) = -2\rho + 2$	$L_0^3(2\rho) = 1$	
$n = 3$	$L_2^1(2\rho) = 1/2[4\rho^2 - 12\rho + 6]$	$L_1^3(2\rho) = -2\rho + 4$	$L_0^5(2\rho) = 1$

From equations (2.114) and (2.105),

$$E = -\frac{me^4}{2\hbar^2 n^2}, \qquad (2.117)$$

which are the discrete energy eigenvalues that explain the spectrum of hydrogen. This equation is often written $E = R/n^2$, where R is the Rydberg constant

$$R = 13.605698eV. \qquad (2.118)$$

The scaling factor for ρ

$$\rho = r/a = \frac{\sqrt{-2mE_1}}{\hbar}r = \frac{me^2}{\hbar^2}r, \qquad (2.119)$$

for the lowest eigenvalue, $n = 1$, is called the Bohr radius

$$a = 0.529177349 \times 10^{-8} \text{cm}. \tag{2.120}$$

This quantity gives a rough measure of the size of a hydrogen atom.

The final form of the eigenfunctions of the hydrogen atom are

$$\psi_{nlm}(\vec{r}) = \kappa^{l+1} e^{-\kappa r} r^l L_{n-l-1}^{2l+1}(2\kappa r) Y_{lm}(\theta, \phi). \tag{2.121}$$

The energy eigenvalues depend only on the quantum number n, as seen in equation (2.117). In accordance with equation (2.115) and the properties of the Laguerre polynomials, the eigenfunctions with a given n and $l = 0, 1, 2, \ldots n - 1$ all correspond to the same eigenvalue. The degeneracy for a given value of l is $2l + 1$, so the degeneracy of an energy eigenvalue E_n

$$\sum_{l=0}^{n-1} (2l + 1) = n^2. \tag{2.122}$$

The eigenvalues for any spherically symmetric potential will have a $(2l + 1)$-fold degeneracy because of the spherical symmetry, but the n^2-fold degeneracy is unique to hydrogen. In a later chapter the effect of symmetry on degeneracy will be discussed using group theory.

2.10 Time-dependent wave equation

The Schrödinger equation (2.58) is the time-independent form of

$$\left[-\frac{\hbar^2}{2m} \nabla^2 + V(\vec{r}) \right] \psi(\vec{r}, t) = i\hbar \frac{\partial \psi(\vec{r}, t)}{\partial t}, \tag{2.123}$$

obtained by using

$$\psi(\vec{r}, t) = \psi(E, \vec{r}) e^{-iEt/\hbar}. \tag{2.124}$$

Equation (2.123) is so familiar that it is easy to overlook just how strange it is. It seems to be equivalent to

$$E\psi(\vec{r}, t) = i\hbar \frac{\partial \psi(\vec{r}, t)}{\partial t}, \tag{2.125}$$

which looks a lot like equation (2.26). This would imply that E and t are conjugate in the same sense as $\mathbf{p}_x$ and $\mathbf{x}$, but this cannot be because E and t are not operators in this form of the theory but are variables. Being numbers, they obviously commute. Another strange feature about equation (2.123) is that it is first order in the time but second order in the space variables.

The wave equation that physicists were familiar with in 1926 described sound waves or light, and had the familiar form

$$\nabla^2 f(\vec{r}, t) = \left(\frac{n(\vec{r})}{c} \right)^2 \frac{\partial^2 f(\vec{r}, t)}{\partial t^2}, \tag{2.126}$$

where $n(\vec{r})$ is a local index of refraction that describes the medium the wave is moving through. The solutions of this equation are real, whereas the solutions of equation (2.123) have to be complex. In addition, since equation (2.126) is second order in both position and time, it can be invariant under a Lorentz transformation, and hence it can be called a relativistic equation. Schrödinger derived a relativistic equation, but when he tried to solve the hydrogen atom problem with it he got nonsense. His relativistic equation was rederived several times, and it is usually called the Klein–Gordon equation. The problems that occur when using the Klein–Gordon equation as the wave equation are discussed in the following chapter.

Schrödinger was forced to conclude that the non-relativistic form of his equation, equation (2.123), works beautifully while the relativistic version does not. It is thus interesting to see how he arrived at that equation. A well trained physicist like Schrödinger knew that their were mathematical similarities between the equations for the paths traveled by light waves and those traveled by particles. In particular, Fermat's principle that the path taken between two points by a ray of light is the path that can be traversed in the least time is closely related to Hamilton's principle of least action in classical mechanics. Starting from Newton's second law, Hamilton derived equations that could be used to solve problems in classical mechanics more efficiently. The most sophisticated of these is the Hamilton–Jacobi (H-J) equation satisfied by Hamilton's principle function $S(\vec{r}, t)$, which is closely related to the action integral. For ease of notation, the one-dimensional version of the H-J equation is used

$$H\left(x, \frac{\partial S}{\partial x}\right) + \frac{\partial S}{\partial t} = \frac{p_x^2}{2m} + U(x) + \frac{\partial S}{\partial t} = \frac{1}{2m}\frac{\partial S}{\partial x}\frac{\partial S}{\partial x} + U(x) + \frac{\partial S}{\partial t} = 0. \quad (2.127)$$

From H-J theory, the momentum of the particle can be obtained from the solution $S(x, t)$

$$p_x = \frac{\partial S}{\partial x}. \quad (2.128)$$

The Schrödinger equation is related to the H-J equation by defining a function

$$\psi(x, t) = ae^{iS(x, t)/\hbar}, \quad (2.129)$$

where a is a constant. The function $\psi(x, t)$ is dimensionless because Planck's constant has the dimension of action (momentum times distance) just as $S(x, t)$ does. Writing the Schrödinger equation and assuming a solution of this form

$$-\frac{\hbar^2}{2m}\frac{\partial^2\psi}{\partial x^2} + V\psi - i\hbar\frac{\partial\psi}{\partial t} = \left[\frac{1}{2m}\frac{\partial S}{\partial x}\frac{\partial S}{\partial x} - \frac{i\hbar}{2m}\frac{\partial^2 S}{\partial x^2} + V + \frac{\partial S}{\partial t}\right]\psi, \quad (2.130)$$

and also assuming $S(x, t)$ is a solution of equation (2.127) leads to

$$-\frac{\hbar^2}{2m}\frac{\partial^2\psi}{\partial x^2} + V\psi - i\hbar\frac{\partial\psi}{\partial t} = -\frac{i\hbar}{2m}\frac{\partial^2 S}{\partial x^2}\psi. \quad (2.131)$$

Schrödinger set the right side of this equation equal to zero, and thus showed the connection between his non-relativistic equation and the H-J equation. This connection does not constitute a derivation because Schrödinger's equation is new physics and it stands on its own. It is justified by its ability to predict the outcome of experiments. However, by staying as close as possible to the H-J equation, he was able to show that the form of his equation is not as strange as it might otherwise seem.

2.11 The time-evolution operator

In the DvN notation, the time-dependent Schrödinger equation, equation (2.123), is

$$\mathbf{H} \mid \psi(t)\rangle = i\hbar \frac{d \mid \psi(t)\rangle}{dt}, \tag{2.132}$$

where the Hamiltonian operator is

$$\mathbf{H} = \frac{\vec{\mathbf{p}}^2}{2m} + V(\vec{\mathbf{r}}). \tag{2.133}$$

It is observed that in wave mechanics the operators have no intrinsic time dependence, but the wave vector $|\psi(t)\rangle$ does. It is useful to define a time-evolution operator

$$\mid \psi(t)\rangle = \mathbf{U}(t, t_0) \mid \psi(t_0)\rangle, \tag{2.134}$$

which satisfies the operator equation

$$\mathbf{H}\mathbf{U}(t, t_0) = i\hbar \frac{d}{dt} \mathbf{U}(t, t_0). \tag{2.135}$$

It is obvious that

$$\mathbf{U}(t, t) = \mathbf{I}, \tag{2.136}$$

and, since the state vectors are normalized for all times t

$$\mathbf{U}^\dagger(t, t_0)\mathbf{U}(t, t_0) = \mathbf{I}. \tag{2.137}$$

This equation means that $\mathbf{U}(t, t_0)$ is a unitary operator. Finally, it's obvious that the time-evolution matrix can be broken down into products

$$\mathbf{U}(t, t_1)\mathbf{U}(t_1, t_0) = \mathbf{U}(t, t_0), \tag{2.138}$$

and

$$\mathbf{U}(t, t_0)^{-1} = \mathbf{U}(t_0, t). \tag{2.139}$$

From equation (2.135),

$$\mathbf{H}\mathbf{U}(t, t_0) = i\hbar \frac{\mathbf{U}(t + \Delta t, t_0) - \mathbf{U}(t, t_0)}{\Delta t}, \tag{2.140}$$

$$\mathbf{H}\Delta t = i\hbar[\mathbf{U}(t + \Delta t, t_0) - \mathbf{U}(t, t_0)]\mathbf{U}(t_0, t)$$

which leads to

$$U(t + \Delta t, t) = I + \frac{H\Delta t}{i\hbar}. \tag{2.141}$$

Iterating this formula, a solution to equation (2.135) is obtained

$$U(t, t_0) = e^{-\frac{i}{\hbar}H(t-t_0)}, \tag{2.142}$$

which holds when **H** has no intrinsic time dependence. Since **H** commutes with any power of **H**, it follows that it commutes with $U(t, t_0)$

$$[U(t, t_0), H] = 0. \tag{2.143}$$

2.12 The time dependence of Heisenberg's operators

In the Heisenberg picture the operator $\hat{A}(t)$ has an explicit time dependence, while in the Schrödinger picture the same operator **A** does not. Heisenberg did not discuss state vectors, but they can be taken to be $|\psi(t_0)\rangle$. In order for the two pictures to give the same results, the expectation of the operator must be the same. The mathematical statement of this fact is

$$\langle A(t)\rangle = \langle \psi(t) \mid A \mid \psi(t)\rangle = \langle \psi(t_0)\mid \hat{A}(t) \mid \psi(t_0)\rangle. \tag{2.144}$$

Using the equations in the preceding section, it is clear that

$$\hat{A}(t) = U^\dagger(t, t_0)AU(t, t_0). \tag{2.145}$$

Taking the derivative with respect to time,

$$\frac{d\hat{A}(t)}{dt} = \frac{dU^\dagger(t, t_0)}{dt}AU(t, t_0) + U^\dagger(t, t_0)A\frac{dU(t, t_0)}{dt}. \tag{2.146}$$

Using equation (2.135) again leads to

$$\frac{d\hat{A}(t)}{dt} = -\frac{1}{i\hbar}[U^\dagger(t, t_0)HAU(t, t_0) - U^\dagger(t, t_0)AHU(t, t_0)]. \tag{2.147}$$

It thus follows that

$$\frac{d\hat{A}(t)}{dt} = \frac{1}{i\hbar}[\hat{A}(t), H], \tag{2.148}$$

which is called Heisenberg's equation of motion.

In classical mechanics, Hamilton's equations are

$$\frac{dq}{dt} = \frac{\delta H}{\delta p}\frac{dp}{dt} = -\frac{\delta H}{\delta q}, \tag{2.149}$$

and this leads to

$$\frac{dA(q,p)}{dt} = \frac{\partial A}{\partial q}\frac{\partial H}{\partial p} - \frac{\partial A}{\partial p}\frac{\partial H}{\partial q} = \{A, H\}, \tag{2.150}$$

assuming the time dependence of $A(q,p)$ arises only from the time dependence of q and p. The quantity $\{A, H\}$ defined above is called the Poisson bracket of A and H. The classical analog of equation (2.148) is thus

$$\frac{1}{i\hbar}[\hat{A}(t), \mathbf{H}] \Leftrightarrow \{A, H\}. \tag{2.151}$$

A number of classical derivations can be transformed to quantum mechanics with the help of this analogy.

For example, using the Hamiltonian operator in equation (2.133) and the lemmas

$$\left[x, \left(p_x^2 + p_y^2 + p_z^2\right)\right] = 2i\hbar p_x, \tag{2.152}$$

and

$$[p_x, x^n] = -i\hbar\frac{dx^n}{dx}, \tag{2.153}$$

leads to

$$\frac{d\hat{\mathbf{r}}(t)}{dt} = \frac{\hat{\mathbf{p}}(t)}{m} \tag{2.154}$$

and

$$\frac{d\hat{\mathbf{p}}(t)}{dt} = -\nabla V(\hat{\mathbf{r}}(t)). \tag{2.155}$$

These last two equations are analogous to the classical ones with the classical variables replaced by Heisenberg operators. They can be combined to obtain the analog to Newton's second law

$$m\frac{d^2\hat{\mathbf{r}}(t)}{dt^2} = -\nabla V(\hat{\mathbf{r}}(t)). \tag{2.156}$$

Taking the expectation values of the operators in the time-independent state $|\psi(t_0)\rangle$ used in equation (2.144) leads to Ehrenfest's theorem

$$m\frac{d^2\langle\hat{\mathbf{r}}(t)\rangle}{dt^2} = -\nabla\langle V(\hat{\mathbf{r}}(t))\rangle. \tag{2.157}$$

Problems

P2.1 What was Heisenberg's connection to Bohr?

P2.2 Why does the harmonic oscillator equation appear so often in physics?

P2.3 What aspect of Schrödinger's background was useful in his development of his famous equation?

P2.4 Find the energy eigenvalues for the two-dimensional harmonic oscillator with the Schrödinger equation

$$-\frac{\hbar^2}{2m}\left(\frac{d^2\psi(x, y)}{dx^2} + \frac{d^2\psi(x, y)}{dy^2}\right) + \frac{m\omega^2}{2}(x^2 + y^2)\psi(x, y) = E\psi(x, y).$$

P2.5 One of the reasons for physicists finding the Schrödinger formulation of quantum mechanics more attractive than the Heisenberg formulation is given above. Give some other reasons.

P2.6 Refer to a book on classical mechanics and solve the Kepler problem, which is a planet moving around the Sun under the influence of gravity.

P2.7 Solve the harmonic oscillator problem using the Hamilton–Jacobi equation.

P2.8 Derive the equations for a Gaussian wave packet evolving in time.

References

[1] Heisenberg W 1925 Über quantentheoretische Umdeutung kinematischer und mechanischer Beziehungen *Zeitschrift für Physik* **33** 879–93

[2] Schrödinger E 1926 An undulatory theory of the mechanics of atoms and molecules *Phys. Rev.* **28** 1049–70

[3] Pauli W 1926 Über das Wasserstoffspektrum vom Standpunkt der neuen Quantenmechanik *Zeitschrift für Physik* **36** 336–63

[4] Born M 1926 Zur Quantenmechanik der Stoßvorgänge *Zeitschrift für Physik* **37** 863–7

Chapter 3

Relativistic quantum mechanics

3.1 The necessity for relativistic quantum mechanics

Not many years ago it was thought that even a graduate text on quantum mechanics could focus entirely on the Schrödinger equation. The assumption was that it would be nice to include a section on the Dirac theory for the sake of completeness, but it was not a necessity. Today, quantum mechanics is widely used to calculate the properties of condensed matter and also the chemistry of aggregates of atoms and molecules. These calculations are extremely complex, but they have become practical due to the development of cluster workstations and massively parallel supercomputers. Condensed matter physicists, chemists, and materials scientists rely on these calculations to provide insights into the wide variety of phenomena that they must understand in order to develop new and useful materials, compounds, devices, and structures. The earliest computer codes contained many approximations, one of which was ignoring relativistic effects. The reason for this was that the computers that were available in the 1960s and 1970s were relatively slow. The speed and number of parallel processors of computers continues to increase at a great rate, so calculations are expected to include fewer approximations. Condensed matter physicists now base their codes on Dirac theory [1] because it must be used for all but the lightest atoms. Modern theories of magnetism make no sense if the calculations are not relativistic [2]. Quantum chemists have likewise found that the errors in their calculations caused by ignoring relativistic effects are unacceptable [3].

In the later chapters of this book, most discussion will be based on the Schrödinger formulation because it contains the essential physics and the Dirac equation requires the inclusion of many subscripts and superscripts that are not necessary for the discussion. It is assumed that the relativistic effects will be added later as needed.

3.2 Klein–Gordon equation

In many ways, the Klein–Gordon (K-G) equation is easier to derive than the Schrödinger equation. As was pointed out in the preceding chapter, Schrödinger derived a relativistic equation that was essentially the K-G equation, but it gave the wrong results for the hydrogen atom. The first version of this equation is obtained by writing the relativistic equation for the energy of a free particle

$$E^2 = p^2 c^2 + m^2 c^4. \tag{3.1}$$

If the momentum is set equal to zero, this equation reduces to the famous mass–energy relation

$$E = mc^2. \tag{3.2}$$

It is hoped that these equations of special relativity will be familiar to the reader. Courses in classical mechanics or modern physics normally contain a discussion of that theory. There are useful discussions on the web, and books on special relativity are very cheap.

The experience of Schrödinger is now invoked to replace E and $\vec{p}$ with the operators,

$$(E^2 - c^2 \mathbf{p}^2 - m^2 c^4)|\,\psi(t)\rangle \tag{3.3}$$

and then transform into the position representation

$$\langle \vec{r} \mid E \mid \psi(t)\rangle \rightarrow i\hbar \frac{\partial \psi(\vec{r},\,t)}{\partial t} \langle \vec{r} \mid \vec{p} \mid \psi(t)\rangle = -i\hbar \nabla \psi(\vec{r},\,t), \tag{3.4}$$

to obtain the differential equation

$$-\hbar^2 \frac{\partial^2 \psi(\vec{r},\,t)}{\partial t^2} + c^2 \hbar^2 \nabla^2 \psi(\vec{r},\,t) - m^2 c^4 \psi(\vec{r},\,t) = 0. \tag{3.5}$$

Dividing this equation by $c^2 \hbar^2$ yields the Klein–Gordon equation for a free particle

$$-\frac{1}{c^2} \frac{\partial^2 \psi(\vec{r},\,t)}{\partial t^2} + \nabla^2 \psi(\vec{r},\,t) - \frac{m^2 c^2}{\hbar^2} \psi(\vec{r},\,t) = 0. \tag{3.6}$$

The position vector in four-dimensional space is

$$x^0 = ct,\; x^1 = x,\; x^2 = y,\; x^3 = z. \tag{3.7}$$

The length of a vector in that space is found with the use of a metric

$$l = \sum_{\mu=0}^{3} \sum_{\nu=0}^{3} x^\mu g_{\mu\nu} x^\nu, \tag{3.8}$$

where

$$g = \begin{pmatrix} 1 & 0 & 0 & 0 \\ 0 & -1 & 0 & 0 \\ 0 & 0 & -1 & 0 \\ 0 & 0 & 0 & -1 \end{pmatrix}. \tag{3.9}$$

Einstein's second postulate of relativity is that the length of a vector in the four-dimensional spacetime is invariant under a Lorentz transformation

$$l = c^2 t'^2 - r'^2 = c^2 t^2 - r^2. \tag{3.10}$$

The four-vector momentum is

$$p^0 = \frac{E}{c}, \, p^1 = p_x, \, p^2 = p_y, \, p^3 = p_z, \tag{3.11}$$

and its length is

$$m^2 c^2 = \sum_{\mu=0}^{3} \sum_{\nu=0}^{3} p^\mu g_{\mu\nu} p^\nu. \tag{3.12}$$

This is identical with equation (3.1) and since the length of the four momentum is invariant under a Lorentz transformation, it follows that the Klein–Gordon is invariant and is a proper relativistic equation.

The next step is to consider a particle in a force field. In special relativity, this is achieved in the same way that it is in classical Hamiltonian theory. The total energy is modified to include a scalar potential $\phi(\vec{r})$ and a vector potential $\vec{A}(\vec{r})$ as shown in this equation

$$(E - e\phi(\vec{r}))^2 = c^2(\vec{p} - e\vec{A}(\vec{r}))^2 + m^2 c^4. \tag{3.13}$$

Converting E, $\vec{p}$, and $\vec{r}$ into operators and transforming into the position representation leads to

$$\left(ih\frac{\partial}{\partial t} - e\phi(\vec{r})\right)^2 \psi(\vec{r}, t) - c^2(-i\hbar\nabla - e\vec{A}(\vec{r}))^2 \psi(\vec{r}, t) = m^2 c^4 \psi(\vec{r}, t). \tag{3.14}$$

It simplifies the analysis to set the vector potential equal to zero and call $e\phi(\mathbf{r}) = V(\mathbf{r})$ and this leads to

$$\left[-h^2\frac{\partial^2}{\partial t^2} - ih2V(\vec{r})\frac{\partial}{\partial t} + V(\vec{r})^2\right]\psi(\vec{r}, t) + c^2\hbar^2\nabla^2\psi(\vec{r}, t) - m^2 c^4 \psi(\vec{r}, t) = 0. \tag{3.15}$$

This equation takes an interesting form in the non-relativistic limit. The first step is to eliminate the rest mass energy, equation (3.2), because that constant is not normally included in non-relativistic equations. This is achieved by replacing the wave function in equation (3.15) with

$$\psi(r, t) = \hat{\psi}(r, t)e^{-imc^2 t/\hbar}. \tag{3.16}$$

Dividing the resulting expression by $2mc^2$ and carrying out some manipulations gives

$$\begin{aligned} &-\frac{\hbar^2}{2m}\nabla^2\hat{\psi}(\vec{r}, t) + V(\vec{r})\hat{\psi}(\vec{r}, t) = i\hbar\frac{\partial\hat{\psi}}{\partial t} \\ &-\frac{\hbar^2}{2mc^2}\frac{\partial^2\hat{\psi}}{\partial t^2} - i\hbar\frac{V(\vec{r})}{mc^2}\frac{\partial\hat{\psi}}{\partial t} + \frac{V(\vec{r})^2}{2mc^2}\hat{\psi}(\vec{r}, t) \end{aligned} \tag{3.17}$$

In the non-relativistic limit, all of the terms in the second line of this equation are ignored, and the result is the non-relativistic Schrödinger equation that revolutionized quantum mechanics

$$-\frac{\hbar^2}{2m}\nabla^2\hat{\psi}(\vec{r}, t) + V(\vec{r})\hat{\psi}(\vec{r}, t) = i\hbar\frac{\partial\hat{\psi}}{\partial t}. \tag{3.18}$$

It is interesting that the second time derivative in the Klein–Gordon equation disappears, leaving only the first derivative.

3.3 Problems with the Klein–Gordon equation

It was discovered almost immediately by the theorists who derived a Klein–Gordon equation, Klein [4], Gordon [5], Schrödinger *et al*, that the eigenfunctions and eigenvalues produced by the equation showed unphysical features. The first point has to do with Born's interpretation of the absolute square of a time independent wave function $\psi(\vec{r})$ as the probability for finding a particle at the point $\vec{r}$. If the wave function depends on time $\psi(\vec{r}, t)$, then the conservation of probability is analogous to the conservation of particles in an incompressible liquid. If the particle density is $\rho(\vec{r}, t)$ and the flux of particles through a small surface is $\vec{j}(\vec{r}, t)$, then the conservation of particles requires that

$$\frac{\partial\rho}{\partial t} = \nabla \cdot \vec{j}. \tag{3.19}$$

Premultiplying the Schrödinger equation, equation (3.18), by $\psi*(\vec{r}, t)$ and premultiplying the complex conjugate of equation (3.18) by $\psi(\vec{r}, t)$ produces two new equations. Subtracting the second from the first and dividing by $-i\hbar$ leads to an equation like equation (3.19) with

$$\rho = \psi*\psi, \tag{3.20}$$

and

$$\vec{j} = \frac{i\hbar}{2m}[\psi*\nabla\psi - \psi\nabla\psi*] = \frac{\hbar}{m}\mathrm{Im}\psi*\nabla\psi. \tag{3.21}$$

These equations fit nicely with Born's interpretation of the wave function.

Premultiplying the Klein–Gordon equation, equation (3.15), by $\psi*(\vec{r}, t)$ and premultiplying the complex conjugate of equation (3.15) by $\psi(\vec{r}, t)$ and then subtracting the second equation from the first and dividing by $-i\hbar 2mc^2$ leads to an equation like equation (3.19) with the same probability flux as is shown in equation (3.21). However, the density $\rho(\vec{r}, t)$ is

$$\rho(\vec{r}, t) = -\frac{i\hbar}{2mc^2}\left[\psi*\frac{\partial\psi}{\partial t} - \psi\frac{\partial\psi*}{\partial t}\right] + \frac{V(\vec{r})}{mc^2}[\psi*\psi]. \tag{3.22}$$

The strange looking time dependence of this density can be interpreted as a relativistic distortion of the infinitesimal cube that is used in its definition.

However the modification of the non-relativistic density by a factor proportional to $V(\vec{r})$ is really disturbing. The potential function is negative more often than not, and there is nothing to prevent the second term from being larger than the first making $\rho(\vec{r}, t)$ negative. A negative probability density has no physical meaning. It can be argued that this is not a conclusive demonstration of the short-comings of the Klein–Gordon equation because there may be other ways to define the probability density and flux.

A more definitive way to illustrate the problems with the Klein–Gordon equation is to apply it to an elementary scattering problem. Consider a one-dimensional system with a potential that is zero for x less than zero and has a value V_0 when x is greater than zero

$$V(x) = 0 \quad x < 0 \qquad V(x) = V_0 \quad x \geqslant 0. \tag{3.23}$$

In classical mechanics, a particle with energy less than V_0 would bounce back from the step at $x = 0$, and if the energy is greater than V_0 the particle would proceed with no effect from the potential. If the non-relativistic Schrödinger equation, equation (3.18), is used to solve this problem, the reflection of the particle for $0 \leqslant E \leqslant V_0$ is not 100% because the particle has a non-zero probability of tunneling into the range where the potential is V_0. The wave function for $x \geqslant 0$ is

$$\psi(x, t) = Te^{-\kappa x}e^{-iEt/\hbar}, \tag{3.24}$$

where

$$\kappa = \frac{\sqrt{2m(V_0 - E)}}{\hbar}. \tag{3.25}$$

It can be seen that κ is real throughout the energy range. When $E \geqslant V_0$, κ becomes imaginary, which means the the particle is propagating. However, even when $E \geqslant V_0$ the reflection coefficient is not zero. It is analogous to light striking a block of glass. Some of the light is transmitted and some is reflected. The classical and Schrödinger results are both reasonable in their way, but the Klein–Gordon equation gives a different answer.

In one dimension, equation (3.14) is

$$\left[ih\frac{\partial}{\partial t} - V(x) \right]^2 \psi(x, t) + c^2\hbar^2 \frac{\partial^2\psi(x, t)}{\partial x^2} - m^2c^4\psi(x, t) = 0. \tag{3.26}$$

For x less than zero, the equation is

$$-h^2\frac{\partial^2\psi(x, t)}{\partial t^2} + c^2\hbar^2\frac{\partial^2\psi(x, t)}{\partial x^2} - m^2c^4\psi(x, t) = 0, \tag{3.27}$$

and the solution is propagating

$$\psi(x, t) = (e^{ikx} + Re^{-ikx})e^{-iEt/\hbar}. \tag{3.28}$$

where

$$p = \hbar k \tag{3.29}$$

and

$$k = \frac{\sqrt{E^2 - m^2c^4}}{\hbar c}. \tag{3.30}$$

For x greater than zero, the solution is as shown in equation (3.24), a function that does not propagate. The difference between the Klein–Gordon and Schrödinger solutions is that for the former, κ is a solution of

$$(E - V_0)^2 = -\hbar^2\kappa^2 c^2 + m^2 c^4, \tag{3.31}$$

or

$$\kappa = \frac{[m^2 c^4 - (E - V_0)^2]^{1/2}}{\hbar c}. \tag{3.32}$$

It is clear from this equation that κ is real when E is such that

$$-mc^2 \leqslant E - V_0 \leqslant mc^2, \tag{3.33}$$

and pure imaginary otherwise.

The real part of κ is plotted in figure 3.1. This figure makes the predictions of the Klein–Gordon theory very clear. For $E \leqslant V_0 - mc^2$, κ is imaginary and the particle will propagate in the $x \geqslant 0$ region. This not only contradicts the Schrödinger result described above, but also it is hard to believe on physical grounds. The Schrödinger equation gives the physically reasonable result that, when E is greater than the height of the step, the wave will propagate. The Klein–Gordon equation, however, predicts that for energies in the range $V_0 \leqslant E \leqslant V_0 + mc^2$, κ is real and the solution is not propagating. Such a result violates physical expectations.

It is very difficult to solve the hydrogen atom problem with the Klein–Gordon equation, but it has been done. The eigenvalues are considerably different from the ones predicted by the Schrödinger equation. The Schrödinger eigenvalues agree very

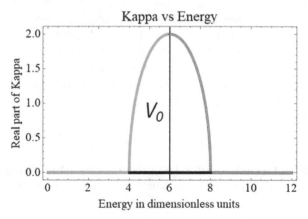

Figure 3.1. The quantity κ from equation (3.32) as a function of energy in dimensionless units. The perpendicular line shows the height of the potential for $x \geqslant 0$. Only the real part of κ is shown.

well with experiments, which means that the Klein–Gordon eigenvalues do not. In addition, heavier atoms could not be treated with the Klein–Gordon equation because the equation for the innermost shell of electrons has no solution for an atomic number Z greater than 69.

For the reasons outlined in this section, most physicists have given up on the Klein–Gordon equation as a wave equation. However, the fact that it is Lorentz invariant means that it serves a useful purpose in field theory.

3.4 Dirac theory

Physicists have often discovered new mathematical principles in the process of developing a theory of some phenomenon. This is particularly true of P A M Dirac. In chapter 1 of this book, the theory of distributions was referenced in order to put into a mathematical context the delta-function that Dirac introduced. Mathematicians have developed new areas of algebra and group theory in order to elucidate the manipulations Dirac performed in his derivation of relativistic quantum theory, in particular, the matrices that he used. His equation led to a number of new concepts in physics, such as spin and antimatter. The theoretical physicist Victor F Weisskopf spent much of his career working with Dirac's equation. He said the following, 'Today it is hard to realize the excitement, the skepticism, and the enthusiasm aroused in the early years by the development of all the new insights that emerged from the Dirac equation. A great deal more was hidden in the Dirac equation than the author had expected when he wrote it down in 1928'. Because the Dirac equation provides the foundation for quantum electrodynamics and field theory, many sophisticated analyses of that equation are found in books on those subjects. Dirac himself remarked in one of his talks that his equation was more intelligent than its author. It should be added, however, that it was Dirac who found many of the additional insights.

Dirac [6] realized that the problem with the Klein–Gordon approach is the second derivative of $\psi(\vec{r}, t)$ with respect to time. The experience with the non-relativistic Schrödinger equation indicated that it would be good to eliminate that feature, but it also made clear that, in order to achieve Lorentz invariance, the derivative with respect to position must also be of first order. He therefore decided to take the square root of both sides of equation (3.1). He was aware that the square root of the quadratic form $c^2(p_x^2 + p_y^2 + p_z^2) + m^2c^4$ doesn't exist within the field of real numbers, so he wrote it as

$$E = c\vec{\alpha} \cdot \vec{p} + \beta mc^2 \tag{3.34}$$

where the elements of $\vec{\alpha}$ and also β have to be matrices. Multiplying the right side of equation (3.34) by itself leads to the quadratic form only if these matrices are such that

$$\alpha_x^2 = \alpha_y^2 = \alpha_z^2 = \beta^2. \tag{3.35}$$

Also

$$\vec{\alpha}\beta + \beta\vec{\alpha} = 0, \tag{3.36}$$

as well as

$$\{\alpha_x, \alpha_y\} = \{\alpha_y, \alpha_z\} = \{\alpha_z, \alpha_x\} = 0, \tag{3.37}$$

where $\{\alpha_a, \alpha_b\}$ is the anticommutator, $\{\alpha_a, \alpha_b\} = \alpha_a\alpha_b + \alpha_b\alpha_a$. These relations are easy to find by simply multiplying the right side of equation (3.34) by itself.

Dirac found a set of matrices that satisfy these conditions. They are 4×4 matrices, but it is more convenient to write them in block matrix form using 2×2 matrices. For example, the 2×2 unit matrix is

$$I_2 = \begin{pmatrix} 1 & 0 \\ 0 & 1 \end{pmatrix}, \tag{3.38}$$

and the matrix β can be written

$$\beta = \begin{pmatrix} I_2 & 0 \\ 0 & -I_2 \end{pmatrix} = \begin{pmatrix} 1 & 0 & 0 & 0 \\ 0 & 1 & 0 & 0 \\ 0 & 0 & -1 & 0 \\ 0 & 0 & 0 & -1 \end{pmatrix}. \tag{3.39}$$

The components of $\vec{\alpha}$ are

$$\alpha_x = \begin{pmatrix} 0 & \sigma_x \\ \sigma_x & 0 \end{pmatrix}, \quad \alpha_y = \begin{pmatrix} 0 & \sigma_y \\ \sigma_y & 0 \end{pmatrix}, \quad \alpha_z = \begin{pmatrix} 0 & \sigma_z \\ \sigma_z & 0 \end{pmatrix}, \tag{3.40}$$

where the sigmas are the Pauli matrices

$$\sigma_x = \begin{pmatrix} 0 & 1 \\ 1 & 0 \end{pmatrix}, \quad \sigma_y = \begin{pmatrix} 0 & -i \\ i & 0 \end{pmatrix}, \quad \sigma_z = \begin{pmatrix} 1 & 0 \\ 0 & -1 \end{pmatrix}. \tag{3.41}$$

They are Hermitean, which means that the α and β matrices are also Hermitean.

The square of a Pauli matrix is a unit matrix

$$\sigma_x^2 = \sigma_y^2 = \sigma_z^2 = -i\sigma_x\sigma_y\sigma_z = I_2, \tag{3.42}$$

and the anticommutators are all zero

$$\{\sigma_x, \sigma_y\} = \{\sigma_y, \sigma_z\} = \{\sigma_z, \sigma_x\} = 0. \tag{3.43}$$

These two properties can be written

$$\{\sigma_\alpha, \sigma_\beta\} = 2\delta_{\alpha\beta}I_2. \tag{3.44}$$

Later, the commutators of these matrices will be needed

$$[\sigma_x, \sigma_y] = 2i\sigma_z, \quad [\sigma_y, \sigma_z] = 2i\sigma_x, \quad [\sigma_z, \sigma_x] = 2i\sigma_y. \tag{3.45}$$

Defining the vector

$$\vec{\sigma} = \sigma_x \hat{x} + \sigma_y \hat{y} + \sigma_z \hat{z}, \tag{3.46}$$

and using the properties outlined above it can be shown that

$$e^{i\theta \hat{n} \cdot \vec{\sigma}} = \cos\theta + i\sin\theta \hat{n} \cdot \vec{\sigma}. \tag{3.47}$$

The commutators in equation (3.45) are obtained by matrix multiplication. It will be seen that multiplying the Pauli matrices by $\hbar/2$ gives operators that have the same commutation rules as the spin operators introduced in chapter 1

$$S_i = \frac{1}{2}\hbar\sigma_i. \tag{3.48}$$

This is proved by comparing equation (3.45) with equation (1.87).

Following Schrödinger, Dirac considered E and $\vec{p}$ as operators and then replaced them with

$$E \Rightarrow i\hbar\frac{\partial}{\partial t}, \vec{p} \Rightarrow -i\hbar\nabla, \tag{3.49}$$

so that equation (3.34) becomes

$$-i\hbar c\vec{\alpha} \cdot \nabla\psi + \beta mc^2\psi = i\hbar\frac{\partial\psi}{\partial t}. \tag{3.50}$$

Since this is a 4×4 matrix equation, the wave function is a column vector

$$\psi(\vec{r}, t) = \begin{pmatrix} \psi^1(\vec{r}, t) \\ \psi^2(\vec{r}, t) \\ \psi^3(\vec{r}, t) \\ \psi^4(\vec{r}, t) \end{pmatrix}. \tag{3.51}$$

It will be explained later why this wave function is more properly called a bispinor. Equation (3.50) is the original, and still a very useful, form of the Dirac equation for a free particle. As described in connection with equation (3.13), the equation can be generalized to treat a particle in a force field by including a scalar potential $\phi(\vec{r})$ and a vector potential $\vec{A}(\vec{r})$ to obtain

$$c\vec{\alpha} \cdot \left(-i\hbar\nabla - \frac{e}{c}\vec{A}(\vec{r})\right)\psi + e\phi(\vec{r})\psi + \beta mc^2\psi = i\hbar\frac{\partial\psi}{\partial t}. \tag{3.52}$$

3.5 Proof of the Lorentz covariance of the Dirac equation

The invariance of the Klein–Gordon equation under a Lorentz transformation was shown in the first section of this chapter. The behavior of the Dirac equation under a Lorentz transformation is not as obvious. In order to investigate this question, it is useful to put the Dirac equation into a different form.

The position and also time was written as a four-vector in equation (3.7). It is useful to first consider a common example of the Lorentz transformation of that vector. According to the rules of special relativity the position of a particle in a coordinate system in which the origin is moving to the right along the x-axis with velocity v is

$$x' = \frac{x - vt}{\sqrt{1 - \beta^2}} = \gamma x - \beta\gamma ct, \tag{3.53}$$

where

$$\beta = \frac{v}{c}, \quad \gamma = \frac{1}{\sqrt{1 - \beta^2}}. \tag{3.54}$$

Obviously, the inverse of this transformation would be

$$x = \frac{x' + vt'}{\sqrt{1 - \beta^2}} = \gamma x' + \beta\gamma ct'. \tag{3.55}$$

The time coordinate also changes in the moving coordinate system

$$ct' = \frac{ct - \beta x}{\sqrt{1 - \beta^2}} = \gamma ct - \beta\gamma x, \tag{3.56}$$

and the inverse is

$$ct = \frac{ct' - \beta x'}{\sqrt{1 - \beta^2}} = \gamma ct' + \beta\gamma x'. \tag{3.57}$$

The Lorentz transformation of the four-vector may thus be written in matrix form

$$\begin{pmatrix} x'^0 \\ x'^1 \\ x'^2 \\ x'^3 \end{pmatrix} = \begin{pmatrix} \gamma & -\beta\gamma & 0 & 0 \\ -\beta\gamma & \gamma & 0 & 0 \\ 0 & 0 & 1 & 0 \\ 0 & 0 & 0 & 1 \end{pmatrix} \begin{pmatrix} x^0 \\ x^1 \\ x^2 \\ x^3 \end{pmatrix}, \tag{3.58}$$

or

$$x'^\mu = \sum_{\nu=0}^{3} \Lambda^\mu_\nu x^\nu. \tag{3.59}$$

Using the fact that

$$\gamma^2(1 - \beta^2) = 1, \tag{3.60}$$

it can be seen algebraically that the inverse of the above equation is

$$\begin{pmatrix} x^0 \\ x^1 \\ x^2 \\ x^3 \end{pmatrix} = \begin{pmatrix} \gamma & \beta\gamma & 0 & 0 \\ \beta\gamma & \gamma & 0 & 0 \\ 0 & 0 & 1 & 0 \\ 0 & 0 & 0 & 1 \end{pmatrix} \begin{pmatrix} x'^0 \\ x'^1 \\ x'^2 \\ x'^3 \end{pmatrix},$$

(3.61)

which is also obvious from the physical arguments above. The compact way to write these matrix equations is

$$x' = \Lambda x.$$

(3.62)

There are other four vectors

$$a_4 = \begin{pmatrix} a^0 \\ a^1 \\ a^2 \\ a^3 \end{pmatrix},$$

(3.63)

that transform under the Lorentz transformation like

$$\begin{pmatrix} a'^0 \\ a'^1 \\ a'^2 \\ a'^3 \end{pmatrix} = \begin{pmatrix} \Lambda^0_0 & \Lambda^0_1 & \Lambda^0_2 & \Lambda^0_3 \\ \Lambda^1_0 & \Lambda^1_1 & \Lambda^1_2 & \Lambda^1_3 \\ \Lambda^2_0 & \Lambda^2_1 & \Lambda^2_2 & \Lambda^2_3 \\ \Lambda^3_0 & \Lambda^3_1 & \Lambda^3_2 & \Lambda^3_3 \end{pmatrix} \begin{pmatrix} a^0 \\ a^1 \\ a^2 \\ a^3 \end{pmatrix},$$

(3.64)

or

$$a' = \Lambda a,$$

(3.65)

in a moving reference frame. The length of the vector requires the use of the metric defined in equation (3.9)

$$\tilde{a}ga = l,$$

(3.66)

where $\tilde{a}$ is the transpose

$$\tilde{a} = (a^0 a^1 a^2 a^3).$$

(3.67)

It follows from the invariance of this length that

$$\tilde{a}'ga' = \tilde{a}\tilde{\Lambda}g\Lambda a = \tilde{a}ga = l,$$

(3.68)

and

$$\tilde{\Lambda}g\Lambda = g.$$

(3.69)

Algebraically, the definition of a proper Lorentz transformation matrix Λ is a real four-by-four matrix that satisfies the preceding equation and has a determinant equal to one.

From the definition in equation (3.7) the components x^1, x^2, and x^3 are the Cartesian coordinates in ordinary three-dimensional space. A rotation by the angle θ around the z axis in that space is also described by a Lorentz transformation

$$\Lambda = \begin{pmatrix} 1 & 0 & 0 & 0 \\ 0 & \cos\theta & \sin\theta & 0 \\ 0 & -\sin\theta & \cos\theta & 0 \\ 0 & 0 & 0 & 1 \end{pmatrix}. \tag{3.70}$$

This relation is familiar from undergraduate physics classes. Obviously, any combination of rotations with translations in time are described by a Lorentz transformation Λ. A transformation in space and time, like the one in equation (3.58), is called a boost.

Defining the matrices

$$\gamma^0 = \beta, \; \gamma^1 = \beta\alpha_x, \; \gamma^2 = \beta\alpha_y, \; \gamma^3 = \beta\alpha_z, \tag{3.71}$$

and multiplying equation by β leads to a new form for the Dirac equation

$$\sum_{\mu=0}^{4} \gamma^\mu \left(i\hbar \frac{\partial}{\partial x^\mu} - eA_\mu \right) \psi - mc\psi = 0 \tag{3.72}$$

where the four-vector potential A is defined by

$$A = \left[\frac{\phi(\vec{r})}{c}, -\vec{A}(\vec{r}) \right]. \tag{3.73}$$

The gamma matrices defined above above and equation (3.72) are similar to the form of his equation carved in Dirac's memorial stone in Westminster Abbey.

Writing the matrices out in detail

$$\gamma^0 = \begin{pmatrix} I_2 & 0 \\ 0 & -I_2 \end{pmatrix},$$

$$\gamma^1 = \begin{pmatrix} 0 & \sigma_x \\ -\sigma_x & 0 \end{pmatrix}, \; \gamma^2 = \begin{pmatrix} 0 & \sigma_y \\ -\sigma_y & 0 \end{pmatrix}, \; \gamma^3 = \begin{pmatrix} 0 & \sigma_z \\ -\sigma_z & 0 \end{pmatrix}, \tag{3.74}$$

it can be seen that γ^0 is Hermitean and the others are anti-Hermitean. The conditions previously worked out for the alphas and beta, the anticommutators of the gamma matrices can be summed up

$$\{\gamma^\mu, \gamma^\nu\} = \gamma^\mu\gamma^\nu + \gamma^\nu\gamma^\mu = 2g_{\mu\nu}I_4, \tag{3.75}$$

where the $g_{\mu\nu}$ are the elements of the metric matrix in equation (3.9).

Using the gamma matrices and the four-vector form of the momentum from equation (3.11), equation (3.34) can be written

$$\sum_{\mu=0}^{4}\sum_{\nu=0}^{4} \gamma^\mu g_{\mu\nu} p^\nu = mc. \tag{3.76}$$

It might appear that it would be useful to define gamma matrices in a moving reference frame as

$$\gamma'^\mu = \sum_{\nu=0}^{3} \Lambda^\mu_\nu \gamma^\nu,$$ (3.77)

and the four-vector momentum in the usual way as

$$p'^\mu = \sum_{\nu=0}^{3} \Lambda^\mu_\nu p^\nu.$$ (3.78)

Then the primed version of equation (3.76) would transform into the unprimed form easily using the condition in equation (3.69). This would lead to time and velocity dependent gamma matrices, and is not the definition of covariance that is desired.

The Lorentz transformed version of the Dirac equation for a free particle is

$$\sum_{\nu=0}^{4} \gamma^\nu \left(i\hbar \frac{\partial}{\partial x'^\nu} \right) \psi'(x') - mc\psi'(x') = 0,$$ (3.79)

where $x' = \Lambda x$ are the coordinates in the moving reference frame. It is assumed that the four-dimensional wave vector must transform according to the equation

$$\psi'(x') = \psi'(\Lambda x) = S\psi(x),$$ (3.80)

with S being a four-by-four matrix that clearly depends on Λ.

It is at this point that it is seen that $\psi(x)$ is not just a 4×1 vector. A function that transforms according to equation (3.80) is called a bispinor. In later chapters, spinors and bispinors will be discussed in more detail.

Using the inverse of equation (3.59)

$$x^\mu = \sum_{\nu=0}^{3} (\Lambda^{-1})^\mu_\nu x'^\nu,$$ (3.81)

and the rule from differential calculus, the derivatives become

$$\frac{\partial}{\partial x'^\nu} = \sum_{\mu=0}^{4} \left[(\Lambda^{-1})^\mu_\nu \frac{\partial}{\partial x^\mu} \right].$$ (3.82)

Multiplying equation (3.79) on the left by S^{-1} and assuming that the differential operator has no effect on S, the equation takes the intermediate form

$$\sum_{\nu=0}^{4} S^{-1} \gamma^\nu S \left(i\hbar \sum_{\mu=0}^{4} \left[(\Lambda^{-1})^\mu_\nu \frac{\partial}{\partial x^\mu} \right] \right) S^{-1} \psi'(x') - mc S^{-1} \psi'(x') = 0.$$ (3.83)

This is the same as the free particle equation in the unprimed variables

$$\sum_{\mu=0}^{4} \gamma^\mu \left(i\hbar \frac{\partial}{\partial x^\mu} \right) \psi(x) - mc\psi(x) = 0,$$ (3.84)

identifying

$$\gamma^{\mu} = \sum_{\nu=0}^{4} S^{-1}\gamma^{\nu}S(\Lambda^{-1})^{\mu}_{\nu}, \tag{3.85}$$

or

$$\sum_{\mu=0}^{4} \gamma^{\mu}\Lambda^{\kappa}_{\mu} = S^{-1}\gamma^{\kappa}S. \tag{3.86}$$

All of the matrices in this equation (3.86) are known with the exception of S. The question is if it exists and can be found. It was proved by Pauli [7] that if there is another set of four four-by-four matrices that satisfy equation (3.75)

$$\{\gamma'^{\mu}, \gamma'^{\nu}\} = \gamma'^{\mu}\gamma'^{\nu} + \gamma'^{\nu}\gamma'^{\mu} = 2g_{\mu\nu}I_4, \tag{3.87}$$

then the two sets of matrices are connected by a similarity transformation

$$\gamma'^{\kappa} = S^{-1}\gamma^{\kappa}S. \tag{3.88}$$

He called this the fundamental theorem on Dirac matrices. It follows from it that the matrices defined in equation (3.86)

$$\gamma'^{\kappa} = \sum_{\mu=0}^{4} \gamma^{\mu}\Lambda^{\kappa}_{\mu}, \tag{3.89}$$

are such a set of matrices and therefore S must exist. Since the gamma matrices are the same for all Dirac equations, S depends only on the Lorentz transformation, Λ.

The algebra required to solve equation (3.85) for S is lengthy and not very illuminating. It has been explained in a number of publications, so the final results will just be quoted. For the case that Λ describes a simple rotation by an angle θ about the z axis, shown in equation (3.70), the matrix S is

$$S = \begin{pmatrix} \cos\frac{\theta}{2}I_2 + i\sigma_3 \sin\frac{\theta}{2} & 0 \\ 0 & \cos\frac{\theta}{2}I_2 + i\sigma_3 \sin\frac{\theta}{2} \end{pmatrix}. \tag{3.90}$$

The Λ for a boost in the x direction is shown in equation (3.58). The corresponding S matrix for this case is

$$S = \begin{pmatrix} I_2 \cosh\frac{\eta}{2} & -\sigma_x \sinh\frac{\eta}{2} \\ -\sigma_x \sinh\frac{\eta}{2} & I_2 \cosh\frac{\eta}{2} \end{pmatrix}, \tag{3.91}$$

where

$$\tanh \eta = \frac{v}{c}. \tag{3.92}$$

It turns out that the four-vector potential in equation (3.72) transforms like

$$A'_\nu = \sum_{\mu=0}^{4} \left[(\Lambda^{-1})^\mu_\nu A_\mu \right], \tag{3.93}$$

so the argument for the covariance of the Dirac equation is unchanged for particles moving in a potential. Of course, for that case S will depend on the potential as well as the motion of the coordinate system.

3.6 The fifth gamma matrix

In the preceding section it was shown that the gamma matrices represent a fixed basis of unit vectors in spacetime. The unit volume element on spacetime is, in terms of the gammas,

$$\gamma^5 = i\gamma^0\gamma^1\gamma^2\gamma^3, \tag{3.94}$$

or

$$\gamma^5 = \begin{pmatrix} 0 & I_2 \\ I_2 & 0 \end{pmatrix}. \tag{3.95}$$

This is not one of the gamma matrices, but it is frequently used in connection with them. It anticommutes with the basic four gamma matrices

$$\{\gamma^\mu, \gamma^5\} = 0. \tag{3.96}$$

It takes a leading role when questions of parity arise because it changes sign under a spacetime reflection.

3.7 Free particle solution of the Dirac equation

The Dirac equation for a free particle at rest is

$$\beta mc^2\psi = i\hbar\frac{\partial\psi}{\partial t} \tag{3.97}$$

with $E = \pm mc^2$. Taking β from equation (3.39), leads to

$$\begin{aligned} \left[\frac{\partial}{\partial t} + \frac{imc^2}{\hbar}\right]\psi_i = 0 \quad i = 1, 2 \\ \left[\frac{\partial}{\partial t} - \frac{imc^2}{\hbar}\right]\psi_i = 0 \quad i = 3, 4 \end{aligned}, \tag{3.98}$$

so that

$$\psi_1 = \frac{1}{\sqrt{V}} \begin{pmatrix} 1 \\ 0 \\ 0 \\ 0 \end{pmatrix} e^{-\frac{imc^2}{\hbar}t}, \quad \psi_2 = \frac{1}{\sqrt{V}} \begin{pmatrix} 0 \\ 1 \\ 0 \\ 0 \end{pmatrix} e^{-\frac{imc^2}{\hbar}t}, \tag{3.99}$$

for positive energies and

$$\psi_3 = \frac{1}{\sqrt{V}} \begin{pmatrix} 0 \\ 0 \\ 1 \\ 0 \end{pmatrix} e^{\frac{imc^2}{\hbar}t}, \quad \psi_4 = \frac{1}{\sqrt{V}} \begin{pmatrix} 0 \\ 0 \\ 0 \\ 1 \end{pmatrix} e^{\frac{imc^2}{\hbar}t}, \tag{3.100}$$

for negative energies. Obviously,

$$\int_V \psi_i^\dagger \psi_i d\vec{r} = 1 \tag{3.101}$$

when working in a finite space V.

Solutions of the Dirac equation could be written in many different ways, but the theory derived in the preceding sections requires that they remain bispinors. Thus, it must be possible to obtain the free particle solution for a particle with finite momentum $\vec{p}$ by applying a boost to the preceding equations. Generalizing equation (3.91) for the case of a particle moving in the direction $\hat{n}$, $\vec{p} = p\hat{n}$, is equivalent to giving the origin a boost in the opposite direction

$$S = \begin{pmatrix} I_2 \cosh \frac{\eta}{2} & \vec{\sigma} \cdot \vec{n} \sinh \frac{\eta}{2} \\ \vec{\sigma} \cdot \vec{n} \sinh \frac{\eta}{2} & I_2 \cosh \frac{\eta}{2} \end{pmatrix}. \tag{3.102}$$

The definition of η in equation (3.92) is not useful for quantum mechanics. It is better to define

$$\tanh \eta = \frac{cp}{E}, \tag{3.103}$$

which leads to

$$\cosh \eta = \frac{E}{mc^2}. \tag{3.104}$$

Trigonometric identities lead to

$$\cosh \frac{\eta}{2} = \sqrt{\frac{1 + \cosh \eta}{2}} = \sqrt{\frac{E + mc^2}{2mc^2}}, \tag{3.105}$$

and

$$\sinh \frac{\eta}{2} = \sqrt{\frac{\cosh \eta - 1}{2}} = \sqrt{\frac{E - mc^2}{2mc^2}}. \tag{3.106}$$

3-16

These relations make it possible to write the boost matrix S in terms of the energy, and applying that matrix to ψ_1 leads to

$$\psi_1(x, p) = S\psi_1(x) = \frac{1}{\sqrt{V}}\cosh\frac{\eta}{2}\begin{pmatrix}\begin{pmatrix}1\\0\end{pmatrix}\\ \tanh\frac{\eta}{2}\vec{\sigma}\cdot\vec{n}\begin{pmatrix}1\\0\end{pmatrix}\end{pmatrix}e^{-i\frac{mc^2}{\hbar}t}. \tag{3.107}$$

The four momentum is $\mathbf{p} = \left(\frac{E}{c}, p_x, p_y, p_z\right)$ and the four-vector position is $\mathbf{x} = (ct, x, y, z)$. Since both of these vectors transform according the the same Lorentz transformation, Λ, it follows that their inner product is invariant. Thus,

$$(\mathbf{p}, \mathbf{x}) = Et - \vec{p}\cdot\vec{r} = mc^2 t, \tag{3.108}$$

the last form being obtained by evaluating the inner product at $t = 0$. It is also useful to write the hyperbolic tangent in a different way

$$\tanh\frac{\eta}{2} = \sqrt{\frac{E - mc^2}{E + mc^2}} = \frac{\sqrt{E^2 - m^2 c^4}}{E + mc^2} = \frac{cp}{E + mc^2}. \tag{3.109}$$

Using these relations, the final form of $\psi_1(x, p)$ is

$$\psi_1(x, p) = \frac{1}{\sqrt{V}}\sqrt{\frac{E + mc^2}{2mc^2}}\begin{pmatrix}\begin{pmatrix}1\\0\end{pmatrix}\\ \frac{c\vec{\sigma}\cdot\vec{p}}{E + mc^2}\begin{pmatrix}1\\0\end{pmatrix}\end{pmatrix}e^{i\frac{(\vec{p}\cdot\vec{r}-Et)}{\hbar}}. \tag{3.110}$$

Similar manipulations lead to

$$\psi_2(x, p) = \frac{1}{\sqrt{V}}\sqrt{\frac{E + mc^2}{2mc^2}}\begin{pmatrix}\begin{pmatrix}0\\1\end{pmatrix}\\ \frac{c\vec{\sigma}\cdot\vec{p}}{E + mc^2}\begin{pmatrix}0\\1\end{pmatrix}\end{pmatrix}e^{i\frac{(\vec{p}\cdot\vec{r}-Et)}{\hbar}}. \tag{3.111}$$

These are the positive energy states, $E > mc^2$. The top two components make up the large part of the wave function and the bottom two are generally smaller because of $1/\sqrt{E + mc^2}$.

$$\psi_3(x, p) = \frac{1}{\sqrt{V}}\sqrt{\frac{E + mc^2}{2mc^2}}\begin{pmatrix}-\frac{c\vec{\sigma}\cdot\vec{p}}{E + mc^2}\begin{pmatrix}1\\0\end{pmatrix}\\ \begin{pmatrix}1\\0\end{pmatrix}\end{pmatrix}e^{i\frac{(\vec{p}\cdot\vec{r}+Et)}{\hbar}}, \tag{3.112}$$

and

$$\psi_3(x, p) = \frac{1}{\sqrt{V}}\sqrt{\frac{E + mc^2}{2mc^2}}\begin{pmatrix}-\frac{c\vec{\sigma}\cdot\vec{p}}{E + mc^2}\begin{pmatrix}0\\1\end{pmatrix}\\ \begin{pmatrix}0\\1\end{pmatrix}\end{pmatrix}e^{i\frac{(\vec{p}\cdot\vec{r}+Et)}{\hbar}}. \tag{3.113}$$

These are the negative energy states, $E < -mc^2$. The bottom two components make up the large part of the wave function and the top two are generally smaller. Naming the parts of the wave function as to large and small parts and positive and negative energy is brought up frequently in discussions of the solutions of the Dirac equation.

A rotation is normally represented mathematically by a three-by-three matrix operating on a three-dimensional vector. It can also be represented by a two-by-two matrix operating on a two-dimensional object called a spinor. For the case of a rotation through the angle θ about an axis that is pointing in the direction of a unit vector $\vec{n}$, the two-by-two matrix is

$$s(\theta) = e^{i\vec{n}\cdot\vec{\sigma}\frac{\theta}{2}}, \tag{3.114}$$

where $\vec{\sigma}$ is the matrix defined in equation (3.46). The simplest way to write the complete set of spinors is

$$\chi_1 = \begin{pmatrix} 1 \\ 0 \end{pmatrix} \quad \chi_2 = \begin{pmatrix} 0 \\ 1 \end{pmatrix}. \tag{3.115}$$

These objects will appear often in succeeding chapters. Comparing them with the solutions of the Dirac equation in equations (3.110) through (3.113) makes it obvious why these solutions are called bispinors.

The free particle solutions found by the boost method are exactly the same as those obtained from a conventional solution of the Dirac equation

$$(c\vec{\alpha} \cdot \vec{p} + \beta mc^2)\psi = i\hbar\frac{\partial\psi}{\partial t}. \tag{3.116}$$

There are two advantages to the boost method. First, it is easier. Second, it demonstrates more clearly the theoretical point that the bispinors transform as expected under the Lorentz transformation.

3.8 Angular momentum and spin

Using equation (2.148) the rate of change of the angular momentum $\vec{L} = \vec{r} \times \vec{p}$ can be found for the case of a free particle by calculating the commutator with the Hamiltonian

$$\frac{d\vec{L}}{dt} = \frac{i}{\hbar}[H, \vec{L}]. \tag{3.117}$$

The math is easier when the focus is on one component

$$L_x = yp_z - zp_y, \tag{3.118}$$

and

$$H = c\alpha_x p_x + c\alpha_y p_y + c\alpha_z p_z + \beta mc^2, \tag{3.119}$$

so

$$[H, L_x] = c\alpha_y[p_y, y]p_z - c\alpha_z[p_z, z]p_y = -i\hbar c(\alpha_y p_z - \alpha_z p_y), \tag{3.120}$$

using the standard position-momentum commutation rule. With a few more manipulations, the result is

$$\frac{d\vec{L}}{dt} = c(\vec{\alpha} \times \vec{p}). \tag{3.121}$$

This result shows that angular momentum is not conserved for a Dirac particle, while it is conserved for a particle described by the Schrödinger equation.

The spin of a Dirac particle is represented as a four-by-four matrix by

$$\vec{S} = \frac{\hbar}{2}\vec{\Sigma}, \tag{3.122}$$

where

$$\vec{\Sigma} = \hat{x}\Sigma_x + \hat{y}\Sigma_y + \hat{z}\Sigma_z, \tag{3.123}$$

and

$$\Sigma_i = \begin{pmatrix} \sigma_i & 0 \\ 0 & \sigma_i \end{pmatrix}. \tag{3.124}$$

The time derivative of Σ_z is given by

$$\frac{d\Sigma_z}{dt} = \frac{i}{\hbar}[H, \Sigma_z] \tag{3.125}$$

where the useful form of H is in equation (3.119). Using the commutation rules in equation (3.45),

$$\frac{i}{\hbar}[H, \Sigma_z] = \frac{c}{\hbar}[\alpha_x, \Sigma_z]p_x + \frac{c}{\hbar}[\alpha_y, \Sigma_z]p_y, \tag{3.126}$$

and hence

$$\frac{d\Sigma_z}{dt} = -\frac{2c}{\hbar}(\vec{\alpha} \times \vec{p})_z, \tag{3.127}$$

and finally

$$\frac{d\vec{S}}{dt} = -c(\vec{\alpha} \times \vec{p}), \tag{3.128}$$

which means that the spin of the Dirac particle is not conserved.

It can be seen from the preceding equations that neither $\vec{L}$ nor $\vec{S}$ is conserved for a Dirac particle, but the total angular momentum $\vec{L} + \vec{S}$ is conserved

$$\frac{d(\vec{L} + \vec{S})}{dt} = 0. \tag{3.129}$$

The spin of an electron was used before Dirac published his equation. It appeared in the Pauli equation, and was used in the explanation of atomic spectra. However, the appearance of the spin in those equations was completely ad hoc. The first equation in which the spin and its behavior appeared naturally is the Dirac equation. This first-principles theory of spin is one of Dirac's great achievements.

3.9 The magnetic moment of the electron

The Dirac equation for a system with non-zero electric and magnetic fields is

$$\left(c\vec{\alpha} \cdot \left(\vec{p} - \frac{e}{c}\vec{A}\right) + \beta mc^2\right)\psi = (E - e\phi)\psi, \tag{3.130}$$

where $\vec{A}$ and ϕ are the scalar and vector potentials. This is obtained in the usual way by replacing

$$\vec{p} \to \vec{p} - \frac{e}{c}\vec{A}, E \to E - E\phi. \tag{3.131}$$

Defining

$$\vec{\pi} = \vec{p} - \frac{e}{c}\vec{A}, \tag{3.132}$$

and writing the wave function in terms of two component spinors

$$\psi = \begin{pmatrix} \psi_A \\ \psi_B \end{pmatrix} \tag{3.133}$$

the Dirac equation can be written

$$\begin{aligned} c\vec{\sigma} \cdot \vec{\pi}\psi_B &= (E - mc^2 - e\phi)\psi_A \\ c\vec{\sigma} \cdot \vec{\pi}\psi_A &= (E + mc^2 - e\phi)\psi_B \end{aligned}. \tag{3.134}$$

Solving the bottom equation for ψ_B in terms of ψ_A and inserting that into the top equation leads to

$$\frac{-i\hbar e c^2(\vec{\sigma} \cdot \nabla\phi)\vec{\sigma} \cdot \vec{\pi}}{(2mc^2 + W - e\phi)^2}\psi_A + \frac{c^2(\vec{\sigma} \cdot \vec{\pi})(\vec{\sigma} \cdot \vec{\pi})}{2mc^2 + W - e\phi}\psi_A = (W - e\phi)\psi_A, \tag{3.135}$$

where the energy that appears in classical mechanics has been introduced

$$W = E - mc^2. \tag{3.136}$$

The first term in equation (3.135) is small relative to the others and will be ignored for the rest of this section. It will play a large role in the following section.

In the low-energy limit

$$W - E\phi \ll mc^2, \tag{3.137}$$

so

$$\frac{1}{2m}(\vec{\sigma} \cdot \vec{\pi})(\vec{\sigma} \cdot \vec{\pi})\psi_A = (W - e\phi)\psi_A. \tag{3.138}$$

Using the properties of the Pauli matrices described previously, the identity

$$(\vec{\sigma} \cdot \vec{\pi})(\vec{\sigma} \cdot \vec{\pi}) = \vec{\pi} \cdot \vec{\pi} + i\sigma \cdot (\vec{\pi} \times \vec{\pi}) \tag{3.139}$$

can be proved. Pursuing the argument further, it is shown that

$$\vec{\pi} \times \vec{\pi} = \left(-i\hbar\nabla - \frac{e}{c}\vec{A}\right) \times \left(-i\hbar\nabla - \frac{e}{c}\vec{A}\right) = i\frac{e\hbar}{c}\vec{B} \tag{3.140}$$

so the low-energy limit of the Dirac equation is

$$\frac{1}{2m}\left(-i\hbar\nabla - \frac{e}{c}\vec{A}\right)^2 \psi_A - \frac{e\hbar}{2mc}\vec{\sigma} \cdot \vec{B}\psi_A + e\phi\psi_A = W\psi_A. \tag{3.141}$$

The preceding equation is usually called the Pauli equation because it was written down by him before Dirac's equation was available. The difference is that Pauli's version of the spin magnetic moment was ad hoc. Dirac's equation gives a first principles derivation of it

$$\vec{\mu} = \frac{e\hbar}{2mc}\vec{\sigma} = 2 \times \frac{e}{2mc}\vec{S}. \tag{3.142}$$

The reason for writing this moment as the above is that a classical derivation for an angular momentum $\vec{L}$ would give

$$\vec{\mu}_L = \frac{e}{2mc}\vec{L}. \tag{3.143}$$

The factor of 2 in equation (3.142) is called a g-factor. It was understood from experiments on atomic spectra that it had to have this value, but it arises naturally from Dirac's equation.

Feynman was extremely proud that his quantum electrodynamics (QED) predicts the radiative corrections to the g-factor

$$\frac{g - 2}{2} = 0.0011596521564, \tag{3.144}$$

which agrees with experiment to the first eleven decimal places. He liked to point out that this was the most accurate theoretical prediction of a physical constant ever.

3.10 Scalar relativistic approximation

Starting in the 1960s, quantum mechanical calculations of the electronic structure of atoms, molecules, and solids became increasingly common. These calculations allowed the fields of atomic physics, condensed matter physics and quantum chemistry to become analytical. The rapid increase in the understanding of materials that arose from the ability to calculate properties from first principles is the basis for the technological advances in communications, computers, and consumer products that are the basis for the present economic development.

The earliest calculations were based on the Schrödinger equation, but it was soon realized that relativistic corrections are significant. With the computers that were available at the time, it was thought that solving the Dirac equation for large systems would be too difficult. Therefore, approximations were made, the most common being the scalar relativistic approximation [8].

Let us assume that there is no magnetic field. The Dirac equation is written as in equation (3.134), but with $E\phi = V$

$$
\begin{aligned}
c\vec{\sigma} \cdot \vec{p}\,\psi_B &= (W - V)\psi_A \\
c\vec{\sigma} \cdot \vec{p}\,\psi_A &= (W + 2mc^2 - V)\psi_B
\end{aligned} \tag{3.145}
$$

The spinor ψ_B from the lower equation is inserted into the upper to obtain

$$
\begin{aligned}
&- i\hbar c^2(W + 2mc^2 - V)^{-2}\vec{\sigma} \cdot \nabla V \vec{\sigma} \cdot \vec{p}\,\psi_A + (W + 2mc^2 - V)^{-1}c^2(\vec{\sigma} \cdot \vec{p})^2\psi_A \\
&= (W - V)\psi_A
\end{aligned} \tag{3.146}
$$

Manipulating this equation leads to

$$
\begin{aligned}
&\frac{\vec{p}^2\psi_A}{2m} + V\psi_A \\
&- \frac{(W - V)^2}{2mc^2}\psi_A + \frac{i\hbar}{2m(W + 2mc^2 - V)}[\nabla V \cdot \vec{p} + i\vec{\sigma} \cdot (\nabla V \times \vec{p})]\psi_A \\
&= W\psi_A
\end{aligned} \tag{3.147}
$$

which looks like the Schrödinger equation but with a more complicated potential function. The first addition to the potential,

$$
-\frac{(W - V)^2}{2mc^2}, \tag{3.148}
$$

is called the mass–velocity correction. The second,

$$
\frac{i\hbar\nabla V \cdot \vec{p}}{2m(W + 2mc^2 - V)}, \tag{3.149}
$$

is called the Darwin term. The third,

$$
\frac{-\hbar\vec{\sigma} \cdot (\nabla V \times \vec{p})}{2m(W + 2mc^2 - V)}, \tag{3.150}
$$

is the spin–orbit correction. In scalar relativistic calculations, only the mass–velocity and Darwin terms are used. If it is desirable to see the effect of the spin–orbit interactions, they can be calculated later using perturbation theory.

The scalar relativistic approximation reproduces relativistic effects reasonably well, and is still used in calculations where great accuracy is not required. However, the gold standard for calculations of atoms, molecules, and solids is to use the full Dirac theory.

3.11 The Dirac theory of the hydrogen atom

The measure of success for a quantum theory in 1928 was how well it treated the hydrogen atom. Schrödinger abandoned his first effort at a quantum equation, which was relativistic, because it gave a very bad result for hydrogen. His non-relativistic equation successfully explained the main features of the hydrogen spectrum, The Dirac equation does the same, but it then goes on to predict fine structure that Schrödinger's equation can not explain.

Since 1928 when the Dirac equation for the hydrogen atom was first solved correctly, other approaches have been used that emphasize different aspects of the solutions. The one shown here was used by Martin and Glauber in a study of orbital electron capture [9, 10]. It requires that the Dirac equation is written in a form that, like the Klein–Gordon equation, is second order in time and space derivatives. Starting with equation (3.52) and defining a Hamiltonian operator leads to

$$c\vec{\alpha} \cdot \left(-i\hbar\nabla - \frac{e}{c}\vec{A}(\vec{r})\right)\psi + e\phi(\vec{r})\psi + \beta mc^2\psi = i\hbar\frac{\partial\psi}{\partial t} = H\psi. \tag{3.151}$$

An operator P is defined by

$$P\beta = \frac{1}{2mc^2}\left[\beta\left(i\hbar\frac{\partial}{\partial t} - H\right) + 2mc^2\right]\beta. \tag{3.152}$$

Applying this operator to equation (3.151)

$$P\beta\left(i\hbar\frac{\partial}{\partial t} - H\right) = 0 \tag{3.153}$$

leads, after a lot of manipulations, to

$$\left[\frac{1}{c^2}\left(i\hbar\frac{\partial\psi}{\partial t} - e\phi\right)^2 - \left(-i\hbar\nabla - \frac{e}{c}\vec{A}\right)^2 - m^2c^2 + \frac{e\hbar}{c}(\vec{\sigma}\cdot\vec{H} - i\vec{\alpha}\cdot\vec{E})\right]\psi = 0. \tag{3.154}$$

This is the Klein–Gordon equation (3.14) with the addition of the term $(\vec{\sigma}\cdot\vec{H} - i\alpha\cdot\vec{E})$ that couples the electron to the magnetic field $\vec{H}$ and the electric field $\vec{E}$.

A solution of this second order equation doesn't have to be a solution of the Dirac equation, although a solution of the Dirac equation is always a solution of the second order equation. The operator $\beta(i\hbar\frac{\partial}{\partial t} - H)$ commutes with the P operator. It follows that a spinor obtained by operating P on any solution of the second order equation will be such that

$$P\beta\left(i\hbar\frac{\partial}{\partial t} - H\right)\psi = \beta\left(i\hbar\frac{\partial}{\partial t} - H\right)P\psi = 0. \tag{3.155}$$

Therefore, if ψ is any solution of the second order equation, $P\psi$ is a solution to the Dirac equation. If the original ψ is already a solution of the Dirac equation, then $P\psi = \psi$.

For the hydrogen atom or a nucleus with charge Ze and one electron,

$$\vec{A} = 0 \quad \vec{H} = 0 \quad e\phi = -\frac{Ze^2}{r}. \tag{3.156}$$

For a stationary state with energy E, equation (3.154) becomes

$$\left[\frac{1}{c^2}\left(E + \frac{Ze^2}{r}\right)^2 + \hbar^2\nabla^2 - m^2c^2 + \frac{i\hbar}{c}\frac{Ze^2}{r^2}\vec{\alpha}\cdot\hat{r}\right]\psi = 0 \tag{3.157}$$

and $\hat{r}$ is the unit vector in the direction of $\vec{r}$. It was observed in equation (2.100) that

$$-\hbar^2\nabla^2 = -\hbar^2\frac{1}{r^2}\frac{\partial^2}{\partial r^2}r^2 + \frac{L^2}{r^2}. \tag{3.158}$$

Equation (3.157) then becomes

$$\left[\frac{E^2 - m^2c^4}{c^2} + \frac{2EZe^2}{c^2r} + \hbar^2\frac{1}{r^2}\frac{\partial}{\partial r}r^2\frac{\partial}{\partial r} + \frac{L^2 - (Ze^2/c)^2 - (i\hbar Ze^2/c)\vec{\alpha}\cdot\hat{r}}{r^2}\right]\psi = 0. \tag{3.159}$$

Let us now define the scalar operator

$$K = \beta\left(1 + \frac{1}{\hbar}\vec{\sigma}\cdot\vec{L}\right). \tag{3.160}$$

This operator commutes with

$$\vec{\alpha}\cdot\vec{p}\,\vec{\alpha}\cdot\hat{r}r^2\vec{J} = \vec{L} + \vec{S}. \tag{3.161}$$

From the equations in the preceding sections, it can be seen that these relations imply that K commutes with the usual Dirac equation for the hydrogen atom

$$H = c\vec{\alpha}\cdot\vec{p} + \beta mc^2 - \frac{Ze^2}{r}, \tag{3.162}$$

and is a constant of the motion for that system. Since J^2 and J_z are also constants of motion, the energy levels of the relativistic hydrogen atom can be labeled with the eigenvalues of K, J^2, and J_z.

The eigenvalues of J^2, and J_z are $\hbar^2 j(j + 1)$ and $\hbar j$ where $j = 1/2, 3/2, 5/2, ...$ The square of K is

$$K^2 = \frac{1}{\hbar^2}J^2 + \frac{1}{4}, \tag{3.163}$$

and the eigenvalues of K^2 are thus $(j + 1/2)^2$. It can be shown that K anticommutes with the fifth gamma matrix discussed in section 3.6 of this chapter

$$\{K, \gamma^5\} = 0. \tag{3.164}$$

It follows that, if there is an eigenstate of K corresponding to an eigenvalue k,

$$K \mid k\rangle = k \mid k\rangle, \tag{3.165}$$

then

$$K\gamma_5 \mid k\rangle = -\gamma_5 K \mid k\rangle = -k\gamma_5 \mid k\rangle. \tag{3.166}$$

This means that, for every positive eigenvalue and eigenvector, there is a negative eigenvalue and an eigenvector that has the opposite parity. From the preceding analysis it is concluded that the eigenvalues of K are

$$k = \pm1, \pm2, \pm3, \dots. \tag{3.167}$$

If an operator Λ can be found such that

$$\hbar^2\Lambda(\Lambda + 1) = \mathbf{L}^2 - (Ze^2/c)^2 - (i\hbar Ze^2/c)\vec{\alpha} \cdot \hat{r}, \tag{3.168}$$

then equation (3.159) can be written

$$\left\{-\frac{\hbar^2}{2m}\left[\frac{1}{r^2}\frac{d}{dr}\left(r^2\frac{d}{dr}\right) + \frac{\Lambda(\Lambda + 1)}{r^2}\right] - \frac{Z'e^2}{r}\right\}R_{nl}(r) = E'R_{nl}(r). \tag{3.169}$$

This equation looks just like the radial part of the non-relativistic Schrödinger equation for the hydrogen atom, equation (2.102), except that

$$Z' = \frac{EZ}{mc^2}, \tag{3.170}$$

and

$$E' = \frac{E^2 - m^2c^4}{2mc^2}. \tag{3.171}$$

Of course the angular momentum quantum number l has been replaced by the operator Λ. The advantage of this manipulation is that, assuming the solution of the relativistic equation is also an eigenstate of Λ, $\Lambda\psi = l'\psi$, then it follows immediately that the solutions of the equation are the generalized Laguerre polynomials and the energy eigenvalue is

$$E' = -\frac{mZ'^2e^4}{2\hbar^2n'^2}, \tag{3.172}$$

where n' is to be determined. This analogy also defines the relation between the eigenvalues of Λ, l', and n'. It turns out that the operator

$$\Lambda = -\beta K - (iZe^2/\hbar c)\vec{\alpha} \cdot \hat{r}, \tag{3.173}$$

satisfies all the requirements outlined above. It can also be shown that

$$n' = n - j - 1/2 + \sqrt{(j + 1/2)^2 - \alpha^2 Z^2}, \tag{3.174}$$

where n is the non-relativistic principle quantum number. The constant α is the well known fine structure constant

$$\alpha = \frac{e^2}{\hbar c}, \tag{3.175}$$

which has the numerical value

$$\alpha^{-1} = 137.035999084. \tag{3.176}$$

It should not be confused with the matrices in $\vec{\alpha}$.

Eliminating the primed quantities from equations (3.170), (3.171) and (3.172) leads to the expression for the energy eigenvalue of the relativistic equation

$$E = \frac{mc^2}{\left[1 + \dfrac{\alpha^2 Z^2}{n'^2}\right]^{1/2}}. \tag{3.177}$$

This is perhaps the easiest way to find the energy eigenvalues of the Dirac equation. It also has the advantage that it is easy to relate these eigenvalues to the ones for the non-relativistic Schrödinger equation. For example, Taylor's expansion of the energy in equation (3.177) in powers of $(\alpha Z)^2$ is

$$E = mc^2 - 1/2mc^2\left(\frac{\alpha Z}{n'}\right)^2 + 3/8mc^2\left(\frac{\alpha Z}{n'}\right)^4 + \dots. \tag{3.178}$$

The expansion of n' in powers of $(\alpha Z)^2$ is

$$n' = n - \frac{(\alpha Z)^2}{2j + 1} + \dots. \tag{3.179}$$

Ignoring the second term in this equation, the second term in equation (3.178) is just the non-relativistic Schrödinger energy. The terms of order $(\alpha Z)^4$ and higher are the relativistic corrections. They clearly depend on j as well as n.

One of the first successes of the Dirac equation was the calculation of the splitting of the energy levels of the hydrogen atom. The Schrödinger equation predicts that the 2s and 2p levels are degenerate. They are split by the spin–orbit effects in the Dirac equation. The splitting of the Lyman-alpha spectral lines that is the consequence of this effect was seen experimentally years before the Dirac equation was proposed.

3.12 Advantages and disadvantages

The energy levels and total energy of atoms with the atomic number $Z>1$ can be calculated with relativistic Hartree–Fock codes based on the Dirac equation. This technique will be discussed in more detail later. The results of these calculations

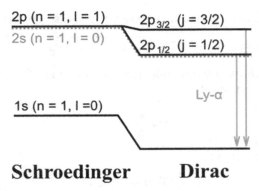

Figure 3.2. The lowest few energy levels of the hydrogen atom calculated with the Schrödinger and the Dirac equations.

shown in figure 3.2 agree very well with experiment. The difference between them and non-relativistic calculations increases with increasing Z. However, the solution of the Dirac equation for the 1S electron is oscillatory and unbounded for $Z \geqslant 137$. This can be seen from equation (3.174) because the square root becomes imaginary for such values of $j + 1/2 = 1$ and $Z \geqslant 137$. It could be argued that there is no atom on the periodic table with a Z that large, but this still indicates that Dirac equation is fundamentally limited as a theory of electrons.

Another feature of the Dirac equation that has led to great debate over the years is the occurrence of negative energy solutions with $E < -mc^2$. Dirac suggested that, in the equilibrium state, these negative energy levels are filled with two electrons each. In an excited state, an electron goes into a positive energy state, leaving a hole in the negative energy sea. This hole behaves like a particle of positive charge. The only positive charged particle known at the time was a proton, so Dirac gave the 1930 paper in which he put forward this theory 'A Theory of Electrons and Protons'. It was shown theoretically that Dirac's theory could not work unless the positively charged particle had the same mass as an electron. Dirac's prediction seemed to be vindicated in 1932 when Carl D Anderson discovered the positron in a cloud chamber. This discovery added to the enthusiasm for Dirac's theory, and he shared the 1933 Nobel prize with Schrödinger. Dirac's idea of an infinite sea of electrons with negative energies is not only illogical, but efforts to use it to make predictions of specific phenomena were not successful.

It is now generally agreed that the explanation for the difficulties in Dirac's theory is that his theory starts with ordinary mechanics in which potential functions are used to describe interactions between particles. With the development of quantum electrodynamics and field theory, this picture has been replaced with one in which the force between particles is caused by the exchange of virtual particles. For example, the Coulomb force between electrons is mediated by the exchange of photons. The forces between protons and neutrons are mediated by pi and rho mesons. From this point of view, the surprise is that Dirac's theory works so well for atoms with reasonable atomic numbers.

The pair production of a positron and an electron, or the reverse interaction in which the positron and electron are annihilated with the resulting energy being carried away by photons is also best explained by field theory. In 1973 Julian Schwinger remarked, 'The picture of an infinite sea of negative energy electrons is now best regarded as a historical curiosity, and forgotten'.

Schwinger's remark does not mean the Dirac equation should be forgotten. The field theory for spin one half particles is based on the Dirac equation. Also, as mentioned before, practical calculations of the energetics, chemical behavior, cohesion, and magnetic properties of atoms, molecules, and solids require the solution of the Dirac equation with the help of high speed supercomputers.

Problems

P3.1 Find in the literature the relative size of relativistic corrections on the total energy of atoms as a function of atomic number.

P3.2 Derive equation (3.22) from equation (3.15) putting in all the steps.

P3.3 What is the difference between the ground state of the hydrogen atom as obtained from the Schröedinger equation and the Klein–Gordon equation?

P3.4 Use the properties of the Pauli spin matrices to derive equations (3.35), (3.36), and (3.37).

P3.5 Derive an equation similar to equations (3.21) and (3.22) for the Dirac equation.

P3.6 What is the difference between a 4×1 matrix and a bispinor?

P3.7 A transformation in space and time, like the one in equation (3.58), is called a boost. Use such a transformation to find the wave function for a particle moving in the y direction.

P3.8 Extend the preceding proofs by calculating $[H, L_y], [H, L_z], [H, \Sigma_x]$, and $[H, \Sigma_y]$.

P3.9 Use Maxwell's equations to show that the classical magnetic moment is proportional to the angular momentum $\vec{m} = \frac{e}{2m}\vec{L}$.

P3.10 Derive an expression for the zitterbewegung of a moving electron.

P3.11 Use the expansion of n' to get the correct expression for the lowest relativistic corrections to E from equation (3.178).

References

[1] Strange P 1998 *Relativistic Quantum Mechanics: With Applications in Condensed Matter and Atomic Physics* (Cambridge: Cambridge University Press)

[2] Mohn P 2003 *Magnetism in the Solid State: An Introduction* (*Springer Series in Solid-State Sciences*) (Berlin: Springer)

[3] Reiher M and Wolf A 2009 *Relativistic Quantum Chemistry* (New York: Wiley)

[4] Klein O 1926 *Z Phys.* **37** 895

[5] Gordon W 1926 *Z Phys.* **40** 117

[6] Dirac P A M 1928 *Proc. R. Soc. Lond.* **A117** 610 ibid **A118** 351

[7] Pauli W 1936 *Ann. Inst. Henri Poincare* **6** 109

[8] Koelling D D and Harmon B N 2001 *J. Phys.* C **10** 3107

[9] Martin P C and Glauber R J 1958 *Phys. Rev.* **109** 1307

[10] Baym G 1969 Lectures on quantum mechanics *Lecture Notes and Supplements in Physics* (New York: The Benjamin/Cummings Publishing Company)

Chapter 4

Symmetry

4.1 The importance of symmetry in physics

People have always been fascinated by symmetry, as illustrated by the ancient Buddhist mandala shown in figure 4.1. Clearly four lines could be drawn through this object, and it would look the same if it is reflected through any one of those lines. It would also appear unchanged if the center is held fixed and it is rotated it through 90, 180, and 270 degrees.

Figure 4.1. A Buddhist mandala. Credit: Twitter: https://twitter.com/claudmang/status/1255670446034284544.

The mathematical description of symmetry is called group theory, a field that was developed in the late nineteenth and early twentieth century. The importance of symmetry and group theory in physics, particularly quantum mechanics, was recognized very early. The German mathematician Herman Weyl wrote a book entitled '*Gruppentheorie und Quantenmechanik*' in 1928. The famous Hungarian physicist Eugene Wigner was awarded the Nobel prize in 1963 'for his contributions to the theory of the atomic nucleus and the elementary particles, particularly through the discovery and application of fundamental symmetry principles'.

The atomic structure of crystals shows a fundamental symmetry, which can be studied by x-ray and neutron diffraction. The atoms in most molecules are likewise arranged symmetrically. The wave function for electrons in an atom has features that are best described using group theory. These manifestations of geometrical symmetry are not surprising, but group theory turns out to have applications in the purely theoretical universe of fundamental particle theory and the standard model. As first noted in the 1930s by Eugene Wigner, there is a natural connection between particle physics and representation theory. The properties of elementary particles are linked to the structure of Lie groups and Lie algebras. According to this connection, the different quantum states of an elementary particle give rise to an irreducible representation of the Poincaré group. Moreover, the properties of the various particles, including their spectra, can be related to representations of Lie algebras, corresponding to 'approximate symmetries' of the Universe. Lie groups are named after Norwegian mathematician Sophus Lie, who laid the foundations of the theory of continuous transformation groups.

No attempt will be made in this chapter to survey the massive literature on symmetries in physics. The aspects that are fundamental for understanding the other material in this book will be explained, however.

4.2 A simple example

The simplest example of a highly symmetrical geometrical form is the equilateral triangle shown in figure 4.2. It is highly colored to make the symmetry operations

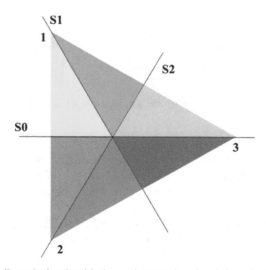

Figure 4.2. An equilateral triangle with the vertices numbered and the reflection lines indicated.

more obvious. For example, counter clockwise rotations of 120° and 240° about the center map the triangle back on itself as shown in figure 4.3.

r0   r1   r2

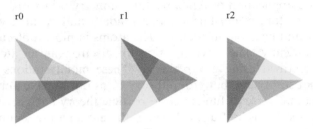

Figure 4.3. Rotations.

The rotation r0 is called the identity because it doesn't do anything. The rotation of 120° about the center is called r1, while the rotation of 240° is called r2. The product of two operations is defined. For example, a rotation of r1 followed by another r1*r1 has the same effect as the rotation r2, r1*r1 = r2. Using the same definition, it can be seen that r1*r2 = r0. It should be clear that rotation operations are commutative, ri*rj = rj*ri.

The triangle is also mapped on itself when it is reflected through one of the diagonals shown by the lines S_0, S_1, or S_2 in figure 4.2. The example of a reflection through S_0 is shown in the second panel of figure 4.4. In the third panel the result of a reflection through S_0 followed by a rotation through 120° is shown to be equivalent to a reflection through S_1.

r0 S_0 r1*S_0 = S_2

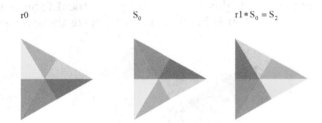

Figure 4.4. Reflection through S_0 and reflection followed by rotation r1*S_0 = S_2.

The third panel if figure 4.5 shows the result of first rotating through 120° and then reflecting through S_0, which is equivalent to reflecting through S_2. From figures 4.4 and 4.5 it can be seen that the operations do not commute, r1*S_0 ≠ S_0*r1.

The table 4.1 shows the complete resume of all of the multiplications. An element in this table is obtained by multiplying the element in a column by the one in a row.

r0 r1 $S_0 * r1 = S_1$

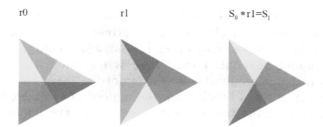

Figure 4.5. The third panel shows the result of a rotation through 120° followed by a reflection through S_0 which is equivalent to a reflection through S_1.

Table 4.1. The multiplication table for the dihedral group D_3.

	r0	r1	r2	S_0	S_1	S_2
r0	r0	r1	r2	S_0	S_1	S_2
r1	r1	r2	r0	S_2	S_0	S_1
r2	r2	r0	r1	S_1	S_2	S_0
S_0	S_0	S_1	S_2	r0	r1	r2
S_1	S_1	S_2	S_0	r2	r0	r1
S_2	S_2	S_0	S_1	r1	r2	r0

4.3 Theory of finite groups

The set of operations that map an equilateral triangle onto itself has some special features that cause it to be called a group, and more precisely a finite group. The mathematical definition of a finite group is a finite set of objects that have a single operation, usually called multiplication, such that two elements **a** and **b** can be multiplied to obtain a third element that is also in the group **a** • **b** = **c**. Using this definition, the elements and their products satisfy the following axioms:

1. **CLOSURE**: If **a** and **b** are in the group then **a** • **b** is also in the group.
2. **ASSOCIATIVITY**: If **a, b** and **c** are in the group then (**a** • **b**) • **c** = **a** • (**b** • **c**).
3. **IDENTITY**: There is an element **e** of the group such that for any element **a** of the group **a** • **e** = **e** • **a** = **a**. This is called the identity element.
4. **INVERSES**: For any element **a** of the group there is an element $\mathbf{a}^{-1}$ such that
   - $\mathbf{a} \cdot \mathbf{a}^{-1} = \mathbf{e}$ and
   - $\mathbf{a}^{-1} \cdot \mathbf{a} = \mathbf{e}$.

From table 4.1 it can be seen that the six actions that represent the operations that map the equilateral triangle onto itself, with multiplication being defined as the repetition of the operations, satisfy the four axioms listed above. The axioms lead to the result that every element of the group appears in each row of the multiplication table once and only once. The same is true for all the columns in the table. It should be noticed that there is no axiom that would require multiplication to be commutative, and it is not for the present example. The inverse of the element **a** is found from the multiplication table by finding the identity operator **e** in the row corresponding to **a.** The column that **e** is in corresponds to the inverse of **a.**

The group discussed in the preceding section belongs to a class called geometric groups. Geometric group theory is an area in mathematics devoted to the study of finite groups by exploring the connections between the algebraic properties of a group and the topological and geometric properties of the object on which it acts. A dihedral group D_n is the group of symmetries of a polygon. It includes rotations and reflections. The group of the equilateral triangle is the dihedral group D_3.

The order of a group g is the number of elements in the group. The order of D_3 is 6.

The mathematical objects that are used to define a group are not unique. For example, the permutations of the three numbers 1, 2, and 3 also constitute the group D_3. The rotations are equivalent to cyclic permutations, r0 = 123, r1 = 231, and r2 = 312. These are called even permutations because they can be achieved by an even number of interchanges of pairs of numbers, r1 → 123 → 213 → 231. The reflections through a line are equivalent to an odd permutation S_0 = 213, S_1 = 132, and S_2 = 321. These can be achieved by an odd number of interchanges S_0 → 123 → 213. The product of two elements is the operation of carrying out the permutations successively. For example

$$r2 \times S_1 = (312) \times (132) = (321) = S_2. \tag{4.1}$$

These are called permutation groups, and were the focus of the first text on group theory published in 1897 by William Burnside [1].

It is also possible to find a set of matrices that satisfy the axioms for a group listed above. Such a set of matrices is called a matrix group. Matrix groups will play a major roll in the next section.

A group $G_A = \{A_1, A_2, ...A_{g_A}\}$ is said to be isomorphic to a group $G_B = \{B_1, B_2, ...B_{g_B}\}$ if there is a one-to-one correspondence between the elements of the groups $A_i \leftrightarrow B_i$ that preserves the multiplication rules $A_i A_j \leftrightarrow B_i B_j$. This implies that the orders must be the same, $g_A = g_B$, and G_A has the same multiplication table as G_B. If these conditions hold, there is said to be a isomorphism between G_A and G_B.

A group $G_A = \{A_1, A_2, ...A_{g_A}\}$ is said to be homomorphic to a group $G_B = \{B_1, B_2, ...B_{g_B}\}$ if there is a many-to-one correspondence between the elements of the groups $A_{i_1}, A_{i_2}, ... \leftrightarrow B_i$ that preserves the multiplication rules $A_{i_\alpha} A_{j_\beta} \leftrightarrow B_i B_j$. The term many to one includes one to one, so a homomorphism can be an isomorphism. If $g_A \geqslant g_B$, G_B is said to be the homomorph of G_A.

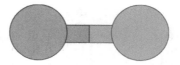

Figure 4.6. An illustration of the dihedral group D_1.

The permutation group described above is isomorphic to D_3. The two element permutation group $G_2 = \{1, -1\}$ is a homomorph to D_3 with the correspondences $r0$, $r1$, $r2 \leftrightarrow 1$ and S_0, S_1, $S_2 \leftrightarrow -1$.

The dihedral group, D_1, is illustrated in figure 4.6. It has only two operations, the identity operation, E, and a reflection through a vertical line, R. This simple group has a remarkable number of uses in physics. The multiplication table for D_1 is shown in table 4.2 the permutation group G_2 introduced above is isomorphic to D_1.

Table 4.2. The multiplication table for D_1.

	E	R
E	E	R
R	R	E

4.4 Representations of finite groups

A representation Γ of a group G is a matrix group homomorph together with a specification of the homomorphism. A faithful representation is one that is isomorphic to G.

For example, considering integers to be a one-by-one matrix, the group $\{1, -1\}$ is a representation of D_3.

One faithful representation of a group is called the regular representation. It is defined as follows. Let $G_A = \{A_1, A_2, ...A_{g_A}\}$ and A_i an element of G_A. The set of elements $\{A_iA_1, A_iA_2, ...A_iA_{g_A}\}$ are simply a rearrangement of the set $\{A_1, A_2, ...A_{g_A}\}$. If both of these sets are written as column vectors, the matrix $D(A_i)$ that will rearrange the first set to obtain the second can be used to represent A_i in the regular representation.

For example, the matrix $D(r2)$ that corresponds to the element $r2$ in D_3 is

$$\begin{pmatrix} r0 \\ r1 \\ r2 \\ s0 \\ s1 \\ s2 \end{pmatrix} = \begin{pmatrix} 0 & 1 & 0 & 0 & 0 & 0 \\ 0 & 0 & 1 & 0 & 0 & 0 \\ 1 & 0 & 0 & 0 & 0 & 0 \\ 0 & 0 & 0 & 0 & 0 & 1 \\ 0 & 0 & 0 & 1 & 0 & 0 \\ 0 & 0 & 0 & 0 & 1 & 0 \end{pmatrix} \begin{pmatrix} r2 \\ r0 \\ r1 \\ s1 \\ s2 \\ s0 \end{pmatrix} = D(r2) \begin{pmatrix} r2 \\ r0 \\ r1 \\ s1 \\ s2 \\ s0 \end{pmatrix}. \qquad (4.2)$$

The matrices for the rotation operations are

$$
D(r0) = \begin{pmatrix} 1 & 0 & 0 & 0 & 0 & 0 \\ 0 & 1 & 0 & 0 & 0 & 0 \\ 0 & 0 & 1 & 0 & 0 & 0 \\ 0 & 0 & 0 & 1 & 0 & 0 \\ 0 & 0 & 0 & 0 & 1 & 0 \\ 0 & 0 & 0 & 0 & 0 & 1 \end{pmatrix} \quad
D(r1) = \begin{pmatrix} 0 & 0 & 1 & 0 & 0 & 0 \\ 1 & 0 & 0 & 0 & 0 & 0 \\ 0 & 1 & 0 & 0 & 0 & 0 \\ 0 & 0 & 0 & 0 & 1 & 0 \\ 0 & 0 & 0 & 0 & 0 & 1 \\ 0 & 0 & 0 & 1 & 0 & 0 \end{pmatrix}
$$

$$
D(r2) = \begin{pmatrix} 0 & 1 & 0 & 0 & 0 & 0 \\ 0 & 0 & 1 & 0 & 0 & 0 \\ 1 & 0 & 0 & 0 & 0 & 0 \\ 0 & 0 & 0 & 0 & 0 & 1 \\ 0 & 0 & 0 & 1 & 0 & 0 \\ 0 & 0 & 0 & 0 & 1 & 0 \end{pmatrix},
$$

(4.3)

and the matrices for the reflections are

$$
D(S_0) = \begin{pmatrix} 0 & 0 & 0 & 1 & 0 & 0 \\ 0 & 0 & 0 & 0 & 1 & 0 \\ 0 & 0 & 0 & 0 & 0 & 1 \\ 1 & 0 & 0 & 0 & 0 & 0 \\ 0 & 1 & 0 & 0 & 0 & 0 \\ 0 & 0 & 1 & 0 & 0 & 0 \end{pmatrix} \quad
D(S_1) = \begin{pmatrix} 0 & 0 & 0 & 0 & 1 & 0 \\ 0 & 0 & 0 & 0 & 0 & 1 \\ 0 & 0 & 0 & 1 & 0 & 0 \\ 0 & 0 & 1 & 0 & 0 & 0 \\ 1 & 0 & 0 & 0 & 0 & 0 \\ 0 & 1 & 0 & 0 & 0 & 0 \end{pmatrix}
$$

$$
D(S_2) = \begin{pmatrix} 0 & 0 & 0 & 0 & 1 & 0 \\ 0 & 0 & 0 & 0 & 0 & 1 \\ 0 & 0 & 0 & 1 & 0 & 0 \\ 0 & 0 & 1 & 0 & 0 & 0 \\ 1 & 0 & 0 & 0 & 0 & 0 \\ 0 & 1 & 0 & 0 & 0 & 0 \end{pmatrix}.
$$

(4.4)

By doing matrix multiplications, it can be shown that these matrices lead to the same multiplication table as the one in table 4.1.

The theory of group representations is a major area of study in mathematics, and the language used in the definitions and theorems can quickly become very formal. In the following, a description that emphasizes applicability to physics over mathematical precision will be chosen. However, it must be recognized that a set of matrices can be discussed without specifying that they are a matrix group because there are definitions and lemmas that do not depend on the matrices satisfying all the requirements of a group.

Two square matrices A and B are said to be equivalent if they have the same dimension and there exists a nonsingular matrix P such that

$$
P^{-1}AP = B.
$$

(4.5)

If A and B are equivalent, then they have the same trace $\text{Tr}(A) = \text{Tr}(B)$ and the same determinant $\text{Det}(A) = \text{Det}(B)$.

A matrix is block diagonal if it has the form

$$
\begin{pmatrix}
a_{11} & a_{12} & a_{13} & 0 & 0 & 0 & 0 & 0 & 0 & 0 \\
a_{21} & a_{22} & a_{23} & 0 & 0 & 0 & 0 & 0 & 0 & 0 \\
a_{31} & a_{32} & a_{33} & 0 & 0 & 0 & 0 & 0 & 0 & 0 \\
0 & 0 & 0 & b_{11} & b_{12} & b_{13} & b_{14} & 0 & 0 & 0 \\
0 & 0 & 0 & b_{21} & b_{22} & b_{23} & b_{24} & 0 & 0 & 0 \\
0 & 0 & 0 & b_{31} & b_{32} & b_{33} & b_{34} & 0 & 0 & 0 \\
0 & 0 & 0 & b_{41} & b_{42} & b_{43} & b_{44} & 0 & 0 & 0 \\
0 & 0 & 0 & 0 & 0 & 0 & 0 & c_{11} & 0 & 0 \\
0 & 0 & 0 & 0 & 0 & 0 & 0 & 0 & d_{11} & d_{12} \\
0 & 0 & 0 & 0 & 0 & 0 & 0 & 0 & d_{21} & d_{22}
\end{pmatrix}. \tag{4.6}
$$

A block diagonal matrix is said to be reduced.

Two sets of matrices $[C_1, C_2, C_3, \ldots C_n]$ and $[D_1, D_2, D_3, \ldots D_n]$ are equivalent if there exists a matrix P such that $P^{-1}C_iP = D_i$ for all i. A set of matrices is reducible if it is equivalent to a set of reduced matrices, all of which have the same shape. A set of matrices is irreducible if it is not reducible. These definitions are not useful for proving the reducibility of irreducibility of a given set because they rely on being clever enough to find P. Other approaches to this question will be put forward.

Schur's lemma and its first corollary are extremely important in group representation theory and the application of group theory to physics. Schur's lemma is: let $[C_1, C_2, C_3, \ldots C_n]$ be an irreducible set of d_C dimensional matrices and $[D_1, D_2, D_3, \ldots D_n]$ be an irreducible set of d_D dimensional matrices. If there exists a matrix S such that $C_iS = SD_i$ for some ordering of the second set, then either S is a matrix with every element zero or $d_C = d_D = d$ and S is a square nonsingular matrix of dimension d. The corollary is: a matrix S that commutes with an irreducible set of matrices $[C_1, C_2, C_3, \ldots C_n]$ must be a scalar matrix. That is, if $SC_i = C_iS$ for all i, then S is a number times the unit matrix.

If the set of matrices is a representation of a group, then the irreducible set is an irreducible representation of the group. There are entire books written to explain the methods for finding the irreducible representations of groups. Since many of these books tabulate the irreducible representations of the common groups, it is easier to look one up than to deduce it yourself. For example, a web search shows that the irreducible representations of D_3 are those shown in table 4.3.

Table 4.3. The irreducible representations of D_3.

	r0	r1	r2	S_0	S_1	S_2
Γ_1	1	1	1	1	1	1
Γ_2	1	1	1	-1	-1	-1
Γ_3	$\begin{pmatrix} 1 & 0 \\ 0 & 1 \end{pmatrix}$	$\begin{pmatrix} -1/2 & -\sqrt{3}/2 \\ \sqrt{3}/2 & -1/2 \end{pmatrix}$	$\begin{pmatrix} -1/2 & +\sqrt{3}/2 \\ \sqrt{3}/2 & -1/2 \end{pmatrix}$	$\begin{pmatrix} 1 & 0 \\ 0 & -1 \end{pmatrix}$	$\begin{pmatrix} -1/2 & -\sqrt{3}/2 \\ -\sqrt{3}/2 & +1/2 \end{pmatrix}$	$\begin{pmatrix} -1/2 & +\sqrt{3}/2 \\ \sqrt{3}/2 & 1/2 \end{pmatrix}$

Table 4.4. The irreducible representations of D_1.

	E	R
Γ_1	1	1
Γ_2	1	-1

The first representation Γ_1 in which every element is mapped on 1 is called the trivial representation. The representation Γ_2 is called the sign representation. The representation Γ_3 is a faithful representation of D_3. It is obtained from the matrices that are commonly used to describe rotations in two dimensions.

The regular representation is faithful but it also reducible. There is an interesting theorem that states that the regular representation can be reduced to a representation that has all of the irreducible representation of the group arranged in blocks down the diagonal. A given representation appears a number of times that is equal to its dimension. Thus, the matrix from the regular representation in equation (4.2) is equivalent to

$$P^{-1}D(r2)P = \begin{pmatrix} 1 & 0 & 0 & 0 & 0 & 0 \\ 0 & 1 & 0 & 0 & 0 & 0 \\ 0 & 0 & -\frac{1}{2} & \frac{\sqrt{3}}{2} & 0 & 0 \\ 0 & 0 & -\frac{\sqrt{3}}{2} & -\frac{1}{2} & 0 & 0 \\ 0 & 0 & 0 & 0 & -\frac{1}{2} & \frac{\sqrt{3}}{2} \\ 0 & 0 & 0 & 0 & -\frac{\sqrt{3}}{2} & -\frac{1}{2} \end{pmatrix}. \tag{4.7}$$

The group D_1 has only two representations.

4.5 Theory of infinite groups and Lie groups

Essentially all of the groups that have an infinite number of elements and that are of interest in physics fall into the category of Lie groups [2]. As with finite groups, most of the results that are needed are available in the literature, so it is only necessary to understand their meaning and to know where to look for them. It was recognized very early by Weyl that group theory, and more specifically Lie group theory, provides a useful mathematical underpinning for quantum mechanics [3]. Weyl was close with Schrödinger, but many other physicists were unenthusiastic about the intrusion of an unfamiliar form of mathematics into their theory, some even referring to it as a 'gruppenpest'. They developed their own methods for treating certain problems, although in some cases these methods are a mere change in notation from Lie's theory. It is only in relatively recent years that this prejudice has receded.

The first thing that should be understood is that there is no group that goes by the name 'Lie group'. Infinite groups are defined and studied without any reference to Lie, and they have their own names. The question to be addressed is, can a given group be treated using Lie's theory? There is another mathematical structure that must be used in connection with such a study, and that is called a Lie algebra. That name is reasonable because the concept has no independent meaning outside of Lie group theory.

4.6 Continuous groups in physics

Hermann Weyl said 'You can not apply mathematics as long as words still becloud reality'. Most modern discussions of Lie group theory follow this axiom literally, and are replete with symbolic statements that contain few if any words. The following discussion contains a high percentage of words because the target audience has not had the time or motivation to complete a full mathematics curriculum. The groups to be considered are of infinite order because they are continuous functions of a parameter, but they consist of finite dimensional $N \times N$ matrices.

The first example is the special orthogonal group in two dimensions SO(2). This group is made up of all real 2×2 orthogonal matrices that have a determinant equal to one

$$R(\theta) \in \text{SO}(2), \ \tilde{R}(\theta)R(\theta) = I, \ \det R(\theta) = 1. \tag{4.8}$$

The group operation is matrix multiplication $R(\theta' + \theta) = R(\theta')R(\theta)$. The unit element in the group is $R(0) = I$, and the inverse is $R(\theta)^{-1} = R(-\theta)$. This group can be identified with the rotations of a two-dimensional rigid object. If the coordinates of a point on the object are written as a column vector

$$v = \begin{pmatrix} x_1 \\ x_2 \end{pmatrix}, \tag{4.9}$$

it is learned in elementary physics that counter clockwise rotation of the object through the angle θ changes them to $v' = R(\theta)v$ where

$$R(\theta) = \begin{pmatrix} \cos\theta & -\sin\theta \\ \sin\theta & \cos\theta \end{pmatrix}. \tag{4.10}$$

The orthogonality of $R(\theta)$ assures that angles and distances are preserved

$$\tilde{v}'v' = x_1'^2 + x_2'^2 = x_1^2 + x_2^2 = \tilde{v}v. \tag{4.11}$$

This can be written in abstract vector form by defining the two-dimensional Euclidean vector $|x_1, x_2\rangle$. When the abstract operator also called $R(\theta)$ is applied

$$R(\theta)|x_1, x_2\rangle = |\cos\theta x_1 - \sin\theta x_2, \ \sin\theta x_1 + \cos\theta x_2\rangle. \tag{4.12}$$

The three-dimensional rotation group SO(3) is pictured as the change in the coordinates of a point on a three-dimensional rigid body that is rotated. The rotation can be described as a two-dimensional rotation in a plane perpendicular to a unit vector $\hat{n}$. It is also easy to describe the rotations around the unit vectors of a Cartesian coordinate system $\hat{e}_1$, $\hat{e}_2$, and $\hat{e}_3$

$$R_x(\theta, \hat{e}_1) = \begin{pmatrix} 1 & 0 & 0 \\ 0 & \cos\theta & -\sin\theta \\ 0 & \sin\theta & \cos\theta \end{pmatrix}, R_y(\theta, \hat{e}_2) = \begin{pmatrix} \cos\theta & 0 & \sin\theta \\ 0 & 1 & 0 \\ -\sin\theta & 0 & \cos\theta \end{pmatrix},$$
$$R_z(\theta, \hat{e}_3) = \begin{pmatrix} \cos\theta & -\sin\theta & 0 \\ \sin\theta & \cos\theta & 0 \\ 0 & 0 & 1 \end{pmatrix}. \tag{4.13}$$

It is occasionally convenient to switch from the subscripts 1, 2, and 3 to x, y, and z. This will become more obvious as the discussion evolves. The three-dimensional rotation is conventionally described by the Euler angles

$$R(\phi, \theta, \psi) = R''_z(\psi, \hat{e}''_3)R'_x(\theta, \hat{e}''_1)R_z(\phi, \hat{e}_3). \tag{4.14}$$

The object is first rotated counter clockwise through the angle ϕ about the z-axis. It is then rotated through ϑ about the new x-axis. Finally, it is rotated through ψ about the newer z-axis. The resulting matrix $R(\phi, \theta, \psi)$ is complicated, but it is factored into three simpler matrices. The elements are orthogonal, their inverse is

$$R(\phi, \theta, \psi)^{-1} = R(-\phi, -\theta, -\psi), \tag{4.15}$$

and their determinant is one. It is possible to write these equations in terms of abstract operators operating on three-dimensional Euclidean vectors $|x_1, x_2, x_3\rangle$ in a manner analogous to equation (4.12), but it is rather messy.

Another continuous group that operates on the Euclidean vector is the group of translations

$$T(x'_1, x'_2, x'_3)|x_1, x_2, x_3\rangle = |x_1 + x'_1, x_2 + x'_2, x_3 + x'_3\rangle. \tag{4.16}$$

Because of the orthogonality, the operations can be broken up into three subgroups

$$T(x'_1, x'_2, x'_3) = T_x(x'_1)T_y(x'_2)T_z(x'_3). \tag{4.17}$$

This group is Abelian in that the operations can be applied in any order

$$T(x'_1, x'_2, x'_3)T(x''_1, x''_2, x''_3) = T(x''_1, x''_2, x''_3)T(x'_1, x'_2, x'_3). \tag{4.18}$$

The elements of the Euclidean group are combinations of rotations and translations.

Another group of great interest to physicists is the two-dimensional special unitary group SU(2). To picture this group it is necessary to start with a vector with complex components

$$v = \begin{pmatrix} z_1 \\ z_2 \end{pmatrix}. \tag{4.19}$$

The length of this vector is

$$|v|^2 = z_1^* z_1 + z_2^* z_2. \tag{4.20}$$

The vector is transformed by a 2×2 matrix that is a continuous function of a real parameter t

$$v' = g(t)v. \tag{4.21}$$

In order to preserve distances and angles in a generalized sense and also have only proper transformations, $U(t)$ is unitary with determinant equal one

$$g(t) \in \text{SU}(2), \quad g^{\dagger}(t)g(t) = I, \quad \det g(t) = 1. \tag{4.22}$$

The inverse operator is $g(t)^{-1} = g(-t)$ and the unit operator is $g(0) = I$. The elements of SU(2) are operators in a two-dimensional Hilbert space, which was discussed in chapter 1.

Consider a group G with elements $g(t_1, t_2, t_3, \ldots t_m)$ that depend on m parameters and a vector space $\mathfrak{g}$ with m linearly independent elements X_i. If every element of G can be obtained by exponentiation,

$$g(t_1, t_2, t_3, \ldots t_m) = e^{t_1 X_1 + t_2 X_2 + t_3 X_3 + \ldots t_m X_m}, \tag{4.23}$$

with t_i real numbers, then G is a Lie group and $\mathfrak{g}$ is a Lie algebra. This deceptively simple statement has many consequences.

From the rules of exponentiation, $g(t_1, t_2, t_3, \ldots t_m) = g_1(t_1)g_2(t_2)g_3(t_3)\ldots g_m(t_m)$. The expression obtained by differentiating equation (4.23)

$$X_i = \left(\frac{dg_i(t_i)}{dt_i}\right)_{t_i=0}, \tag{4.24}$$

turns out to be a very useful way to find X_i. It is obvious that

$$g_i g_j g_i^{-1} = g_i e^{tX_j} g_i^{-1} \in G. \tag{4.25}$$

Expanding the exponent in the previous equation in a Taylor's series and using the fact that $g_i X_j^n g_i^{-1} = g_i X_j g_i^{-1} g_i X_j g_i^{-1} g_i X_j g_i^{-1}\ldots$

$$g_i e^{tX_j} g_i^{-1} = e^{t g_i X_j g_i^{-1}}. \tag{4.26}$$

Since the left-side of this equation is an element of G, it follows that

$$g_i X_j g_i^{-1} \in \mathfrak{g} \tag{4.27}$$

Differentiating the expression in equation (4.25) in the form

$$\frac{d}{dt}(e^{tX_i} X_j e^{-tX_i}) = (X_i X_j - X_j X_i)(e^{tX_i} X_j e^{-tX_i}) \in \mathfrak{g}, \tag{4.28}$$

proves that the Lie bracket

$$[X_j, X_k] = X_j X_k - X_k X_j \tag{4.29}$$

is an element of $\mathfrak{g}$ and hence some combination of the X_j

$$[X_j, X_k] = \sum_{l=1}^{n} \epsilon_{jkl} X_l. \tag{4.30}$$

The constants ϵ_{jkl} are called structure constants. It should be noted that the Lie bracket $[X_j, X_k]$ was introduced fifty years before the same structure was used by Heisenberg under the name 'commutator' in his matrix mechanics as described in chapter 2.

The number of independent parameters that are required to specify the elements in a group, m, is an important thing to know. Obviously the number of parameters required for a group of $n \times n$ real matrices is n^2 and for complex matrices it is $2n^2$. If the real matrices have the condition that they are orthogonal the number of parameters becomes $n(n-1)/2$. If the complex matrices are unitary and have determinant 1, the number of parameters is $n^2 - 1$.

The general rules outlined in the previous paragraph show that the elements in the group SO(2) depend on only one independent real parameter. For all orthogonal groups,

$$\tilde{g}g = e^{t\tilde{X}}e^{tX} = e^{t(\tilde{X}+X)} = I, \tag{4.31}$$

so

$$\tilde{X} = -X, \tag{4.32}$$

the matrices in the Lie algebra are skew-symmetric. The elements in SO(2) have determinant one. From the general rule

$$\det(e^A) = e^{\text{Trace}A}, \tag{4.33}$$

the matrices in the Lie algebra have zero trace

$$\text{Trace}X = 0. \tag{4.34}$$

The only real two-dimensional matrix that satisfies these requirements is

$$X = \begin{pmatrix} 0 & -1 \\ 1 & 0 \end{pmatrix}, \tag{4.35}$$

or some constant times it. It is easy to calculate the powers of this matrix, $X^2 = -I$, $X^3 = -X$, $X^4 = X^0 = I$. Considering the Taylor's series expansions it can be shown that

$$R(\theta) = e^{\theta X} = \cos\theta I + X \sin\theta. \tag{4.36}$$

Inserting X from equation (4.35) into this equation gives the $R(\theta)$ in equation (4.10).

The Lie group theory for the three-dimensional rotation group SO(3) is a straightforward extension of that for SO(2). Using the general rule, a group of 3×3 real orthogonal matrices depends on 3 real parameters. They are conventionally chosen to be the Euler angles. The elements of SO(3), $R(\phi, \theta, \psi)$, are described in equation (4.14) as the product of three two-dimensional rotations. Those rotations are orthogonal with determinant one, so the corresponding elements of the Lie algebra are

$$X_1(\hat{e}_1) = \begin{pmatrix} 0 & 0 & 0 \\ 0 & 0 & -1 \\ 0 & 1 & 0 \end{pmatrix}, \; X_2(\hat{e}_2) = \begin{pmatrix} 0 & 0 & 1 \\ 0 & 0 & 0 \\ -1 & 0 & 0 \end{pmatrix}, \; X_3(\hat{e}_3) = \begin{pmatrix} 0 & -1 & 0 \\ 1 & 0 & 0 \\ 0 & 0 & 0 \end{pmatrix}. \tag{4.37}$$

The elements $R(\phi, \theta, \psi) \in SO(3)$ are thus

$$R(\phi, \theta, \psi) = \exp\left[\psi X''_3(\hat{e}''_3) + \theta X'_1(\hat{e}'_1) + \phi X_3(\hat{e}_3)\right]. \tag{4.38}$$

These exponentials can be written in terms of sines and cosines as in equation (4.36) with one caveat. The unit matrix in that equation is replaced by a matrix with a zero on the diagonal. For example

$$R_x(\theta, \hat{e}_1) = e^{\theta X_1(\hat{e}_1)} = (I - I_1) + \cos \theta I_1 + X_1(\hat{e}_1) \sin \theta, \tag{4.39}$$

with

$$I_1 = \begin{pmatrix} 0 & 0 & 0 \\ 0 & 1 & 0 \\ 0 & 0 & 1 \end{pmatrix}. \tag{4.40}$$

When $X_2(\hat{e}_2)$ is used the zero is in the 2,2 position, and its in the 3,3 position when $X_3(\hat{e}_3)$ is used. The term $(I - I_a)$ appears in the equation because there is an $(I - I_a)e^0$ in the exponentiation.

A feature that arises with SO(3) is that there are three elements in the Lie algebra. By simply multiplying them together, the Lie brackets are calculated

$$\begin{aligned} [X_1(\hat{e}_1), X_2(\hat{e}_2)] &= X_3(\hat{e}_3) \\ [X_2(\hat{e}_2), X_3(\hat{e}_3)] &= X_1(\hat{e}_1). \\ [X_3(\hat{e}_3), X_1(\hat{e}_1)] &= X_2(\hat{e}_2) \end{aligned} \tag{4.41}$$

It follows that, for SO(3), the structure constants ε_{jkl} is one if jkl is an even permutation of 123, and minus one if it is an odd permutation.

The use of Euler angles for describing the rotation of an object has a historical basis, but there are many other ways that fit into the above analysis. The only requirements are that they use the Lie algebra and they contain three parameters. For example,

$$R(\phi, \theta, \psi) = \exp\left[\psi X_1(\hat{e}_1) + \theta X_2(\hat{e}_2) + \phi X_3(\hat{e}_3)\right] \tag{4.42}$$

would work just as well. Two angles can be used to define the parameters

$$n_1 = \sin \alpha \cos \beta n_2 = \sin \alpha \sin \beta n_3 = \cos \alpha, \tag{4.43}$$

which are the components of a unit vector pointing in the direction defined by the angles α and β. The elements of the Lie algebra can be combined into a kind of vector

$$\vec{X} = \hat{e}_1 X_1 + \hat{e}_2 X_2 + \hat{e}_3 X_3, \tag{4.44}$$

and the rotation operator becomes

$$R(\phi, \alpha, \beta) = R(\phi, \hat{n}) = \exp[\hat{n} \cdot \vec{X}\phi]. \tag{4.45}$$

The interpretation of this formula is that two angles are used to define the direction of $\hat{n}$ and the third angle gives the magnitude of the rotation around that unit vector. It should be pointed out that the elements of the Lie algebra in equation (4.37) are not unique. It may be possible to find other three-dimensional real orthogonal matrices that satisfy the Lie brackets.

The special unitary group SU(2) is related to spin in physics, and is made up of unitary complex 2×2 matrices with determinant one. From the general rules, the elements of SU(2) depend on three independent real parameters. Therefore, it is written

$$g(t_1, t_2, t_3) \in \mathrm{SU}(2) g(t_1, t_2, t_3) = e^{(t_1 X_1 + t_2 X_2 + t_3 X_3)}. \tag{4.46}$$

Because the elements are unitary

$$g^{\dagger}g = e^{(t_1 X_1 + t_2 X_2 + t_3 X_3)^{\dagger}} e^{(t_1 X_1 + t_2 X_2 + t_3 X_3)} = e^{\left(t_1\left(X_1^{\dagger}+X_1\right)+t_2\left(X_2^{\dagger}+X_2\right)+t_3\left(X_3^{\dagger}+X_3\right)\right)} = I, \tag{4.47}$$

and the corresponding elements of the Lie algebra $X_i \in \mathbf{su}(2)$ are skew Hermitean

$$X_i^{\dagger} = -X_i. \tag{4.48}$$

Since $\det g = 1$ it follows that $\mathrm{Trace} X_i = 0$. If the skew Hermitean matrix is written as $X_i = iY_i$, then Y_i is Hermitean. Three linearly independent two-dimensional Hermitean matrices with zero trace, the Pauli matrices, were introduced in chapter 1

$$\sigma_1 = \begin{pmatrix} 0 & 1 \\ 1 & 0 \end{pmatrix}, \quad \sigma_2 = \begin{pmatrix} 0 & -i \\ i & 0 \end{pmatrix}, \quad \sigma_3 = \begin{pmatrix} 1 & 0 \\ 0 & -1 \end{pmatrix}. \tag{4.49}$$

By matrix multiplication it can be proved that the square of each of these matrices is the two-dimensional unit matrix, and the commutation rules are

$$[\sigma_1, \sigma_2] = 2i\sigma_3, \ [\sigma_2, \sigma_3] = 2i\sigma_1, \ [\sigma_3, \sigma_1] = 2i\sigma_2. \tag{4.50}$$

Defining

$$X_i = -\frac{i}{2}\sigma_i, \tag{4.51}$$

the Lie bracket relations for SO(3) in equation (4.41) become identical with the commutation rules in equation (4.50). From the above relations, it can be shown that

$$(t_1\sigma_1 + t_2\sigma_2 + t_3\sigma_3)^2 = (t_1^2 + t_3^2 + t_3^2)I = \theta I. \tag{4.52}$$

The exponentiation in equation (4.46) then becomes

$$g(\theta, \vec{n}) = e^{-i\frac{\vec{\sigma}\cdot\hat{n}\theta}{2}} = \cos\frac{\theta}{2} - i\vec{\sigma}\cdot\hat{n}\sin\frac{\theta}{2}, \qquad (4.53)$$

where $\hat{n}$ is a unit vector. The relation between SU(2) and SO(3) is close. First, the two groups have the same Lie algebra. Second, the preceding equation relates the elements of SU(2) to rotations through the angle θ about the vector $\hat{n}$. Of course, the factor of two must be noted, these elements have to go through 720° in order to describe one complete rotation. For this reason, SU(2) is called a double cover group of SO(3).

There are other Lie groups that are of interest in physics. Although it was not described in that way, the Lorentz group of all transformations of Minkowski space-time that leaves the origin unchanged is a Lie group. The Dirac theory produces a Hamiltonian that is required to be invariant under the Lorentz group. The group is called O(1,3) because the elements are not rotations in the same sense as SO(3). Instead of being orthogonal, they satisfy

$$\Lambda \in O(1, 3) \leftrightarrow \tilde{\Lambda}g\Lambda, \qquad (4.54)$$

where g is the metric $g_{11} = 1$ and $g_{22} = g_{33} = g_{44} = -1$, all other elements being zero.

The group U(1) is the unitary group of all the points on a unit circle in complex space

$$g(\theta) \in U(1) \quad g(\theta) = e^{i\theta}. \qquad (4.55)$$

Many phenomena in physics are caused by an invariance under this group.

4.7 Conservation laws from Noether's theorem

Physicists learn about the conservation of energy, momentum, and angular momentum in their first classes. They may understand these laws as facts, like so many other facts they were inundated with, or they might have been given simple proofs based on Newton's laws. Credit for the mathematical proof that conservation laws are the consequences of Lie group symmetries of the system is usually given to Emmy Noether. [4]

The statement of classical mechanics that was in use when she was doing her work in 1915 was the Lagrange formulation. It differs from the Hamiltonian formulation in that the Lagrangian $L(q_i(t), \dot{q}_i(t), t)$ does not treat the momentum as an independent parameter. As time evolves, the points $(q_i(t), \dot{q}_i(t))$ trace out a path in a $6N$ dimensional phase space. The integral

$$S = \int_{t_1}^{t_2} L(q_i(t), \dot{q}_i(t), t)dt. \qquad (4.56)$$

is called the action. The Lagrange theory states that the path in phase space between $(q_i(t_1), \dot{q}_i(t_1))$ and $(q_i(t_2), \dot{q}_i(t_2))$ must be such that the action will take a minimum value. This is called the principle of least action. From the calculus of variations, it requires that the Lagrangian satisfies the Euler–Lagrange equations

$$\frac{\partial \boldsymbol{L}}{\partial q_i} - \frac{d}{dt}\left(\frac{\partial \boldsymbol{L}}{\partial \dot{q}_i}\right) = 0. \tag{4.57}$$

Taking the Lagrangian to be the difference between the kinetic energy and the potential energy, these equations are equivalent to Newton's laws.

Noether's first theorem states that every differentiable symmetry of the action, S, of a physical system has a corresponding conservation law. This theorem only applies to continuous and smooth symmetries over physical space, which is the definition of a Lie group. This point was made by Noether, who quoted Lie frequently in her paper.

She went on to prove that a system invariant under a unitary group U(1), with the parameter being time, will have one element in the Lie algebra, the Hamiltonian. It follows that the Hamiltonian, which is to say the energy, is conserved. The invariance under U(1) simply means that the Lagrangian does not depend on time explicitly. If the Lagrangian doesn't depend on the position coordinates, the action is invariant under translations and the momentum is conserved. Finally if the action is invariant under SO(3), the group of rotations, the angular momentum is conserved.

The quantum theory of Heisenberg, Schrödinger, and Dirac is based on the Hamiltonian formulation. Those publications started a trend, and the Lagrange theory was not thought to have any relevance to the new mechanics. This changed in 1948 when Feynman introduced his path-integral approach to quantum mechanics, which uses Lagrangians. The Lagrange formulation has also turned out to be very useful in quantum field theory, and Noether's theorem has been rederived using that approach.

4.8 Conservation laws from quantum mechanics

Max Born was very happy that he was able to prove the conservation of energy and angular momentum from Heisenberg's new laws of quantum mechanics. In his quantum mechanics book, Sakurai [5] attributes the following arguments to the famous Dreimännerarbeit of Born, Heisenberg, and Jordan [6] but it is hard to find that in the original manuscripts. Born proves the conservation of energy using the arguments of elementary classical mechanics but with the new quantum laws replacing some of the calculus. His study of angular momentum is essentially the same as in chapter 2, deriving the commutation rules for angular momentum from Heisenberg's commutation rules for position and momentum. The words Lie, Noether, and symmetry do not appear in Born's papers of 1925 and 1926, even though Noether was greatly applauded by the likes of Hilbert and Einstein before that time.

All transformation operators in quantum mechanics are unitary. Groups to be considered are isomorphic to the rotation and translation groups discussed above, but they operate in a different space. The Schrödinger approach to quantum mechanics leads to wave functions $\psi(x_1, x_2, x_3) = \langle x_1, x_2, x_3 | \psi \rangle$. For the sake of this argument it will be assumed that the Hilbert space of interest is of finite or countably infinite dimension

$$\psi(x_1, x_2, x_3) = \sum_{i=1}^{N} c_i \psi_i(x_1, x_2, x_3), \tag{4.58}$$

where the $\psi_i(x_1, x_2, x_3)$ are basis functions. A transformation g of the wave function, $g\psi(x_1, x_2, x_3)$, leads to another function in the same Hilbert space, so the only change is to the coefficients c_i

$$g\psi(x_1, x_2, x_3) = \sum_{i=1}^{N} c_i g \psi_i(x_1, x_2, x_3) = \sum_{i=1}^{N} c'_i \psi_i(x_1, x_2, x_3). \tag{4.59}$$

Assuming the standard situation in which the $\psi_i(x_1, x_2, x_3)$ form a complete orthonormal set, it can be seen that

$$c'_j = \sum_{i=1}^{N} c_i \int \psi_j^*(x_1, x_2, x_3) g \psi_i(x_1, x_2, x_3) dx_1 dx_2 dx_3 = \sum_{i=1}^{N} D_{ji}(g) c_i. \tag{4.60}$$

It follows that the matrices $D(g)$ form a $N \times N$ matrix group called D. If $g \in G$ and $D(g) \in D$, the isometry of G and D is obvious, with $g \rightarrow D(g)$. The matrix element $D_{ji}(g)$ can be written in abstract notation

$$D_{ji}(g) = \langle \psi_j | g | \psi_i \rangle. \tag{4.61}$$

Some people call D the group action.

The elements of D can be found by studying

$$g\psi(x_1, x_2, x_3) = \psi(x_1', x_2', x_3') = \langle x_1, x_2, x_3 | g \mid \psi \rangle. \tag{4.62}$$

Using the manipulations described in chapter 1, the new argument is

$$\langle x_1, x_2, x_3 | g = (g^\dagger | x_1, x_2, x_3 \rangle)^\dagger = (g^{-1} | x_1, x_2, x_3 \rangle)^\dagger = \langle x_1', x_2', x_3' |, \tag{4.63}$$

so the Euclidean components are obtained by applying the inverse of g to them. As an example, consider the rotation about the z-axis in three-dimensional space $R_z(\phi) \in SO(3)$. From the equations above,

$$R_z(\phi) | x_1, x_2, x_3 \rangle = | \cos \phi x_1 - \sin \phi x_2, \ \sin \phi x_2 + \cos \phi x_1, x_3 \rangle. \tag{4.64}$$

It follows that a rotation in Hilbert space leads to

$$R_z(\phi)\psi(x_1, x_2, x_3) = \psi(\cos \phi x_1 + \sin \phi x_2, -\sin \phi x_2 + \cos \phi x_1, x_3), \tag{4.65}$$

and

$$
\begin{aligned}
&D_{ji}(R_z(\phi)) \\
&= \int_\infty \psi_j^*(x_1, x_2, x_3) \psi_i(\cos \phi x_1 + \sin \phi x_2, -\sin \phi x_2 + \cos \phi x_1, x_3) dx_1 dx_2 dx_3
\end{aligned} \tag{4.66}
$$

The group of rotation operators in three-dimensional space $SO(3)$ is usually thought to be isomorphic to a group of 3×3 matrices, and that three is a 'natural' dimension. The matrices in the group action $D(R(\phi, \theta, \psi))$ can have any dimension from one to infinity.

The time displacement operator $U(t, t_0)$ takes a vector for a state at t_0 and converts it into one at time t. Writing the unitary infinitesimal displacement as

$$U(t_0 + dt, t_0) = (1 - iXdt),\qquad(4.67)$$

the operator X must be Hermitean. The operator $U(t, t_0)$ can be obtained by repeating the infinitesimal displacements an infinite number of times using the definition

$$e^{-x} = \lim_{N\to\infty}\left(1 - \frac{x}{N}\right)^N \qquad(4.68)$$

to obtain the general expression for the time displacement operator

$$U(t, t_0) = e^{-iX(t-t_0)}.\qquad(4.69)$$

Schrödinger's equation can be written as an equation for the time displacement operator

$$HU(t, t_0) = i\hbar\frac{\partial U(t, t_0)}{\partial t}.\qquad(4.70)$$

Comparing the last two equations the operator X is the Hamiltonian divided by $\hbar$ and

$$U(t, t_0) = e^{-i\frac{H(t-t_0)}{\hbar}}.\qquad(4.71)$$

If the Hamiltonian does not depend on time, when operating on an eigenfunction of that operator the time displacement operator becomes

$$U(t, t_0)|\psi(E)\rangle = e^{-i\frac{E(t-t_0)}{\hbar}}|\psi(E)\rangle,\qquad(4.72)$$

which says that if the energy for the state is initially E, it will remain that for all time.

The conservation of momentum argument is similar. Start with a unitary translation operator $T_x(x')$ that translates a vector describing a particle at x_1 to one for the particle being at $x_1 + x'$

$$T_x(x')|x_1, x_2, x_3\rangle = |x_1 + x', x_2, x_3\rangle.\qquad(4.73)$$

The infinitesimal translation operator can be written

$$T_x(dx) = (I - iYdx),\qquad(4.74)$$

where Y is a Hermitean operator. An infinite number of infinitesimal leads to

$$T_x(x') = e^{-iYx}.\qquad(4.75)$$

From the arguments in chapter 2, it follows that the Hermitean operator must be

$$Y = \frac{p_x}{\hbar}, \tag{4.76}$$

where p_x is the x component of the momentum operator

$$T_x(x') = e^{-i\frac{p_x}{\hbar}x'}. \tag{4.77}$$

If a particle is in an eigenstate of the momentum operator, operating the translation operator on that state gives

$$T_x(x')|P_x'\rangle = e^{-i\frac{p_x'}{\hbar}x'}|P_x'\rangle, \tag{4.78}$$

so it will remain in that state even after a translation.

As has been shown, the three-dimensional rotation group $R(\phi, \theta, \psi) \in SO(3)$ can be broken down into a series of rotations around the orthogonal axes. Infinitesimal rotations are

$$\begin{aligned}
R_x(d\phi) &= (I - iXd\phi) \\
R_y(d\theta) &= (I - iYd\theta) \ , \\
R_z(d\psi) &= (I - iZd\psi)
\end{aligned} \tag{4.79}$$

and an infinite number of infinitesimal rotations allow the rotations by a finite angle to be expressed as exponentials. Physicists replace the Hermitean operators with abstract angular momentum operators

$$X = \frac{J_x}{\hbar} \quad Y = \frac{J_y}{\hbar} \quad Z = \frac{J_z}{\hbar}. \tag{4.80}$$

to obtain

$$R_x(d\phi) = e^{-i\frac{J_x}{\hbar}\phi} \quad R_y(d\theta) = e^{-i\frac{J_y}{\hbar}\theta} \quad R_z(d\psi) = e^{-i\frac{J_z}{\hbar}\psi}. \tag{4.81}$$

As was the case before, operating a rotation operator on an eigenstate will not change the angular momentum. The commutation rules are obtained by a method that is essentially the same as the ones leading to equation (4.41) with $X_\alpha(\hat{e}_\alpha) = -i\frac{J_\alpha}{\hbar}$, which leads to

$$[J_x, J_y] = i\hbar J_z[J_y, J_z] = i\hbar J_x[J_z, J_x] = i\hbar J_y. \tag{4.82}$$

The operators in equation (4.81) can be used with the Euler angles as in equation (4.38), and they can also be used as in equation (4.45) in the form

$$R(\phi, \vec{n}) = \exp\left[-\frac{i\hat{n} \cdot \vec{J}\phi}{\hbar}\right]. \tag{4.83}$$

Although it was not emphasized in the preceding chapter, the Lorentz transformations discussed there form a group. There is a Lie group description of the Lorentz group, but it is not necessary to discuss that at the present time.

Comparing the results of this section with those of the preceding section, it is clear that the physicists reproduced to a large extent mathematics that had been around for decades. Leaving that aside, the Hamiltonian, momentum, and angular momentum are important in analyzing physics problems because they all have the attribute that they are conserved. The student of physics should know about the analyses of Lie and Noether as well as that of Born, Heisenberg, and Jordan.

4.9 Continuous group representations

In the discussion of Lie groups above it was noted that the groups SU(2) and SO(3) have the same Lie algebra. It will be seen below that these groups are closely linked when the physics approach to finding representations is us

The abstract operators J_x, J_y, and J_z that satisfy the commutation rules in equation (4.82) can be considered the components of a vector

$$\vec{J} = J_x \hat{e}_1 + J_y \hat{e}_2 + J_z \hat{e}_3, \tag{4.84}$$

where the $\hat{e}_\alpha$ are the ordinary Euclidean unit vectors. This operator is used in the general expression for a rotation operator in equation (4.83). The square of the vector is

$$J^2 = J_x^2 + J_y^2 + J_z^2. \tag{4.85}$$

Using equation (4.82) it can be shown that J^2 commutes with any of the operators J_α. Since those operators don't commute with each other, a complete set of commuting observables is J^2 and one of the J_α that is conventionally chosen to be J_z

$$[J^2, J_z] = 0, \tag{4.86}$$

so a generic eigenvector is

$$\begin{aligned} J^2 \,|\, a, b\rangle &= a \,|\, a, b\rangle \\ J_z \,|\, a, b\rangle &= b \,|\, a, b\rangle \end{aligned} \tag{4.87}$$

The eigenvalues of these operators are found by defining raising and lowering operators

$$J_\pm = J_x \pm iJ_y. \tag{4.88}$$

These operators obviously commute with J^2, and their commutation relations with J_z are

$$[J_z, J_\pm] = \pm\hbar J_\pm. \tag{4.89}$$

Operating $J_\pm$ followed or preceded with J_z on the generic eigenvector leads to

$$J_z J_\pm \,|\, a, b\rangle = (b \pm \hbar)J_\pm \,|\, a, b\rangle, \tag{4.90}$$

which means that operating the raising operator on the eigenvector $|a, b\rangle$ produces another eigenvector of J_z with the eigenvalue increased by $\hbar$

$$J_+ \mid a, b \rangle = c_b^+ \mid a, b + \hbar \rangle. \tag{4.91}$$

The lowering operator leads to

$$J_- \mid a, b \rangle = c_b^- \mid a, b - \hbar \rangle. \tag{4.92}$$

All of the eigenvectors are assumed to be normalized to unity, so the constants $c_b^\pm$ will have to be found.

The next step in the analysis starts from the algebra

$$\begin{aligned} J_+ J_- &= J^2 - J_z^2 + \hbar J_z \\ J_- J_+ &= J^2 - J_z^2 - \hbar J_z \end{aligned}, \tag{4.93}$$

and hence

$$\langle a, b \mid J^2 - J_z^2 \mid a, b \rangle = \tfrac{1}{2} \langle a, b \mid J_+ J_- + J_- J_+ \mid a, b \rangle = \tfrac{1}{2}(|c_b^-|^2 + |c_b^+|^2) \geqslant 0. \tag{4.94}$$

The conclusion that can be drawn from these equations is that there is some $b_{\max}$ and some $b_{\min}$ such that $b_{\max}^2 \leqslant a$ and $b_{\min}^2 \leqslant a$, and $b_{\max}$ and $b_{\min}$ must be separated by an integer times $\hbar$. Any effort to create a state with J_z greater than $b_{\max}$ will lead to a null vector

$$J_- J_+ \mid a, b_{\max} \rangle = a - b_{\max}(b_{\max} + \hbar) = 0, \tag{4.95}$$

and an effort to create a state with J_z will have the same result

$$J_+ J_- \mid a, b_{\min} \rangle = a - b_{\min}(b_{\min} - \hbar) = 0, \tag{4.96}$$

so

$$b_{\max}(b_{\max} + \hbar) = b_{\min}(b_{\min} - \hbar) = a \tag{4.97}$$

These conditions can be satisfied if $b_{\max} = j\hbar$ and $b_{\min} = -j\hbar$, with j an integer.

At this point it is convenient to write the eigenvector as $\mid j, m \rangle$. The eigenvalue of J^2 is

$$J^2 \mid j, m \rangle = \hbar^2 j(j + 1) \mid j, m \rangle, \tag{4.98}$$

and of J_z

$$J_z \mid j, m \rangle = m\hbar \mid j, m \rangle. \tag{4.99}$$

The quantity m starts at $-j$ and increases in integer steps to j. The number of steps is $2j$ and must be an integer n, so $j = n/2$. This proves that j must be an integer or half integer. Since $m = -j, j + 1, j + 2, \ldots, j$, it to must be an integer or half integer. If $j = 0$ then $m = 0$.

The abstract operator in equation (4.83) can be used with the vectors $\mid j, m \rangle$ to create an action

$$\langle j, m \mid R(\phi, \hat{n}) \mid j', m' \rangle = \boldsymbol{D}_{jj'mm'}(\phi, \hat{n}). \tag{4.100}$$

It is easy to show that J^2 commutes with $R(\phi, \hat{n})$, so there are no elements for which $j \neq j'$. The action of the elements of SO(3) on the vectors $|j, m\rangle$ creates a set of $2j + 1$ dimensional matrices, $\boldsymbol{D}^{(j)}(\phi, \hat{n})$, for $j = 0, 1/2, 1, 3/2, 2, \ldots$. The matrices for each j are a matrix group that is homomorphic (or double covering homomorphic) to SO(3), and hence are representations of that group. It is easy to see that the representations $\boldsymbol{D}^{(j)}(\phi, \hat{n})$ are irreducible.

Particular angular momentum operators that satisfy the commutation rules in equation (4.82) but have the additional feature that they can be related to position and momentum operators by a formula that is analogous to the classical one, $\vec{L} = \vec{r} \times \vec{p}$, were discussed in chapter 2. In the position representation of Hilbert space, the operators L^2 and L_z became first and second order differential operators. The solutions of the eigenvalue equations

$$\begin{aligned} L^2 Y_{lm}(\theta, \phi) &= l(l + 1)\hbar^2 Y_{lm}(\theta, \phi) \\ L_z Y_{lm}(\theta, \phi) &= m\hbar Y_{lm}(\theta, \phi) \end{aligned}, \tag{4.101}$$

are called spherical harmonics. The abstract vectors that correspond to the $Y_{lm}(\theta, \phi)$, $|l, m\rangle$, differ from the more general ones discussed above in that the quantum numbers l can only take on integer values $l = 0, 1, 2, \ldots$. The important thing about the spherical harmonics is that they are a complete set in the sense that any smooth function of position can be expanded in terms of them

$$f(x, y, z) = \sum_{l,m}^{\infty} f_{lm}(r) Y_{lm}(\theta, \phi). \tag{4.102}$$

It is easy to express the action of a rotation about the z-axis $R_z(\phi, \hat{e}_3)$ on the functions

$$Y_l^m(\theta, \phi) = (-1)^m \sqrt{\frac{(2l + 1)}{4\pi} \frac{(l - m)!}{(l + m)!}} P_{lm}(\cos \theta) e^{im\phi}. \tag{4.103}$$

For example, for $l = 0$, $\boldsymbol{D}^{(0)}(\phi, \hat{e}_3) = 1$. For $l = 1$,

$$\boldsymbol{D}^{(1)}(\theta, \hat{e}_3) = \begin{pmatrix} e^{-i\phi} & 0 & 0 \\ 0 & 1 & 0 \\ 0 & 0 & e^{i\phi} \end{pmatrix}. \tag{4.104}$$

It has been noted that

$$Y_{10}(\theta, \phi) \propto z \, Y_{1,\pm1}(\theta, \phi) \propto \frac{x \pm iy}{\sqrt{2}}. \tag{4.105}$$

These relations make it possible to rearrange the formula for $R_z(\theta, \hat{e}_3)$ in equation (4.13) to obtain equation (4.104). The 3×3 representations for $l = 1$ are called the defining representations.

For more general rotations it is necessary to invoke the Euler angles

$$R(\phi, \theta, \psi) = e^{-i\frac{L_z}{\hbar}\psi} e^{-i\frac{L_x}{\hbar}\theta} e^{-i\frac{L_z}{\hbar}\phi}, \tag{4.106}$$

and

$$D_{mm'}^{(l)}(\phi, \theta, \psi) = e^{-im\psi}\langle l, m \mid e^{-i\frac{L_x}{\hbar}\theta} \mid l, m'\rangle e^{-im'\phi} = e^{-i(m\psi+m'\phi)}d_{mm'}^{(l)}(\theta). \quad (4.107)$$

The development of efficient algorithms for calculating the matrices $d_{mm'}^{(l)}(\theta)$ was a prime goal in the early days of quantum mechanics. This is now considered to be a solved problem. [7] With these formulae a series of $2l + 1$ dimensional irreducible representations for SO(3) can be obtained for every integer l from zero to infinity.

The defining representation for the special unitary group SU(2) is the two-dimensional representation in equation (4.53). Using the relation between the spin operator and the Pauli spin matrices,

$$\vec{\sigma} = \frac{2}{\hbar}\vec{S}, \quad (4.108)$$

the rotation operator can be written in the form being used in this section

$$R(\theta) = e^{-i\frac{\vec{S}\cdot\hat{n}}{\hbar}\theta}. \quad (4.109)$$

There is an irreducible representation of SU(2), $D^{(j)}(\theta)$, created by the action of $R(\theta)$ on the vectors $|j, m\rangle$ for all integer and half integer values of j from zero to infinity. The two-dimensional defining representation

$$D^{(1/2)}(\theta) = \cos\frac{\theta}{2} - i\vec{\sigma} \cdot \hat{n} \sin\frac{\theta}{2}, \quad (4.110)$$

is obtained from the action of $R(\theta)$ on the two-dimensional Hilbert space spanned by the vectors $|\frac{1}{2}, \frac{1}{2}\rangle$ and $|\frac{1}{2}, -\frac{1}{2}\rangle$. In physics this is the spin space for a single Fermion.

The actions of SU(2) on spaces corresponding to $j = 0$ and $j = 1$ were actually worked out in chapter 2. The eigenvectors for two Fermions, in the present language are

$$|0, 0\rangle = \frac{1}{\sqrt{2}}(|\frac{1}{2}, \frac{1}{2}\rangle|\frac{1}{2}, -\frac{1}{2}\rangle - |\frac{1}{2}, -\frac{1}{2}\rangle|\frac{1}{2}, \frac{1}{2}\rangle), \quad (4.111)$$

and

$$\begin{aligned}
|1, 1\rangle &= |\frac{1}{2}, \frac{1}{2}\rangle|\frac{1}{2}, \frac{1}{2}\rangle \\
|1, 0\rangle &= \frac{1}{\sqrt{2}}(|\frac{1}{2}, \frac{1}{2}\rangle|\frac{1}{2}, -\frac{1}{2}\rangle + |\frac{1}{2}, -\frac{1}{2}\rangle|\frac{1}{2}, \frac{1}{2}\rangle). \\
|1, -1\rangle &= |\frac{1}{2}, -\frac{1}{2}\rangle|\frac{1}{2}, -\frac{1}{2}\rangle
\end{aligned} \quad (4.112)$$

The Pauli operator in this two particle space Σ_z can be written

$$\Sigma_z = \sigma_z^{(1)} + \sigma_z^{(2)}. \quad (4.113)$$

The eigenvalues of S_z operating on $|\frac{1}{2}, \pm\frac{1}{2}\rangle$ are $\pm\hbar/2$, so the eigenvalues of σ_z are

$$\sigma_z|{}^1\!/_2, \pm{}^1\!/_2\rangle = \pm|{}^1\!/_2, \pm{}^1\!/_2\rangle. \tag{4.114}$$

Writing the rotation operator as $R(\theta) = e^{i\Sigma_z\frac{\theta}{2}}$ leads to

$$\boldsymbol{D}^{(0)} = 1 \tag{4.115}$$

and

$$\boldsymbol{D}^{(1)} = \begin{pmatrix} e^{-i\theta/2} & 0 & 0 \\ 0 & 1 & 0 \\ 0 & 0 & e^{i\theta/2} \end{pmatrix}. \tag{4.116}$$

Adding more Fermions and diagonalizing can be used to obtain the matrices for larger values of j, but there are more efficient methods.

4.10 Groups of a Hamiltonian

Suppose a Hamiltonian H is invariant under a group G,

$$HG = GH. \tag{4.117}$$

This means that the group is made up of n operations $G = \{X_1, X_2, X_3, ...X_n\}$, and the Hamiltonian is invariant under all of the operations

$$HX_i = X_iHi = 1, 2, ...n. \tag{4.118}$$

In addition, suppose that there is a set of m vectors $\{|j_1\rangle, |j_2\rangle, |j_3\rangle, ...|j_m\rangle\}$ that span a subspace of the Hilbert space. The matrix elements

$$\boldsymbol{D}^{(m)}(X_i) = \begin{pmatrix} \langle j_1|X_i|j_1\rangle & \langle j_1|X_i|j_2\rangle & \cdots & \langle j_1|X_i|j_m\rangle \\ \langle j_2|X_i|j_1\rangle & \langle j_2|X_i|j_1\rangle & \cdots & \langle j_2|X_i|j_m\rangle \\ \cdots & \cdots & \cdots & \cdots \\ \langle j_m|X_i|j_1\rangle & \langle j_m|X_i|j_2\rangle & \cdots & \langle j_m|X_i|j_m\rangle \end{pmatrix}, \tag{4.119}$$

arising from the action of G on this subspace obviously are a representation of G. Let us use the same set of vectors to create a matrix

$$H^{(m)} = \begin{pmatrix} \langle j_1|H|j_1\rangle & \langle j_1|H|j_2\rangle & \cdots & \langle j_1|H|j_m\rangle \\ \langle j_2|H|j_1\rangle & \langle j_2|H_i|j_1\rangle & \cdots & \langle j_2|H|j_m\rangle \\ \cdots & \cdots & \cdots & \cdots \\ \langle j_m|H|j_1\rangle & \langle j_m|H|j_2\rangle & \cdots & \langle j_m|H|j_m\rangle \end{pmatrix}. \tag{4.120}$$

Since G is a group of H, it follows from equation (4.118) that the matrix $H^{(m)}$ commutes with all the matrices $\boldsymbol{D}^{(m)}(X_i)$

$$H^{(m)}\boldsymbol{D}^{(m)}(X_i) = \boldsymbol{D}^{(m)}(X_i)H^{(m)}. \tag{4.121}$$

This straightforward analysis leads to one of the most important applications of group theory to quantum mechanics. A lemma that follows from Schur's theorem takes the following form. *A matrix that S that commutes with an irreducible set of matrices* $[C_1, C_2, C_3, ...C_n]$ *must be a scalar matrix. That is, if $SC_i = C_iS$ for all i,* *then S is a number times the unit matrix.* From this lemma it follows that, if the representation $\boldsymbol{D}^{(m)}(X_i)$ is irreducible, then the matrix $H^{(m)}$ must be a constant times a m dimensional unit matrix

$$H^{(m)} = E_m I^{(m)}. \tag{4.122}$$

There are several conclusions that can be derived from this equation. It is clear that E_m is an eigenvalue of H, and it is m-fold degenerate. Group theory does not give a value for E_m. It does identify the part of Hilbert space in which the eigenfunction can be found. If a subspace of Hilbert space is spanned by the set of vectors $\{|j_1\rangle, |j_2\rangle, |j_3\rangle, \dots |j_m\rangle\}$ that form the basis set for an irreducible representation for the group of the Hamiltonian G, then

$$E_m = \langle j_i|H|j_i\rangle. \tag{4.123}$$

The conclusion in the preceding paragraph can be made more clear by reversing the argument. Assume that a Hamiltonian H is invariant under a group G. Then the eigenvectors of H are basis vectors for an irreducible representation of the group G.

The easiest example of this application of Schur's lemma is the one-dimensional square well potential. Take the middle of the well as the origin of the x-axis, the potential function is symmetrical, $V(-x) = V(x)$. The Hamiltonian is therefore invariant under the dihedral group D_1. This group has two one-dimensional representations. The basis set for the trivial representation is any normalizable even function $\psi_g^i(-x) = \psi_g^i(x)$. The basis set for the other representation is any normalizable odd function $\psi_u^i(-x) = -\psi_u^i(x)$. Group theory does not give the eigenvalues for this problem, but it proves that they will be brought up into two sets, $\{E_g^1, E_g^2, E_g^2, \dots\}$ with symmetric eigenfunctions and $\{E_u^1, E_u^2, E_u^2, \dots\}$ with antisymmetric eigenfunctions. It is not necessary to consider functions that are neither symmetric nor antisymmetric.

A somewhat more complicated case is a three-dimensional Hamiltonian with a potential function that is spherically symmetric, $V(\vec{r}) = V(r)$. The isotropic harmonic oscillator or the hydrogen atom have such potentials. A Hamiltonian with a spherically symmetric potential commutes with the rotation group SO(3) with elements $R(\phi, \theta, \psi)$. The basic functions for the irreducible representations of this group are the spherical harmonics from equation (4.103), and the representations $\boldsymbol{D}_{mm'}^{(l)}(\phi, \theta, \psi)$ are given in equation (4.107). Group theory predicts that the energy eigenvalues of H depend only on l and have degeneracies of $2l + 1$. This result was illustrated by the direct calculation of the eigenvalues and eigenfunctions for the hydrogen atom in chapter 2.

This last example demonstrates the fact that results that could be predicted by group theory will also be found by direct calculations. However, it is almost always easier to use group theory to reduce the labor of such calculations.

4.11 Conclusions

Most Lie groups describe symmetries of abstract quantities rather than physical objects. Following the discussion of SU(2), it is straightforward to proceed to SU(3).

From the rules given above, there are eight elements in the Lie algebra **su**(3), they are skew Hermitean, and they have trace zero. The theory of fundamental particles and their non-gravitational interactions is based on an extremely successful quantum field theory known as the Standard Model. This quantum field theory is determined by a set of quantum gauge fields and a Hamiltonian. The gauge fields are invariant under the symmetry $U(1) \times SU(2) \times SU(3)$.

Problems

P4.1 Find the multiplication table of the group that leaves a rectangle invariant, the dihedral group D_2.

P4.2 Find a group of permutations that is isomorphic to the geometrical group D_2.

P4.3 Find the irreducible representations of the group of the rectangle D_2.

P4.4 Find the X in equation (4.35) using equation (4.24).

P4.5 Find the elements of the Lie algebra SO(3) using equation (4.24).

P4.6 Prove the conservation of momentum for a free particle using Noether's theorem.

P4.7 Describe the X operator in equation (4.69), the X, Y, and Z operators in equation (4.80), and the operator Y in equation (4.76) in the language of Lie groups.

P4.8 A transition metal ion with an outer electron that has angular momentum $l = 2$ is placed as an impurity on a crystal site with cubic symmetry. How will the degeneracies of the energy levels change?

P4.9 Did the standard model predict the existence of the Higgs boson?

References

[1] Burnside W 1897 *Theory of Groups of Finite Order* (Cambridge: Cambridge University Press)
[2] Lie S 1888 *Theorie der Transformationsgruppen I* (Leipzig: B G Teubner Verlag)
[3] Weyl H 1928 *Gruppentheorie und Quantenmechanik* (Leipzig: S Hirzel)
[4] Noether E 1918 *Invariante Variationsprobleme* (Nachrichten von der Gesellschaft der Wissenschaften zu Göttingen. Mathematisch-Physikalische Klasse)
[5] Sakurai J J 1994 *Modern Quantum Mechanics* (New York: Addison Wesley Publishing Co)
[6] Born M, Heisenberg W and Jordan P 1926 Zur Quantenmechanik II *Z. Phys.* **35** 557–615
[7] Biedenharn L C and Van Dam H 1965 *Quantum Theory of Angular Momentum* (New York: Academic)

Chapter 5

Approximate methods

5.1 Rayleigh–Ritz variational method

Physicists and chemists have played a major role in the new economy that has boomed since the end of World War II. Much of their effort has been in the field of applied quantum mechanics. The understanding of properties of solids and molecules that are of interest in the development of high-tech devices like smart phones, computers, and smart televisions starts from calculations of the electronic states using the Schrödinger or Dirac equations. The exactly solvable problems that are the focus in undergraduate quantum mechanics e.g. the harmonic oscillator, the hydrogen atom, or the particle in a box are of no interest in the modern context where scientists are collaborating with engineers and materials developers. What is needed are accurate but less than exact solutions for complex systems.

The numerical methods that have been developed for studying the electronic structure of materials have many names, but they are usually based on the Rayleigh–Ritz variational method [1] first proposed by Lord Rayleigh in 1870. It will be seen that this method lends itself to calculations that can be carried out on a digital computer. The widespread availability of high speed supercomputers has made this the method of choice for finding approximate solutions of the Schrödinger equation.

The derivation of the Rayleigh–Ritz (R-R) variational method starts with the observation that the number

$$E(\psi) = \frac{\langle \psi \mid H \mid \psi \rangle}{\langle \psi | \psi \rangle} \tag{5.1}$$

is a functional of the state vector $|\psi\rangle$. It is assumed that the Hamiltonian H is Hermitean, so $E(\psi)$ is real. If the lowest eigenvalue of H, E_0, is a finite number, then H is said to be bounded from below.

Theorem: If H is bounded from below, an upper bound to its lowest eigenvalue can be found by minimizing $E(\psi)$ with respect to all of the vectors in the allowed Hilbert space.

doi:10.1088/978-0-7503-2167-9ch5

Proof: Expand $|\psi\rangle$ in the eigenvalues of H,

$$|\psi\rangle = \sum_i c_i |i\rangle. \tag{5.2}$$

Then,

$$E(\psi) = \frac{\sum_i |c_i|^2 E_i}{\sum_i |c_i|^2} = E_0 + \frac{\sum_i |c_i|^2 (E_i - E_0)}{\sum_i |c_i|^2} \geqslant E_0, \tag{5.3}$$

because the sum in the second term is obviously positive.

Let us approximate the state vector $|\psi\rangle$ by a set of trial vectors

$$|\psi\rangle \approx \sum_{i=1}^{N} c_i |f_i\rangle. \tag{5.4}$$

The functions, $f_i(\vec{r}) = \langle \vec{r} | f_i \rangle$, do not have to be a solution of any differential equation. They are chosen because they make certain integrals easy and because they can be used efficiently in numerical calculations. The theorem guarantees that

$$E(\psi) = \frac{\sum_{i,j=1}^{N} c_i^* H_{ij} c_j}{\sum_{i,j=1}^{N} c_i^* \Delta_{ij} c_j} \geqslant E_0, \tag{5.5}$$

is an upper bound to E_0. The matrix elements in this expression are

$$\begin{aligned} H_{ij} &= \langle f_i | H | f_j \rangle \\ \Delta_{ij} &= \langle f_i | f_j \rangle \end{aligned} \tag{5.6}$$

The accuracy with which E_0 can be approximated depends on the completeness of the set of trial vectors and the choice of the coefficients. The coefficients are found using the calculus of variations. They minimize the numerator while keeping the denominator fixed. Using a Lagrange multiplier λ, this leads to the equations

$$\frac{\partial \left[\sum_{i,j=1}^{N} c_i^* H_{ij} c_j - \lambda \sum_{i,j=1}^{N} c_i^* \Delta_{ij} c_j \right]}{\partial c_i^*} = \sum_{j=1}^{N} H_{ij} c_j - \lambda \sum_{j=1}^{N} \Delta_{ij} c_j = 0, \tag{5.7}$$

which is a set of N simultaneous homogeneous equations for the unknowns c_i and λ.

It is well known that there are no non-trivial solutions unless the determinant of the coefficients is zero

$$\det(\mathbf{H} - \lambda \mathbf{\Delta}) = 0. \tag{5.8}$$

This equation can be solved of up to N eigenvalues λ_m. The lowest eigenvalue λ_0 is the best approximation to the upper bound of E_0 that can be obtained with the set of trial functions that have been chosen, $\lambda_0 \geqslant E_0$. It is an observation, not a theorem, that, assuming a good set of trial functions is chosen, the other eigenvalues λ_m yield reasonable approximations for the other eigenvalues of H, $\lambda_m \approx E_m$.

It is equally important to find the eigenfunctions corresponding to a given energy eigenvalue using equation (5.4). The coefficients c_i^m corresponding to a given $\lambda_m \approx E_m$ are found by throwing away one of the simultaneous equations and solving the inhomogeneous equations

$$
\begin{pmatrix}
(H_{11} - \lambda_m \Delta_{11})c_1^m & (H_{12} - \lambda_m \Delta_{12})c_2^m & \cdots & \left(H_{1(N-1)} - \lambda_m \right. \\
 & & & \left. \Delta_{1(N-1)}\right)c_{(N-1)}^m \\
(H_{21} - \lambda_m \Delta_{21})c_1^m & (H_{22} - \lambda_m \Delta_{22})c_2^m & \cdots & \left(H_{1(N-1)} - \lambda_m \right. \\
 & & & \left. \Delta_{1(N-1)}\right)c_{(N-1)}^m \\
\cdots & \cdots & \cdots & \cdots \\
\left(H_{(N-1)1} - \lambda_m \Delta_{(N-1)1}\right)c_1^m & \left(H_{(N-1)2} - \lambda_m \right. & \cdots & \cdots \\
 & \left. \Delta_{(N-1)1}\right)c_3^m &
\end{pmatrix}
$$
$$
=
\begin{pmatrix}
(H_{1N} - \lambda_m \Delta_{1N})c_N^m \\
(H_{2N} - \lambda_m \Delta_{2N})c_N^m \\
\cdots \\
\left(H_{(N-1)N} - \lambda_m \Delta_{(N-1)N}\right)c_N^m
\end{pmatrix} \tag{5.9}
$$

It might seem strange that the derivative used in this derivation was only with respect to c_i^*. The coefficients c_i and c_i^* are not independent. Let us write $c_i = x_i + iy_i$. Then $c_i^* = x_i - iy_i$, and

$$
\begin{aligned}
\frac{\partial F}{\partial x_i} &= \frac{\partial F}{\partial c_i}\frac{\partial c_i}{\partial x_i} + \frac{\partial F}{\partial c_i^*}\frac{\partial c_i^*}{\partial x_i} = \frac{\partial F}{\partial c_i} + \frac{\partial F}{\partial c_i^*} = 0 \\
\frac{\partial F}{\partial y_i} &= \frac{\partial F}{\partial c_i}\frac{\partial c_i}{\partial y_i} + \frac{\partial F}{\partial c_i^*}\frac{\partial c_i^*}{\partial y_i} = \frac{\partial F}{\partial c_i} - \frac{\partial F}{\partial c_i^*} = 0
\end{aligned} \tag{5.10}
$$

Adding and subtracting these equations leads to

$$\frac{\partial F}{\partial c_i} = \frac{\partial F}{\partial c_i^*} = 0, \tag{5.11}$$

so either derivative can be used.

Example : One-dimensional molecule

The Schrödinger equation for the one-dimensional δ-function atom is

$$\left[-\frac{\hbar}{2m}\frac{d^2}{dx^2} - \frac{\hbar}{2m}p\delta(x) \right]\psi = E_0\psi. \tag{5.12}$$

The constant p is assumed to be positive and $E_0 \leqslant 0$, so the above equation can be rewritten,

$$\left[\frac{d^2}{dx^2} + p\delta(x) \right]\psi = \varepsilon_0\psi\varepsilon_0 = -\frac{2mE_0}{\hbar^2}. \tag{5.13}$$

The simplest solutions to this equation that are continuous and can be normalized are

$$\begin{align} \psi(x) &= Ae^{-\sqrt{\varepsilon_0}\,x} \quad x > 0 \\ \psi(x) &= Ae^{+\sqrt{\varepsilon_0}\,x} \quad x < 0 \end{align}. \tag{5.14}$$

Integrating this equation from $x = -\delta$ to δ leads to the condition

$$\left(\frac{d\psi}{dx}\right)_\delta - \left(\frac{d\psi}{dx}\right)_{-\delta} = -p\psi(0) = -A\sqrt{\varepsilon_0} - A\sqrt{\varepsilon_0} = -pA, \tag{5.15}$$

and hence

$$\sqrt{\varepsilon_0} = \frac{p}{2}\varepsilon_0 = \frac{p^2}{4}. \tag{5.16}$$

The wave function $\psi(x)$ is symmetric about $x = 0$, so it is normalized by requiring

$$\int_{-\infty}^{\infty} |\psi(x)|^2 dx = \frac{2A^2}{p} = 1, \tag{5.17}$$

which leads to

$$\psi(x) = \sqrt{p/2}\, e^{-\frac{p|x|}{2}}. \tag{5.18}$$

The Schrödinger equation for a δ-function molecule with an 'atom' at $x = \pm a$ is

$$\left[\frac{d^2}{dx^2} + p\delta(x - a) + p\delta(x + a) \right]\psi(x) = \varepsilon\psi(x). \tag{5.19}$$

The R-R variational method will be used to find the energies and wave functions of this equation using the one-atom wave functions as trial functions

$$\psi_1(x) = \sqrt{p/2}\, e^{-\frac{p|x-a|}{2}} \quad \psi_2(x) = \sqrt{p/2}\, e^{-\frac{p|x+a|}{2}}. \tag{5.20}$$

Clearly, $\Delta_{11} = \Delta_{22} = 1$. The integral $\Delta_{12} = \int_{-\infty}^{\infty} \psi_1\psi_2 dx$ is called an overlap integral because it measures the overlap of the trial functions. The calculation of these matrix elements leads to

$$\Delta_{12} = \frac{p}{2}e^{-pa}(2a + 1) = \Delta_{21}. \qquad (5.21)$$

The calculation of the Hamiltonian matrix elements starts with the observations,

$$\left[\frac{d^2}{dx^2} + p\delta(x - a) + p\delta(x + a)\right]\psi_1(x) = \varepsilon_0\psi_1(x) + p\delta(x + a)\psi_1(x)$$

$$\left[\frac{d^2}{dx^2} + p\delta(x - a) + p\delta(x + a)\right]\psi_2(x) = \varepsilon_0\psi_2(x) + p\delta(x - a)\psi_2(x) \qquad (5.22)$$

From these it can be seen that

$$\begin{aligned} H_{11} &= \varepsilon_0 + p\,|\psi_1(-a)|^2 \approx \varepsilon_0 \\ H_{22} &= \varepsilon_0 + p\,|\psi_2(a)|^2 \approx \varepsilon_0 \end{aligned}, \qquad (5.23)$$

where it is assumed that a is large enough that the wave function centered on one site is small on the other site. The off diagonal elements are

$$H_{12} = \varepsilon_0\Delta_{12} + \frac{p^2}{2}e^{-pa} = H_{21}. \qquad (5.24)$$

The R-R matrix equation is

$$\begin{pmatrix} \varepsilon_0 - \lambda & (\varepsilon_0 - \lambda)\Delta_{12} + X \\ (\varepsilon_0 - \lambda)\Delta_{12} + X & \varepsilon_0 - \lambda \end{pmatrix}\begin{pmatrix} c_1 \\ c_2 \end{pmatrix} = 0 \qquad (5.25)$$

where $X = \frac{p^2}{2}e^{-pa}$. Setting $\lambda = \varepsilon_0 + \delta\varepsilon$ and $\delta\varepsilon\Delta_{12} \approx 0$, The determinant of the coefficients is

$$\begin{vmatrix} -\delta\varepsilon & X \\ X & -\delta\varepsilon \end{vmatrix} = \delta\varepsilon^2 - X^2 = 0 \qquad (5.26)$$

with solutions $\delta\varepsilon = \pm X$. The eigenvalues $\lambda = \varepsilon_g$ and $\lambda = \varepsilon_u$ and therefore the energy eigenvalues of equation (5.19) are

$$\varepsilon_g = \frac{p^2}{4} + \frac{p^2}{2}e^{-pa}$$

$$\varepsilon_u = \frac{p^2}{4} - \frac{p^2}{2}e^{-pa}. \qquad (5.27)$$

From equation (5.25) the coefficients satisfy the equations

$$\begin{aligned} -\delta\varepsilon c_1 + Xc_2 &= 0 \\ Xc_1 - \delta\varepsilon c_2 &= 0 \end{aligned}. \qquad (5.28)$$

Inserting $\delta\varepsilon = X$ into equation (5.28), the coefficients corresponding to this energy are $c_1 = c_2 = 1/\sqrt{2}$, and hence

$$\psi_g(x) = 1/\sqrt{2}\,[\psi_1(x) + \psi_2(x)]. \tag{5.29}$$

Since this wave function has the property $\psi_g(x) = \psi_g(-x)$, it is called the symmetric (gerade) state. When $\delta\varepsilon = -X$, the coefficients are $c_2 = -c_1 = 1/\sqrt{2}$, so the wave function is

$$\psi_u(x) = 1/\sqrt{2}\,[\psi_1(x) - \psi_2(x)]. \tag{5.30}$$

This function has the property that $\psi_u(-x) = -\psi_u(x)$, and is the anti-symmetric (ungerade) state. The use of German words for even and odd is a tribute to the early history of quantum mechanics.

This problem is sufficiently simple that it would not be of great help, but the group theoretical analysis in the preceding chapter based on Schur's lemma gives part of this answer immediately. It tells us that, as soon as it is realized that the potential function in equation (5.19) is symmetric, the eigenfunctions must be symmetric and anti-symmetric.

5.2 Time-independent perturbation theory

The most widely used method for finding approximate solutions of the Schrödinger equation is the Rayleigh–Ritz variational method described above. However, the Rayleigh–Schrödinger (R-S) perturbation theory [2] can be used if the system of interest is 'close' to a simple system that can be solved exactly. The simple system is described by a Hamiltonian H_0, and it is assumed that all the eigenvalues $E_n^{(0)}$ and eigenvectors $|n^{(0)}\rangle$ are known

$$H_0|n^{(0)}\rangle = E_n^{(0)}|n^{(0)}\rangle, \quad n = 1, 2, 3, \cdots \tag{5.31}$$

It is perturbed by adding a potential V multiplied by a smallness parameter λ

$$H = H_0 + \lambda V. \tag{5.32}$$

The eigenvalues and eigenvectors of the perturbed system are found by solving the equation

$$(H_0 + \lambda V)|n\rangle = E_n|n\rangle. \tag{5.33}$$

It is assumed that the exact eigenvalues and eigenvectors are analytic in λ, and that λ can be set equal to one at the end of the derivation. It follows that the energy

$$E_n = E_n^{(0)} + \lambda E_n^{(1)} + \lambda^2 E_n^{(2)} + \cdots, \tag{5.34}$$

and the eigenket

$$|n\rangle = |n^{(0)}\rangle + \lambda|n^{(1)}\rangle + \lambda^2|n^{(2)}\rangle + \cdots, \tag{5.35}$$

can be expanded in powers of λ. Inserting these quantities into equation (5.33) and equating the coefficients of λ leads to the equation

$$H_0|n^{(1)}\rangle + V|n^{(0)}\rangle = E_n^{(0)}|n^{(1)}\rangle + E_n^{(1)}|n^{(0)}\rangle. \tag{5.36}$$

Premultiplying this equation by $\langle n^{(0)}|$ gives

$$E_n^{(1)} = \langle n^{(0)}|V|n^{(0)}\rangle. \tag{5.37}$$

The result that the first order correction of the energy requires only the unperturbed wave function is one of the most useful results of perturbation theory.

To find the first order correction to the wave function, rewrite equation (5.36)

$$\left(E_n^{(0)} - H_0\right)|n^{(1)}\rangle = V|n^{(0)}\rangle - E_n^{(1)}|n^{(0)}\rangle. \tag{5.38}$$

Inserting the resolution of the identity operator

$$\sum_k |k^{(0)}\rangle\langle k^{(0)}| = I, \tag{5.39}$$

and premultiplying with $\langle k^{(0)}|$ yields

$$\left(E_n^{(0)} - E_k^{(0)}\right)\langle k^{(0)}|n^{(1)}\rangle = \langle k^{(0)}|V|n^{(0)}\rangle \tag{5.40}$$

which obviously only makes sense for $k \neq n$. Assuming that the eigenvectors of H_0 are normalized, $\langle n^{(0)}|n^{(0)}\rangle = 1$, the eigenvectors of H can only be normalized if $\langle n^{(0)}|n^{(m)}\rangle = 0$ for m $\geqslant$ 1. From the preceding equation,

$$\sum_{k \neq n}\langle k^{(0)}|n^{(1)}\rangle = \sum_{k \neq n}\frac{\langle k^{(0)}|V|n^{(0)}\rangle}{E_n^{(0)} - E_k^{(0)}}|k^{(0)}\rangle. \tag{5.41}$$

To find the second order contribution to the energy, equate the coefficients of λ^2,

$$H_0|n^{(2)}\rangle + V|n^{(1)}\rangle = E_n^{(0)}|n^{(2)}\rangle + E_n^{(1)}|n^{(1)}\rangle + E_n^{(2)}|n^{(0)}\rangle. \tag{5.42}$$

Premultiplying this equation by $\langle n^{(0)}|$ yields

$$E_n^{(2)} = \langle n^{(0)}|V|n^{(1)}\rangle. \tag{5.43}$$

Combining this with the preceding equations and setting $\lambda = 1$ leads to the expression for the energy through second order

$$E_n = E_n^{(0)} + \langle n^{(0)}|V|n^{(0)}\rangle + \sum_{k \neq n}\frac{|\langle k^{(0)}|V|n^{(0)}\rangle|^2}{E_n^{(0)} - E_k^{(0)}}. \tag{5.44}$$

There are almost no applications of physical interest for which the R-S perturbation theory provides a practical means for calculating the desired answer. It does, however, provide some general insights into quantum systems that are useful. It has been pointed out that energy shifts to a given order can be found using wave functions that are accurate to one order lower. Another insight is that the wave function will be perturbed a lot no matter how small V is if $E_n^{(0)} \cong E_k^{(0)}$.

The preceding observation can be used to explain qualitatively the anomalous scattering of x-rays from atoms. Assume that the electronic states of copper are similar to those found from the hydrogen atom. There are 29 electrons in a neutral copper atom, and the configuration of the atom is $(1s)^2(2s)^2(2p)^6(3s)^2(3p)^6(3d)^{10}(4s)^1$. If the energy of the incoming (and scattered) x-ray $E_n^{(0)}$ is about equal to the energy of the $2s$ and $2p$ electrons $E_k^{(0)}$ the energy denominator will be very small and the wave function will be greatly perturbed. This shows up in the scattering of x-rays as a great increase in the scattering cross section. This is very useful in x-ray studies of condensed matter because it makes it possible to 'see' a specific atom.

If there is more than one eigenvector $|n_i^{(0)}\rangle$ corresponding to the eigenvalue $E_n^{(0)}$ of the simple system described by a Hamiltonian H_0

$$H_0|n_i^{(0)}\rangle = E_n^{(0)}|n_i^{(0)}\rangle, \quad n = 1, 2, 3, \cdots, \quad i = 1, 2, 3, \cdots, d_n, \quad (5.45)$$

the integer d_n is called the degeneracy of the level n. It can be shown that, if H_0 commutes with the elements of a group G and H does not, then the perturbation $\lambda V = H - H_0$ will remove the degeneracy, which is obvious from group theory.

5.3 Time-dependent perturbation theory

The Schrödinger equation in the abstract operator and vector notation is

$$H(t) \,|\, \psi(t)\rangle = i\hbar \frac{d \,|\, \psi(t)\rangle}{dt}, \quad (5.46)$$

where the possibility that the Hamiltonian depends on time has been included. A time displacement operator can be defined with the property $|\psi(t)\rangle = U(t, t_0)|\psi(t_0)\rangle$, and it satisfies the equation

$$H(t)U(t, t_0) = i\hbar \frac{dU(t, t_0)}{dt}. \quad (5.47)$$

Simple integration gives

$$U(t, t_0) = I + \left(\frac{-i}{\hbar}\right)\int_{t_0}^{t} H(t')U(t', t_0)dt' \quad (5.48)$$

since $U(t_0, t_0) = I$. This equation can be solved by iteration

$$U(t, t_0) = I + \left(\frac{-i}{\hbar}\right)\int_{t_0}^{t} H(t_1)dt_1 + \left(\frac{-i}{\hbar}\right)^2 \int_{t_0}^{t} H(t_1) \int_{t_0}^{t_1} H(t_2)dt_2dt_1$$
$$+ \left(\frac{-i}{\hbar}\right)^3 \int_{t_0}^{t} H(t_1) \int_{t_0}^{t_1} H(t_2) \int_{t_0}^{t_2} H(t_3)dt_3dt_2dt_1 + \ldots \quad (5.49)$$

It is known from integral calculus that

$$\int_{x_0}^{x} f(x_1) \int_{x_0}^{x_1} f(x_2) \int_{x_0}^{x_2} f(x_3)... \int_{x_0}^{x_{n-1}} f(x_n)dx_n...dx_3dx_2dx_1 = \frac{1}{n!}\left[\int_{x_0}^{x} f(x)dx\right]^n.$$

If the Hamiltonian operator for two different times commute with each other then the above formula can be used in the evaluation of $U(t, t_0)$

$$U(t, t_0) = e^{\left(\frac{-i}{\hbar}\right)\int_{t_0}^{t} H(t_1)dt_1}. \tag{5.50}$$

When H is independent of t,

$$U(t, t_0) = e^{\frac{-iH(t-t_0)}{\hbar}}, \tag{5.51}$$

and the time dependence of an energy eigenstate $H|\psi_E\rangle = E|\psi_E\rangle$, is simply

$$|\psi_E(t)\rangle = U(t, t_0)|\psi_E(t_0)\rangle = e^{\frac{-iE(t-t_0)}{\hbar}}|\psi_E(t_0)\rangle. \tag{5.52}$$

When the Hamiltonian depends on time, and $[H(t_i), H(t_j)] \neq 0$, the iterative equation written above is almost impossible to solve. The problem is somewhat easier if the Hamiltonian can be written

$$H = H_0 + V(t), \tag{5.53}$$

where the time-dependent perturbing potential is small compared to the time-independent H_0. In this case it is useful to use the interaction representation.

In the Schrödinger representation, the time dependence of the state vector is given by equation (5.46). Typically, the observables do not depend on time but there is a time dependence of expectation values

$$\langle A(t)\rangle = \langle \psi(t) | A | \psi(t)\rangle. \tag{5.54}$$

As described in chapter 2, in the Heisenberg representation the state vector can be taken to be time independent and

$$\frac{dA(t)}{dt} = \frac{1}{ih}[A(t), H], \tag{5.55}$$

where

$$\langle A(t)\rangle = \langle \psi(0) | A(t) | \psi(0)\rangle. \tag{5.56}$$

Schrödinger showed these formalisms contain the same physics. Using the time translation operator, $U(t) = U(t, 0)$, it can be shown that the time-dependent operator $A(t) = U^\dagger(t)AU(t)$ satisfies the Heisenberg equations.

In the interaction representation, the time translation operator is determined by the time independent part of H

$$H_0U_0(t) = i\hbar\frac{dU_0(t)}{dt}, \tag{5.57}$$

so the operator $A_I(t) = U_0^\dagger(t) A U_0(t)$ satisfies

$$\frac{dA_I(t)}{dt} = \frac{1}{ih}[A_I(t), H_0] \tag{5.58}$$

and the state function in the interaction representation $|\psi_I(t)\rangle = U_0^\dagger(t)|\psi(t)\rangle$ satisfies

$$i\hbar\frac{d|\psi_I\rangle}{dt} = V_I|\psi_I\rangle. \tag{5.59}$$

Since H_0 does not contain t, $U_0(t) = e^{-\frac{iH_0 t}{\hbar}}$, and $V_I(t) = e^{\frac{iH_0 t}{\hbar}} V(t) e^{-\frac{iH_0 t}{\hbar}}$. Note that the time dependence of $V_I(t)$ comes from the intrinsic time dependence of $V(t)$ and the time displacement operator $U_0(t)$. Using the eigenvalues and eigenvectors of H_0, ε_n and $|n\rangle$, $|\psi_I(t)\rangle = \sum_n c_n(t)|n\rangle$ where $c_n(t) = \langle n|\psi_I(t)\rangle$. It follows that

$$i\hbar\dot{c}_n = \sum V_{nm} e^{i\omega_{nm}(t)} c_m, \tag{5.60}$$

where

$$\omega_{nm} = \varepsilon_n - \varepsilon_m. \tag{5.61}$$

5.4 The two-level Hamiltonian

The following generic model explains a surprisingly large number of phenomena that are of interest in the modern high-tech economy. Consider a system in which transitions between two isolated levels is the only thing of interest. Start with a simple model Hamiltonian defined in a Hilbert space spanned by only two vectors

$$H_0 = E_1 |1\rangle\langle 1| + E_2 |2\rangle\langle 2|. \tag{5.62}$$

The time-dependent potential operator is

$$V(t) = \gamma e^{i\omega t} |1\rangle\langle 2| + \gamma e^{-i\omega t} |2\rangle\langle 1|. \tag{5.63}$$

The coefficients in $|\psi_I(t)\rangle = c_1(t)|1\rangle + c_2(t)|2\rangle$ satisfy the equations

$$\begin{aligned}
i\hbar\dot{c}_1 &= \gamma e^{i[\omega - \omega_{21}]t} c_2 \\
i\hbar\dot{c}_2 &= \gamma e^{-i[\omega - \omega_{21}]t} c_1
\end{aligned} \tag{5.64}$$

because

$$V_{I,12} = e^{\frac{iE_1 t}{\hbar}} V_{12}(t) e^{\frac{-iE_2 t}{\hbar}} = \gamma e^{i(\omega - \omega_{21})t}. \tag{5.65}$$

Eliminating c_1 leads to

$$\ddot{c}_2 + i[\omega - \omega_{21}]\dot{c}_2 + \frac{\gamma^2}{\hbar^2}c_2 = 0, \tag{5.66}$$

which is the equation for a harmonic oscillator.

As usual, this equation is solved by substituting $c_2 = e^{\alpha t}$ and solving for α to obtain

$$\alpha^2 + i[\omega - \omega_{21}]\alpha + \frac{\gamma^2}{\hbar^2} = 0. \tag{5.67}$$

Solving this quadratic equation leads to

$$c_2 = e^{-i\frac{[\omega-\omega_{21}]}{2}t}\{Ce^{i\Omega t} + De^{-i\Omega t}\}, \tag{5.68}$$

where

$$\Omega = \sqrt{\frac{[\omega - \omega_{21}]^2}{4} + \frac{\gamma^2}{\hbar^2}}. \tag{5.69}$$

As with all oscillator problems, the coefficients C and D are determined by initial conditional. If the system is initially in state $|1\rangle$, $c_1(0) = 1$ and $c_2(0) = 0$. It follows that

$$c_2(t) = \frac{\gamma}{i\hbar\Omega} \sin \Omega t. \tag{5.70}$$

Taking the absolute square of this solution leads to the formula

$$|c_2|^2 = 1 - |c_1|^2 = \left(\frac{\gamma}{\hbar}\right)^2 \frac{1}{\frac{(\omega - \omega_{21})^2}{4} + \frac{\gamma^2}{\hbar^2}} \left\{\sin\left[\frac{(\omega - \omega_{21})^2}{4} + \frac{\gamma^2}{\hbar^2}\right]^{1/2}t\right\}^2. \tag{5.71}$$

At resonance, $\omega = \omega_{21}$, and $|c_2|^2$ will reach its maximum value of 1. Put another way,

$$|c_2|^2 = 1 - |c_1|^2 = \sin^2\left(\frac{\gamma t}{\hbar}\right) \tag{5.72}$$

and the system goes from being certainly in state $|1\rangle$ to certainly in state $|2\rangle$.

5.5 Spin magnetic resonance

The classical expression for the energy of a magnet with magnetic moment $\vec{\mu}$ in a magnetic field $\vec{B}$ is

$$H = -\vec{\mu} \cdot \vec{B}, \tag{5.73}$$

where the minus sign assures that the lowest energy state is the one in which the magnet is aligned with the field. It has been mentioned before that the spin of an electron acts like a magnet with a moment given by

$$\vec{\mu} = -\frac{|e|}{mc}\vec{S}. \tag{5.74}$$

The spin of a proton acts like a magnet with a factor that is like the above except that the electron mass is replaced by the proton mass. The most common application of magnetic resonance is in the MRI scans that are used to study soft tissue in patients.

The Hamiltonian for a spin in a large uniform field and a circularly polarized oscillating field

$$\mathbf{B} = B_0\hat{\mathbf{z}} + B_1(\hat{\mathbf{x}}\cos\omega t + \hat{\mathbf{y}}\sin\omega t), \tag{5.75}$$

is

$$H = \frac{|e|}{mc}\Big[B_0 S_z + B_1(S_x \cos\omega t + S_y \sin\omega t)\Big]. \tag{5.76}$$

Recall that the spin operators may be written

$$S_z = \frac{\hbar}{2}(|+\rangle\langle+|-|-\rangle\langle-|),$$

$$S_x = \frac{\hbar}{2}(|+\rangle\langle-|+|-\rangle\langle+|), \tag{5.77}$$

$$S_y = \frac{\hbar}{2}(-i|+\rangle\langle-|+i|-\rangle\langle+|)$$

so

$$H = \frac{|e|\hbar}{2mc}[-B_0 \,|\, 1\rangle\langle 1\,| + B_0\,|\, 2\rangle\langle 2\,| + B_1 e^{i\omega t}\,|\, 1\rangle\langle 2\,| + B_1 e^{-i\omega t}\,|\, 2\rangle\langle 1\,|], \tag{5.78}$$

where the notation $|-\rangle \to |1\rangle$, $|+\rangle \to |2\rangle$ has been introduced. This is the same as equations (5.62) and (5.63) when

$$E_1 = -\frac{|e|\hbar B_0}{2mc}, \quad E_2 = \frac{|e|\hbar B_0}{2mc}, \quad \gamma = \frac{|e|\hbar B_1}{2mc}. \tag{5.79}$$

When an electromagnetic wave of frequency ω is shown on the sample, the number of spins in the up position $|2\rangle$ is given by equation (5.71) with a spin resonance frequency

$$\omega_{21} = \frac{E_2 - E_1}{\hbar} = \frac{|e|B_0}{mc}. \tag{5.80}$$

Magnetic resonance imagers modify the alignment of the spins of hydrogen atoms in the body by irradiating it with radiofrequency waves of frequency ω. When the radiofrequency field is turned off, the MRI sensors are able to detect the radiofrequency energy released as the protons realign with the magnetic field. The time it takes for the protons to realign, as well as the amount of energy released, changes depending on the environment of the hydrogen nuclei and the chemical nature of the molecules. Technicians are able to tell the difference between various types of tissues based on the emitted waves.

5.6 The maser

The chemical composition of the ammonia molecule is one nitrogen atom and three hydrogens. The hydrogen atoms form a triangular plane with the nitrogen atom above or below it. Considering the positions of the nitrogen as the variable, the molecule is symmetric under a reflection through the plane of hydrogens. The molecule is thus invariant under the dihedral group D1, and the wave function describing the position of the nitrogen is either symmetric or anti-symmetric. The absolute square of these functions predict that it is equally probable to find the nitrogen above or below the plane, but the energy of the symmetric state E_S is slightly less than that of the anti-symmetric state E_A, so the simplest possible describes the ammonia molecule as a two-level system that can make transitions between the two states

$$H = E_A \mid A\rangle\langle A \mid + E_S \mid S\rangle\langle S \mid. \tag{5.81}$$

By sending a beam of ammonia particles through a microwave cavity that has a natural frequency ω, a device called a maser is created that will generate an intense beam of microwaves. The Hamiltonian, including the microwaves in the cavity, is

$$H = E_A \mid A\rangle\langle A \mid + E_S \mid S\rangle\langle S \mid + \mu E \cos \omega t (\mid A\rangle\langle S \mid + \mid S\rangle\langle A \mid). \tag{5.82}$$

When $\omega = \frac{E_A - E_S}{\hbar}$, the intensity of the microwaves is increased because the molecules make a transition for A to S, dumping the energy in the cavity and creating an intense monochromatic and coherent microwave beam.

5.7 Fermi's golden rule

Consider two eigenstates of H_0 that will be called the initial state $|\phi_i\rangle$ and the final state $|\phi_f\rangle$. What is the probability of ending up in the final state after a time t when the system is originally in $|\phi_i\rangle$? To state this problem in more detail, assume that at $t<0$ a system is in an eigenstate $\|\phi_i\rangle$ of the Hamiltonian H_0. At $t = 0$ the system is perturbed and the Hamiltonian becomes $H = H_0 + V(t)$. Within the interaction picture the probability of finding the system in the eigenstate $|\phi_f\rangle$ of the Hamiltonian H_0 at a later time t is given by

$$P_{if}(t) = |\langle \phi_f | U_I(t) | \phi_i \rangle|^2. \tag{5.83}$$

To first order in the perturbation V

$$P_{if}(t) = \frac{1}{\hbar^2} \left| \int_0^t e^{i\omega_{fi}t'} V_{fi}(t')dt' \right|^2 \tag{5.84}$$

with $\omega_{fi} = \frac{E_f - E_i}{\hbar}$ and $V_{fi}(t) = \langle \phi_f | V(t) | \phi_i \rangle$. The most general harmonic perturbation that is assured of being Hermitian is

$$V(t) = We^{i\omega t} + W^\dagger e^{-i\omega t}, \tag{5.85}$$

and this leads to

$$P_{if}(t) = \frac{1}{\hbar^2} \left| W_{fi} \frac{1 - e^{i(\omega_{fi}+\omega)t}}{\omega_{fi} + \omega} + W_{fi}^\dagger \frac{1 - e^{i(\omega_{fi}-\omega)t}}{\omega_{fi} - \omega} \right|^2, \tag{5.86}$$

where $W_{fi} = \langle \phi_f | W | \phi_i \rangle$ is a time-independent matrix element. The contributions to $P_{if}(t)$ are significant only if the denominators are near zero, so, if $E_f = E_i - \hbar\omega$ corresponding to the emission of a quantum of energy

$$P_{if}(t) = \frac{|W_{fi}|^2}{\hbar^2} \left(\frac{\sin \dfrac{\omega_{fi} + \omega}{2} t}{\dfrac{\omega_{fi} + \omega}{2}} \right)^2. \tag{5.87}$$

The function

$$f(\beta, t) = \left(\frac{\sin \dfrac{\beta}{2} t}{\dfrac{\beta}{2}} \right)^2, \tag{5.88}$$

with $\beta = \omega_{fi} - \omega$, looks like

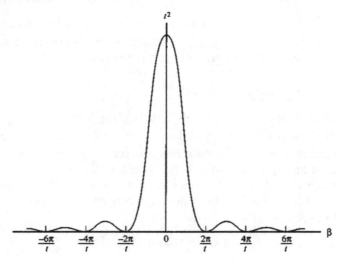

If $\beta = 0$, i.e. $\omega_{fi} = \omega$, then $f(\beta, t) = t^2$ and the probability of finding the system in the state $|\phi_f\rangle$ is

$$P_{if}(t) = \frac{|W_{fi}|^2}{\hbar^2} t^2 \tag{5.89}$$

increasing quadratically with time. The transition probability per unit time $p_{if}(t) = \frac{dP_{if}}{dt}$, increases linearly with t. This answer is incorrect because, in most physical cases, the rate of transitions $p_{if}(t)$ should be independent of time.

In the above figure the height of $f(\beta, t)$ increases proportional to t^2, and the width of the central peak is proportional to $1/t$ so the area under the curve is proportional to t. For a first order approximation to be valid, the condition $t \ll \frac{\hbar}{W_{if}}$ must be satisfied. On the other hand, to justify neglecting the emission term when absorption is being calculated, the separation between the centers of the peaks $\omega + \omega_{fi} - (\omega - \omega_{fi}) = 2\omega_{fi}$ must be larger than the width of the peaks in $P_{if}(t)$, which is $\frac{4\pi}{t}$. From this it follows that $t \gg \frac{1}{\omega_{fi}}$. Combining these two conditions leads to $\frac{1}{|\omega_{fi}|} \ll \frac{\hbar}{|W_{fi}|}$ or $E_f - E_i \gg |W_{if}|$.

If there is a continuum of states $\{|\alpha\rangle\}$, with energies nearly equal to E, and $W_{\alpha i}$ is nearly independent of α for these states then the probability of making a transition to one of these states in a small range $\Delta\alpha$ is $P(\Delta\alpha) = \int_{\Delta\alpha} d\alpha \, |\langle\alpha|\psi(t)\rangle|^2$. If $|\alpha\rangle = |E\rangle$ and $d\alpha = \rho(E)dE$, where $\rho(E)$ is the density of states, the probability of a transition into a state in the range ΔE is

$$P(\Delta E) = \int_{\Delta E} dE\rho(E) \, |\langle E|\psi(t)\rangle|^2. \tag{5.90}$$

For the case of absorption, equation (5.87) leads to

$$P_i(\Delta E) = \int_{\Delta E} dE\rho(E)P_{Ei}(t) = \int_{\Delta E} dE\rho(E)\frac{|W_{Ei}|^2}{\hbar^2}\left(\frac{\sin\frac{\omega_{Ei} - \omega}{2}t}{\frac{\omega_{Ei} - \omega}{2}}\right)^2 \tag{5.91}$$

'where the subscript Ei means that the transition starts from i and ends in a state with energy E in ΔE. The function $P_{Ei}(t)$ peaks at $\omega_{Ei} = \omega$ and has an appreciable amplitude only in a small interval $\Delta\omega_{Ei}$ or ΔE about $\omega_{Ei} = \omega$. Assuming that $\rho(E)$ and $|W_{Ei}|^2$ are nearly constant in that small interval and therefore may be taken out of the integral, it follows that

$$P_i(\Delta E) = \rho(E_f)\frac{|W_{Efi}|^2}{\hbar^2}\int_{E_f - \frac{\Delta E}{2}}^{E_f + \frac{\Delta E}{2}} dE\left(\frac{\sin\frac{\omega_{Ei} - \omega}{2}t}{\frac{\omega_{Ei} - \omega}{2}}\right)^2. \tag{5.92}$$

Some lemmas concerning delta functions are

$$\lim_{t\to\infty}\frac{\sin^2(xt)}{x^2\pi t} = \delta(x)$$
$$\delta(\alpha y) = \frac{1}{\alpha}\delta(y) \tag{5.93}$$

With the help of those lemmas, the function P_{Ei} that appears in equation (5.91) is

$$P_{Ei} = \frac{\pi}{\hbar^2}|W_{Ei}|^2\delta\left(\frac{E - E_i - \hbar\omega}{2\hbar}\right)t = \frac{2\pi}{\hbar}|W_{Ei}|^2\delta(E - E_i - \hbar\omega)t. \tag{5.94}$$

The derivative of this function leads to Fermi's golden rule for the transition probability per unit time

$$p_{fi} = \frac{dP_{fi}}{dt} = \frac{2\pi}{\hbar} |W_{fi}|^2 \delta(E_f - E_i - \hbar\omega). \tag{5.95}$$

This transition probability is time independent as it should be. The delta function guarantees energy conservation because the final energy E is the initial energy plus the quantum of energy that has been absorbed

$$E_f = E_i + \hbar\omega. \tag{5.96}$$

It is attractive to write the golden as in equation (5.95) because its features are most obvious that way, but it should be remembered that the transitions are usually into a continuum of states as in equation (5.91) so

$$P_i(\Delta E) = \int_{\Delta E} dE \rho(E) P_{Ei}(t) = \frac{2\pi}{\hbar} |W_{E_f i}|^2 \rho(E_f) t, \tag{5.97}$$

or

$$\frac{\partial P_i(\Delta E)}{\partial t} = \frac{2\pi}{\hbar} |W_{E_f i}|^2 \rho(E_f), \tag{5.98}$$

which is another useful formula.

The formulae above were derived for the case of absorption of energy into a specific initial state transitioning into one of many possible final states. It would be equally easy to treat the emission case for which the transition is from an initial state into a definite final state. The only changes are $|W_{fi}|^2 \rightarrow |W_{fi}^\dagger|^2$ and $\omega \rightarrow -\omega$. Equation (5.95) is changed very little, but equation (5.97) is changed to

$$P_f(\Delta E) = \int_{\Delta E} dE \rho(E) P_{fE_i}(t) = \frac{2\pi}{\hbar} |W_{fE_i}^\dagger|^2 \rho(E_i) t. \tag{5.99}$$

Since $W_{fi}^\dagger = W_{if}^*$ it follows that $|W_{fi}^\dagger|^2 = |W_{fi}|^2$ and

$$\frac{\text{absorption rate}}{\rho(E_f)} = \frac{\text{emission rate}}{\rho(E_i)}, \tag{5.100}$$

which is called the principle of detailed balancing.

Fermi's golden rule is used to study the absorption of a light wave by an atom in the next section.

5.8 An atom interacting with a plane electromagnetic wave

A way to describe a plane electromagnetic wave propagating in the y direction is to start from a vector potential function

$$\vec{A}(\mathbf{r}, t) = A_0 \widehat{z} e^{i(ky-\omega t)} + A_0^* \widehat{z} e^{-i(ky-\omega t)}. \tag{5.101}$$

The gauge is chosen so that the scalar potential is zero $\phi(\mathbf{r}, t) = 0$ and the divergence of $\mathbf{A}$ is zero $\nabla \cdot \mathbf{A} = 0$. From Maxwell's equations, it is known that the electric and magnetic fields are

$$\vec{E}(\mathbf{r}, t) = \frac{\partial \vec{A}}{\partial t} = i\omega A_0 \hat{z} e^{i(ky-\omega t)} - i\omega A_0^* \hat{z} e^{-i(ky-\omega t)} = E_0 \hat{z} \cos(ky - \omega t), \quad (5.102)$$

and

$$\vec{B}(\mathbf{r}, t) = \nabla \times \vec{A} = ik A_0 \hat{x} e^{i(ky-\omega t)} - ik A_0^* \hat{x} e^{-i(ky-\omega t)} = B_0 \hat{x} \cos(ky - \omega t). \quad (5.103)$$

From these equations

$$\frac{E_0}{B_0} = \frac{\omega}{k} = c, \quad (5.104)$$

where c is the velocity of light. The energy propagated by this wave is given by the Poynting vector

$$\vec{S} = \varepsilon_0 c^2 \vec{E} \times \vec{B} = \varepsilon_0 c \frac{E_0^2}{2} \hat{y}. \quad (5.105)$$

The Hamiltonian of an electron in an atom interacting with this plane wave is

$$H = \frac{1}{3m} (\vec{p} - e\vec{A})^2 + V(\vec{r}) - \frac{e}{m} \vec{S} \cdot \vec{B}, \quad (5.106)$$

where $V(\vec{r})$ is the potential binding the electron in the atom. In this expression only one independent electron is considered, and the spin–orbit interaction is neglected. This leads to

$$H = H_0 + W, \quad (5.107)$$

with H_0 the usual Hamiltonian for the electron in the field V

$$H_0 = \frac{p^2}{2m} + V(\vec{r}), \quad (5.108)$$

and the interaction Hamiltonian

$$W = W_I + W_{II} + W_{III} = -\frac{e}{m} \vec{p} \cdot \vec{A} - \frac{e}{m} \vec{S} \cdot \vec{B} + \frac{e^2}{2m} A^2. \quad (5.109)$$

To find induced transition probabilities, it is necessary to evaluate the matrix elements of $W(t)$ between unperturbed bound states. It turns out that in the optical domain, $W_{II}/W_I \ll I$ and $W_{III}/W_I \ll I$, so

$$W \approx W_I(t) = -\frac{e}{m} p_z [A_0 e^{i(ky-\omega t)} + A_0^* e^{-i(ky-\omega t)}]. \quad (5.110)$$

The matrix elements of W_I will be evaluated between two states of the electron in the atom $|\phi_i\rangle$ and $|\phi_f\rangle$. There is little probability that the electron will be far outside the range of a Bohr radius a_0 in either state. The wavelength of light $\lambda = 2\pi/k$ is much greater than a_0, so $k \approx 0$ is a good approximation and

$$W_I(t) \approx -\frac{e}{m}p_z[A_0e^{i\omega t} + A^*_0e^{-i\omega t}] = \frac{eE_0}{m\omega}p_z \sin \omega t. \qquad (5.111)$$

This is called the electric dipole because it is equivalent to the form that would be obtained starting with an oscillating electric dipole. Comparing with equation (5.85), it is seen that the time-independent operator is

$$W = \frac{eE_0}{2im\omega}p_z. \qquad (5.112)$$

The matrix element that goes into Fermi's golden rule equation (5.95) is

$$W_{fi} = \frac{qE_0}{2im\omega}\langle\phi_f|p_z|\phi_i\rangle. \qquad (5.113)$$

Using Heisenberg's commutation rules with H_0 from equation (5.108) gives

$$p_z = \frac{m}{i\hbar}[z, H_0]. \qquad (5.114)$$

Since $|\phi_i\rangle$ and $|\phi_f\rangle$ are eigenstates of H_0 with eigenvalues E_i and E_f, it follows that

$$\langle\phi_f|p_z|\phi_i\rangle = -\frac{m}{i\hbar}(E_f - E_i)\langle\phi_f|z|\phi_i\rangle \qquad (5.115)$$

and

$$W_{fi} = \frac{qE_0\omega_{fi}}{2\omega}\langle\phi_f|z|\phi_i\rangle. \qquad (5.116)$$

The name 'dipole approximation' appears even more apt when the matrix element is written in this form.

The matrix elements of W_{fi} are proportional to the matrix elements of z, because $\vec{E}$ is in the z direction. The observation that $z = rY_{10}(\vartheta, \varphi)$ has been used before. Writing

$$\left\langle \mathbf{r}|\phi_i\right\rangle = R_{n_il_i}(r)Y_{l_im_i}(\vartheta, \varphi)\left\langle \mathbf{r}|\phi_f\right\rangle = R_{n_fl_f}(r)Y_{l_fm_f}(\vartheta, \varphi) \qquad (5.117)$$

it follows that

$$\langle\phi_f|z|\phi_i\rangle = C\int_\Omega Y^*_{l_fm_f}Y_{10}Y_{l_im_i}d\Omega, \qquad (5.118)$$

where C includes some integrals over r. Integrals over three spherical harmonics have been studied extensively because they appear frequently in atomic physics. It can be shown that $\langle\phi_f|z|\phi_i\rangle$ is zero unless $m_f = m_i$ and $l_f = l_i \pm 1$. An electromagnetic field is most likely to induce a transition between an initial and a final state if these selection rules are satisfied. There can be transitions that violate these selection rules, but they are much weaker. They come from magnetic dipole or electric quadrupole matrix elements that are non-zero even though the electric dipole matrix element is.

5.9 Approximate methods that use computers

As the speed of computers increases, it is frequently easier to study a phenomenon computationally than to use the analytical methods described above. Some of the most common methods fall into a category called Monte Carlo (MC) calculations. The simplest problem that the MC method can be applied to is the order–disorder transition in an Ising model of a magnet. The Hamiltonian for an Ising model is

$$H = -J \sum_i \sum_j s_i s_j, \tag{5.119}$$

where the sum over i is over all sites in the model, and the sum over j is over the nearest neighbors of the site i. The constant J is called the exchange constant. If the spin points upward on the site i, $s_i = 1$. If it points down, $s_i = -1$. This is a highly simplified picture of the ordering process in a real crystal.

The MC calculation proceeds as forward. First choose an initial configuration. Then scan through the sites one by one. For site i, if $s_i = 1$ and n sites on the nearest neighbor shell are spin up, the energy associated with site i is

$$E_i^0 = -J[n - (z - n)], \tag{5.120}$$

where z is the number of nearest neighbors. If $s_i = -1$, E_i^0 is the negative of the above. If the configuration of spins can change, it might be thought that the new configuration is always the one that lowers the total energy. In fact, the way to change the configuration is to use the Metropolis algorithm (Figure 5.1).

All of the sites in the sample are scanned multiple times. On the kth scan, the demon on the ith site (red in this example) calculates the energy difference

$$\Delta E_i^k = E_i^k - E_i^{k-1}. \tag{5.121}$$

According to the Metropolis algorithm, if $\Delta E_i^k \leqslant 0$ the demon switches the site from red to green (spin up to down). If $\Delta E_i^k > 0$, he calculates

$$P_i^k = e^{-\Delta E_i^k / k_B T}. \tag{5.122}$$

He uses a random number generator to obtain a number n such that $0 \leqslant n \leqslant 1$. If $n \leqslant P_i^k$, the demon switches the site from red to green. Otherwise he leaves it alone. After dealing with the ith site, he moves to the $(i + 1)$th and repeats the process. This process is illustrated in figure 5.1.

The Metropolis algorithm introduces a temperature and also a statistical step. The order–disorder temperature for a two-dimensional Ising model is known because the problem was solved exactly by Lars Onsager. He found

$$k_B T_c / J = \frac{2}{\ln\left(1 + \sqrt{2}\right)} = 2.269185314. \tag{5.123}$$

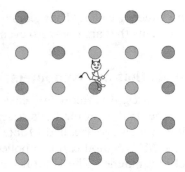

Figure 5.1. An illustration of a sample used in a MC calculation. If the spin is up, the site is red. If the spin is down, the site is green. The demon sitting on site makes the decision if the atom will stay the same or will change.

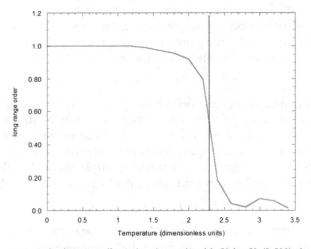

Figure 5.2. The long range order in a two-dimensional sample with 50 by 50 (2,500) sites. One million scans were carried out for each temperature. The temperature is measured in units $T = k_B T / J$, and the long range order disappears at $k_B T_c / J_c \approx 2.3$.

The most common analytical method for solving this problem is the mean field theory in which the atoms surrounding a site have the average moment. Mean field theory leads to

$$k_B T_c / J = 2, \tag{5.124}$$

which is not very good. The calculations described here are quite modest. The two-dimensional sample contains 2500 atoms and there were 1,000,000 scans for each temperature. A laptop computer was used with the Mathematica app. In spite of this, figure 5.2 demonstrates that the results reproduce Onsager's exact number quite well.

Calculations on a work station or a mainframe can reproduce Onsager's result to as many significant figures as required. Almost unlimited accuracy can be obtained using renormalization group theory, a highly sophisticated combination of analytical and numerical methods invented by Kenneth Wilson.

The Monte Carlo calculation was invented by Stanislaw Ulam, John von Neumann, and Nicholas Metropolis when they were working on the atom bomb during World War II. The work was highly classified, so they chose Monte Carlo as a code name because of its statistical aspect. It has been developed greatly over the years and is used for an extremely wide range of problems.

Another type of numerical calculation used instead of analytical approximations is called simulated annealing. This method is closely related to the Monte Carlo. Molecular dynamics also requires large computers, and is useful for certain problems.

Numerical methods have replaced to a large degree the more conventional methods for approximating solutions to physics problems. If they don't replace them, they supplement them. Today it is possible for computers to help solve highly complicated algebraic and analytical problems.

Problems

P5.1 How could group theory be used to simplify certain R-R calculations?

P5.2 Solve this problem exactly and show how the exact result is related to the R-R answer.

P5.3 Why did the experimental observation of anomalous x-ray scattering have to wait on the development of synchrotron x-ray sources?

P5.4 Show that the operator $A(t) = U^\dagger(t) A U(t)$ satisfies Heisenberg's equation of motion.

P5.5 Plot $|c_2|^2$ as a function of ω.

P5.6 Why are cylinders of helium delivered to MRI clinics?

P5.7 How can the ammonia maser be used to make a highly accurate clock?

P5.8 Use Fermi's golden rule to explain the angular resolved photoemission spectroscopy experiment.

P5.9 Show that $\langle \phi_f | z | \phi_i \rangle$ is zero unless $m_f = m_i$ and $l_f = l_i \pm 1$.

P5.10 The calculation described above describes a ferromagnetic transition. What change could be made to the Hamiltonian in equation (5.119) to model an antiferromagnetic transition?

References

[1] Rayleigh J W S 1870 In finding the correction for the open end of an organ-pipe. *Phil. Trans* **161** 77

[2] Rayleigh J W S 1894 *Theory of Sound* 2nd edn vol I (London: Macmillan)

IOP Publishing

Modern Quantum Mechanics and Quantum Information

J S Faulkner

Chapter 6

Scattering and Green's functions

6.1 Potential scattering

The scattering of a particle from a potential is different from other problems in quantum mechanics because the initial state, the free particle, has exactly the same energy as the final state, the particle traveling away from the scatterer. Logically and mathematically the only way to describe this process is for the incoming state to be a wave packet with a large enough distribution in momenta so that the particle can be reasonably well localized. The manipulation of wave packets is laborious, and the outcome of the analysis simply validates the solutions that are obtained from the Lippmann–Schwinger equations, which are much easier to solve.

The one-electron Schrödinger equation may be written in the Dirac notation

$$(E - H_0)| \psi \rangle = V | \psi \rangle, \tag{6.1}$$

where H_0 is the kinetic energy operator. The equation for the free-particle state vector is

$$(E - H_0)| \varphi \rangle = 0. \tag{6.2}$$

The solution of the homogeneous equation, equation (6.2), plus a particular solution of the inhomogeneous equation, equation (6.1), is

$$| \psi \rangle = | \varphi \rangle + (E - H_0)^{-1}V | \psi \rangle. \tag{6.3}$$

The algebra that leads to equation (6.3) is incorrect because E is an eigenvalue of H_0 and thus the inverse $(E - H_0)^{-1}$ doesn't exist. The mathematical trick that is used to get around this difficulty is to add a small imaginary part to the one-electron energy E. This leads to the equation

$$| \psi \rangle = | \varphi \rangle + G_0^+ V | \psi \rangle, \tag{6.4}$$

doi:10.1088/978-0-7503-2167-9ch6

where

$$G_0^+ = \lim_{\varepsilon \to 0}(E - H_0 + i\varepsilon)^{-1}. \qquad (6.5)$$

The imaginary part can approach zero from above or below the real axis. It will be shown that in order to satisfy the physical boundary condition that the second term represents an outgoing wave, must approach zero through positive values.

Equation (6.4) is the Lippmann–Schwinger equation, and it is the starting point for modern treatments of scattering theory. Using the definition

$$(1 - G_0^+V)^{-1} = 1 + G_0^+V + G_0^+VG_0^+V + \ldots = 1 + G_0^+T, \qquad (6.6)$$

where

$$T = V + VG_0^+V + VG_0^+VG_0^+V + \ldots = V(I + G_0^+T), \qquad (6.7)$$

it follows that another way to write the Lippmann–Schwinger equation is

$$|\psi\rangle = (1 + G_{0+}T)|\varphi\rangle. \qquad (6.8)$$

6.2 Position representation

Putting the Lippmann–Schwinger equation in the position representation leads to

$$\psi(\mathbf{r}) = \psi_0(\mathbf{r}) + \iint G_0^+(E, \mathbf{r}, \mathbf{r}')T(\mathbf{r}', \mathbf{r}'')\varphi(\mathbf{r}'')d\mathbf{r}'d\mathbf{r}''. \qquad (6.9)$$

In dimensionless units, the free-particle Hamiltonian H_0 is the negative of the Laplacian

$$H_0 = -\nabla^2. \qquad (6.10)$$

Its eigenvalues are $E = k^2$ and its eigenfunctions are

$$\langle \mathbf{r}|\mathbf{k}\rangle = \frac{1}{(2\pi)^{3/2}}e^{i\mathbf{k}\cdot\mathbf{r}}, \qquad (6.11)$$

so that

$$\langle \mathbf{k} \mid G_0 \mid \mathbf{k}'\rangle = G_o(E, \mathbf{k})\delta(\mathbf{k} - \mathbf{k}') = \frac{1}{E - k^2 + i\varepsilon}\delta(\mathbf{k} - \mathbf{k}'). \qquad (6.12)$$

From this, the free particle Green's function is

$$G_0^+(E, \mathbf{r}, \mathbf{r}') = \iiint \langle \mathbf{r}|\mathbf{k}\rangle\langle \mathbf{k} \mid G_0^+ \mid \mathbf{k}\rangle\langle \mathbf{k}|\mathbf{r}'\rangle d\mathbf{k}$$

$$= \frac{1}{(2\pi)^3}\lim_{\varepsilon \to 0} \iiint \frac{e^{i\mathbf{k}\cdot(\mathbf{r}-\mathbf{r}')}}{(E - k^2 + i\varepsilon)}d\mathbf{k} \qquad (6.13)$$

This integral is evaluated using the contour in complex k-space that is closed in the upper half plane. When ε is positive, the pole in the upper half plane is near $k = \sqrt{E}$, and

$$G_0^+(E, \mathbf{r}, \mathbf{r}') = -\frac{1}{4\pi}\frac{e^{i\alpha|\mathbf{r}-\mathbf{r}'|}}{|\mathbf{r}-\mathbf{r}'|}, \qquad (6.14)$$

where $\alpha = \sqrt{E}$. Inserting it into the Lippmann–Schwinger equation gives an expression for the solution of the one-electron Schrödinger equation

$$\psi_{\mathbf{k}}(\mathbf{r}) = \langle \mathbf{r}|\mathbf{k}\rangle - \frac{1}{4\pi}\int \frac{e^{i\alpha|\mathbf{r}-\mathbf{r}'|}}{|\mathbf{r}-\mathbf{r}'|}V(\mathbf{r}')\psi_{\mathbf{k}}(\mathbf{r}')d\mathbf{r}', \qquad (6.15)$$

or

$$\psi_{\mathbf{k}}(\mathbf{r}) = \langle \mathbf{r}|\mathbf{k}\rangle - \frac{1}{4\pi}\iint \frac{e^{i\alpha|\mathbf{r}-\mathbf{r}'|}}{|\mathbf{r}-\mathbf{r}'|}T(\mathbf{r}', \mathbf{r}'')\langle \mathbf{r}''|\mathbf{k}\rangle d\mathbf{r}'d\mathbf{r}''. \qquad (6.16)$$

In a scattering experiment, the dimensions of the scatterer are small compared with those of the measuring apparatus. Mathematically, this means that $V(\mathbf{r}')$ is zero outside a volume Ω. The wave function is needed only in the neighborhood of the counter where $|\mathbf{r}| \gg |\mathbf{r}'|$. In this limit, $|\mathbf{r} - \mathbf{r}'| \cong r - \frac{\mathbf{r}\cdot\mathbf{r}'}{r}$, and Green's function is

$$G_0^+(E, \mathbf{r}, \mathbf{r}') \cong -\frac{1}{4\pi}\frac{e^{i\alpha r}}{r}e^{-i\mathbf{k}'\cdot\mathbf{r}''}, \qquad (6.17)$$

where $\mathbf{k}' = \frac{k\mathbf{r}}{r}$ is the k-vector pointing toward the outgoing particle. Thus,

$$\psi_{\mathbf{k}}(\mathbf{r}) = \frac{1}{(2\pi)^{3/2}}\left[e^{i\mathbf{k}\cdot\mathbf{r}} + \frac{e^{i\alpha r}}{r}f(\mathbf{k}, \mathbf{k}')\right], \qquad (6.18)$$

where the scattering factor is

$$f(\mathbf{k}, \mathbf{k}') = -2\pi^2\langle \mathbf{k}'|\,V|\psi_{\mathbf{k}}\rangle = -2\pi^2\langle \mathbf{k}'|\,T\,|\mathbf{k}\rangle. \qquad (6.19)$$

The differential scattering cross section is defined by

$$\frac{d\sigma}{d\Omega}\Delta\Omega = \frac{\text{number of particles scattered into } \Delta\Omega \text{ per unit time}}{\text{number of incident particles per unit area and time}}. \qquad (6.20)$$

Using the expression for the current $\mathbf{j} = \frac{\hbar}{m}\mathrm{Im}\psi^*\nabla\psi$, it follows that

$$\frac{d\sigma}{d\Omega} = |f(\mathbf{k}, \mathbf{k}')|^2. \qquad (6.21)$$

The time dependence of the wave function can be factored out

$$\psi_{\mathbf{k}}(\mathbf{r}, t) = \psi_{\mathbf{k}}(\mathbf{r})e^{-i\omega t}, \qquad (6.22)$$

where, in dimensionless units, $\omega = E$. The time-dependent version of equation (6.18) describes a free-particle wave approaching the scatterer, and an outgoing wave leaving it.

The scattering factor $f(\mathbf{k}, \mathbf{k}')$ is proportional to the t-matrix $T(\mathbf{k}, \mathbf{k}')$. Since $|\mathbf{k}| = |\mathbf{k}'| = \sqrt{E}$, this is known as the on-the-energy-shell t-matrix. As mentioned above, the formalism with epsilon approaching zero from above leads to the outgoing wave.

6.3 The spherical scatterer

The plane wave can be expanded in spherical waves

$$\langle \mathbf{r} | \mathbf{k} \rangle = (2\pi)^{-3/2} e^{i\mathbf{k}\cdot\mathbf{r}} = \left(\frac{2}{\pi}\right)^{1/2} \sum_L i^l Y_L^*(\mathbf{k}) \psi_L^0(E, \mathbf{r}), \tag{6.23}$$

where the wave function $\psi_L^0(E, \mathbf{r})$ is written

$$\psi_L^0(E, \mathbf{r}) = Y_L(\mathbf{r}) j_l(\alpha r). \tag{6.24}$$

The free-particle Green's function can be written

$$G_0^+(E, \mathbf{r} - \mathbf{r}') = -\frac{1}{4\pi} \frac{e^{i\alpha|\mathbf{r}-\mathbf{r}'|}}{|\mathbf{r} - \mathbf{r}'|} = -i\alpha \sum_{L'} Y_{L'}(\mathbf{r}) h_{l'}^+(\alpha r) j_{l'}(\alpha r') Y_{L'}^*(\mathbf{r}'), \tag{6.25}$$

so

$$\psi_L(E, \mathbf{r}) = Y_L(\mathbf{r}) j_l(\alpha r) - i\alpha \sum_{L'} Y_{L'}(\mathbf{r}) h_{l'}^+(\alpha r) t_{L',L}(E), \tag{6.26}$$

with

$$t_{L',L} = \iint j_{l'}(\alpha r') Y_{L'}^*(\mathbf{r}') t(\mathbf{r}', \mathbf{r}'') Y_L(\mathbf{r}'') j_l(\alpha r'') d\mathbf{r}' d\mathbf{r}''. \tag{6.27}$$

Scattering theorists prefer to work with the s-matrix, which is related to the t-matrix by

$$\mathbf{S} = \mathbf{I} - i2\alpha\mathbf{T}. \tag{6.28}$$

Using this matrix, the preceding formula can be written

$$\psi_L(E, \mathbf{r}) = \frac{1}{2}\left[Y_L(\mathbf{r}) h_l^-(\alpha r) + \sum_{L'} Y_{L'}(\mathbf{r}) h_{l'}^+(\alpha r) S_{L'L} \right]. \tag{6.29}$$

For a spherical scatterer, $V(\mathbf{r}) = V(r)$, the t-matrix is diagonal in L, and

$$\psi_L(E, \mathbf{r}) = Y_L(\mathbf{r}) \psi_l(E, r), \tag{6.30}$$

where

$$\psi_l(E, r) = j_l(\alpha r) - i\alpha h_l^+(\alpha r) t_l(E). \tag{6.31}$$

It is also possible to find the solution of the Schrödinger equation for a given l, $\phi_l(E, r)$, by direct solution of the radial equation,

$$\left[-\frac{\hbar^2}{2m}\frac{1}{r^2}\frac{d}{dr}\left(r^2\frac{d}{dr} \right) + \frac{\hbar^2 l(l+1)}{2mr^2} + V(r) \right]\phi_l(E, r) = E\phi_l(E, r). \qquad (6.32)$$

The solution of this equation that is regular at the origin is

$$\lim_{r \to 0} \phi_l(E, r) = j_l(\alpha r). \qquad (6.33)$$

When r is outside the range of the potential, the radial solution can be written as a linear combination of the spherical Bessel functions that are the solutions of the radial equation with $V(r) = 0$

$$\phi_l(E, r) = A_l\left[j_l(\alpha r) \cos \delta_l - n_l(\alpha r) \sin \delta_l \right]. \qquad (6.34)$$

Comparing with $\psi_l(E, r)$ gives

$$t_l = -\frac{1}{\alpha}e^{i\delta_l} \sin \delta_l. \qquad (6.35)$$

To find the phase shifts δ_l, the radial equation if first solved for $\phi_l(E, r)$. Assuming that the potential is zero outside a circle of radius a, the continuity of the wave function and its derivative gives the equations

$$\phi_l(a) = A_l\left[j_l(\alpha a) \cos \delta_l - n_l(\alpha a) \sin \delta_l \right]$$

$$\left.\frac{d\phi_l(r)}{dr}\right)_{r=a} = \alpha A_l\left[\frac{dj_l(x)}{d(x)} \cos \delta_l - \frac{dn_l(x)}{d(x)} \sin \delta_l \right]_{x=\alpha a}. \qquad (6.36)$$

This can be looked upon as two equations for the two unknowns A_l and δ_l. They can be solved easily using the identity for Bessel functions

$$j_l(x)\frac{dn_l(x)}{dx} - n_l(x)\frac{dj_l(x)}{dx} = \frac{1}{x^2}. \qquad (6.37)$$

The solution is

$$\tan \delta_l = \frac{\left[\alpha R_l(r)\dfrac{dj_l(x)}{dx} - \dfrac{d\phi_l(r)}{dr}j_l(x) \right]_{x=\alpha r}}{\left[\alpha R_l(r)\dfrac{dn_l(x)}{dx} - \dfrac{d\phi_l(r)}{dr}n_l(x) \right]_{x=\alpha r}}. \qquad (6.38)$$

The rigid sphere is a spherical square well in which the potential is infinite for $r \leqslant a$, and $\phi_l(a) = 0$ for this case. It follows that

$$\tan \delta_l = \frac{j_l(ka)}{n_l(ka)}. \qquad (6.39)$$

For low energies and long wavelengths,

$$j_l(ka) = \frac{(ka)^l}{1 \cdot 3 \cdot 5 \cdot \ldots (2l+1)}$$
$$n_l(ka) = \frac{1 \cdot 3 \cdot 5 \cdot \ldots (2l-1)}{(ka)^{l+1}},$$

(6.40)

so

$$\lim_{k \to 0} \tan \delta_l \to \frac{(ka)^{(2l+1)}}{(2l+1)!!(2l-1)!!}.$$

(6.41)

This is important because it turns out that the scattering from any potential that is bounded in space approaches this limit as the energy approaches zero. From this equation and others, it can be seen that the scattering phase shifts generally become smaller as l becomes bigger.

In many cases, the scattering for all l unequal to zero is considered negligible, leading to

$$f(\theta) = \frac{e^{i\delta_0}}{k} \sin \delta_0.$$

(6.42)

If, for some energy, the scattering phase shift δ_0 is equal to a multiple of π, then there would be no scattered beam. This is called the Ramsauer–Townsend effect, and it has been observed experimentally.

6.4 The optical theorem

From the above arguments it can be seen that, for spherically symmetric potentials,

$$f(\mathbf{k}, \mathbf{k}') = -4\pi \sum_L Y_L(\mathbf{k}) t_l Y_L^*(\mathbf{k}').$$

(6.43)

The addition theorem for spherical harmonics is

$$P_l(\cos \theta) = \frac{4\pi}{2l+1} \sum_{m=-l}^{l} Y_{lm}^*(\mathbf{k}') \, Y_{lm}(\mathbf{k}),$$

(6.44)

where

$$\cos \theta = \frac{\mathbf{k} \cdot \mathbf{k}'}{|\mathbf{k}||\mathbf{k}'|}.$$

(6.45)

Thus

$$f(\theta) = -\sum_l (2l+1) t_l(E) P_l(\cos \theta) = \frac{1}{\alpha} \sum_l (2l+1) e^{i\delta_l} \sin \delta_l P_l(\cos \theta),$$

(6.46)

and, since the differential scattering cross section is the absolute square of $f(\theta)$

$$\frac{d\sigma}{d\Omega} = \frac{1}{\alpha^2} \sum_{ll'} [(2l + 1)e^{i\delta_l} \sin \delta_l][(2l' + 1)e^{-i\delta_{l'}} \sin \delta_{l'}] P_l(\cos \theta) P_{l'}(\cos \theta). \quad (6.47)$$

Using the normalization of the Legendre polynomials,

$$\int_0^\pi P_l(\cos \theta) P_{l'}(\cos \theta) \sin \theta d\theta = \frac{2}{2l + 1} \delta_{ll'}, \quad (6.48)$$

it follows that

$$\sigma = \int_0^{4\pi} \frac{d\sigma}{d\Omega} d\Omega = \frac{4\pi}{k^2} \sum (2l + 1) \sin^2 \delta_l = \frac{4\pi}{k} \mathrm{Im} f(0). \quad (6.49)$$

This proves that the total scattering cross section is proportional to $\mathrm{Im} f(0)$, a result that is called the optical theorem. It has the physical interpretation that an object has a shadow when it is in a beam of incoming rays or particles. The earliest version of this theorem can be traced back to Lord Rayleigh.

6.5 The Born approximation

The approximation proposed by Max Born in the early days of quantum theory is to replace the T operator by its first term, the potential V

$$f(\mathbf{k}, \mathbf{k}') = -2\pi^2 \langle \mathbf{k}' | V | \mathbf{k} \rangle. \quad (6.50)$$

This approximation has been remarkably useful. For example, the theory of x-ray scattering from solids is based on it. For any potential that is spherically symmetric

$$\langle \mathbf{k}' | V | \mathbf{k} \rangle = -\frac{1}{(2\pi)^3} \int_0^\infty \int_\Omega V(r) e^{i(\mathbf{k}-\mathbf{k}')\cdot \mathbf{r}} d\mathbf{r} = \frac{1}{2\pi^2 q} \int_0^\infty V(r) \sin(qr) r dr, \quad (6.51)$$

where $q = |\mathbf{k} - \mathbf{k}'|$.

A spherical square well potential is equal to $-V_0$ inside a sphere of radius a and zero otherwise. For such a potential

$$f(\mathbf{k}, \mathbf{k}') = \frac{1}{q} \int_0^a r V_0 \sin qr dr, \quad (6.52)$$

or

$$f(\mathbf{k}, \mathbf{k}') = -\frac{V_0}{q^3} [qa \cos qa - \sin qa]. \quad (6.53)$$

The angular dependence of this scattering factor is seen by writing

$$q = k\sqrt{2(1 - \cos \theta)} = 2k \sin \frac{\theta}{2}. \quad (6.54)$$

The forward scattering factor, when $\mathbf{k} = \mathbf{k}'$ or $q = 0$, is

$$f(\mathbf{k}, \mathbf{k}) = f(0) = \frac{V_0 a^3}{3}. \tag{6.55}$$

Plugging this result into the optical theorem in equation (6.49) appears to lead to the conclusion that the total scattering cross section is zero. This is of course incorrect, and is explained by the fact that the optical theorem does not apply to the Born approximation. The sum rules that are necessary in the proof of the optical theorem cannot be used when only the first term in the scattering equation is included.

The next example is the Coulomb potential

$$V(r) = \frac{Z_1 Z_2 e^2}{r}. \tag{6.56}$$

The integral in equation (6.51) does not converge for this potential, but it can be made to converge by including a screening term

$$\hat{V}(r) = \frac{Z_1 Z_2 e^2}{r} e^{-\gamma r}, \tag{6.57}$$

which leads to the scattering factor

$$\langle \mathbf{k}' | \ V \ | \ \mathbf{k} \rangle = \frac{Z_1 Z_2 e^2}{2\pi^2 (q^2 + \gamma^2)}. \tag{6.58}$$

Letting $\gamma \to 0$ leads to the Born approximation for the Coulomb potential

$$f(\mathbf{k}, \mathbf{k}') = -\frac{Z_1 Z_2 e^2}{q^2}. \tag{6.59}$$

Using equation (6.54) leads to

$$f(\theta) = -\frac{Z_1 Z_2 e^2}{4k^2 \sin^2 \dfrac{\theta}{2}}, \tag{6.60}$$

which diverges as $\theta \to 0$. The differential scattering cross section obtained by inserting this formula into equation (6.21) gives the formula used by Ernest Rutherford to explain the scattering of alpha particles from gold foil. This famous scattering experiment clarified the structure of atoms.

6.6 Green's function and its adjoint

Since the nineteenth century when the self-trained mathematician George Green began to use the function that bears his name, the primary task of Green's functions has been to convert differential equations with boundary conditions into integral equations. The free-particle Green's function $G_0^+(E, \mathbf{r}, \mathbf{r}')$ was used to convert the Schrödinger equation into the Lippmann–Schwinger equation.

The scattering theory in this chapter has been based on wave functions. For some theoretical studies it is more convenient to focus on Green's functions. Suppose that a system is described by a Hamiltonian H. The equation for Green's function is then

$$\lim_{\varepsilon \downarrow 0}(E - H + i\varepsilon)G = I, \tag{6.61}$$

and the solution is

$$G = \lim_{\varepsilon \downarrow 0}(E - H + i\varepsilon)^{-1}. \tag{6.62}$$

It can be evaluated in the position representation

$$G(E, \mathbf{r}, \mathbf{r}') = \lim_{\varepsilon \downarrow 0} \langle \mathbf{r} \mid (E - H + i\varepsilon)^{-1} \mid \mathbf{r}' \rangle. \tag{6.63}$$

or the momentum representation

$$G(E, \mathbf{k}, \mathbf{k}') = \lim_{\varepsilon \downarrow 0} \langle \mathbf{k} \mid (E - H + i\varepsilon)^{-1} \mid \mathbf{k}' \rangle. \tag{6.64}$$

The charge density of the system is obtained by integrating over the energy of occupied states

$$\rho(\mathbf{r}) = -\frac{2}{\pi}\mathrm{Im} \int_{-\infty}^{E_F} G(E, \mathbf{r}, \mathbf{r})dE. \tag{6.65}$$

The Fermi energy, E_F, is defined by the requirement that the integral of $\rho(\mathbf{r})$ over all space is equal to the total number of electrons, and the 2 accounts for the electron spins. Knowledge of the charge density gives a useful picture of bonding. According to the density functional theory to be discussed in chapter 12, the total energy of the system is a function of the charge density.

The density of energy states is obtained from the equation

$$\rho(E) = -\frac{2}{\pi}\mathrm{Im} \iiint_{\substack{all \\ space}} G(E, \mathbf{r}, \mathbf{r})d\mathbf{r}, \tag{6.66}$$

which is called the trace of Green's function. There are many experiments that measure this quantity. The function

$$\rho(\mathbf{k}) = -\frac{2}{\pi}\mathrm{Im} \int_{-\infty}^{E_F} G(E, \mathbf{k}, \mathbf{k})dE, \tag{6.67}$$

is called the spectral density function, and it is used to analyze the optical properties of condensed matter.

6.7 Green's function with a scatterer

Suppose that the Hamiltonian is the sum of the free particle Hamiltonian, H_0, and the potential of a scatterer

$$H = H_0 + V. \tag{6.68}$$

Using the identity $(A - B)^{-1} = A^{-1} + A^{-1}B(A - B)^{-1}$, Green's function for the system can be written

$$G = G_0 + G_0VG, \tag{6.69}$$

which will later be called a Dyson equation. The free particle wave function is used in the scattering equations above. Iterating G leads to

$$G = G_0 + G_0VG_0 + G_0VG_0VG_0 + G_0VG_0VG_0VG_0 + \ldots, \tag{6.70}$$

which can be written

$$G = G_0 + G_0TG_0, \tag{6.71}$$

with

$$T = V + VG_0V + VG_0VG_0V + VG_0VG_0VG_0V + \ldots. \tag{6.72}$$

Equation (6.71) can be put into the momentum representation

$$\langle \mathbf{k} \mid G \mid \mathbf{k'} \rangle = G_0(E, \mathbf{k})\delta(\mathbf{k} - \mathbf{k'}) + G_0(E, \mathbf{k})\langle \mathbf{k} \mid T \mid \mathbf{k'} \rangle G_0(E, \mathbf{k'}), \tag{6.73}$$

where $G_0(E, \mathbf{k})$ is in equation (6.12). In principle, $\langle \mathbf{k}|T|\mathbf{k'} \rangle$ can be found from

$$\begin{aligned} \langle \mathbf{k} \mid T \mid \mathbf{k'} \rangle = V_{\mathbf{kk'}} + \int V_{\mathbf{kk''}}\frac{1}{E - k''^2 + i\varepsilon}V_{\mathbf{k''k'}}d\mathbf{k''} \\ + \int\int V_{\mathbf{kk''}}\frac{1}{E - k''^2 + i\varepsilon}V_{\mathbf{k''k'''}}\frac{1}{E - k'''^2 + i\varepsilon}V_{\mathbf{k'''k'}}d\mathbf{k''} + \ldots \end{aligned}, \tag{6.74}$$

where the element $V_{\mathbf{kk'}}$ is the scattering matrix in the Born approximation. The techniques for calculating scattering matrices described in the preceding sections are much more efficient.

6.8 The non-spherical scattering potential with bounded domain

Scattering matrices can be found for potential functions that are not spherically symmetric. Extending the arguments of Jost [1], it can be shown by direct substitution that a solution of the equation

$$[-\nabla^2 + v(\mathbf{r}) - E]\phi(E, \mathbf{r}) = 0, \tag{6.75}$$

is

$$\phi_{L_0}(E, \mathbf{r}) = Y_{L_0}(\hat{\mathbf{r}})j_{l_0}(\alpha r) + \iiint_{|\mathbf{r'}|\leqslant r} K(E, \mathbf{r}, \mathbf{r'})v(\mathbf{r'})\phi_{L_0}(E, \mathbf{r'})d\mathbf{r'}, \tag{6.76}$$

where the integration is over all $|\mathbf{r'}| \leqslant |\mathbf{r}| = r$. The kernel of this integral equation is

$$K(E, \mathbf{r}, \mathbf{r'}) = -\alpha\sum Y_L(\hat{\mathbf{r}})\big[j_l(\alpha r)n_l(\alpha r') - n_l(\alpha r)j_l(\alpha r')\big]Y*_L(\hat{\mathbf{r}}). \tag{6.77}$$

This function is regular at the origin, and, in fact,

$$\lim_{|\mathbf{r}|\to 0} \phi_{L_0}^n(E, \mathbf{r}) \to j_{l_0}(ar)\delta_{m,0}. \tag{6.78}$$

It can similarly be shown that

$$f_{L_1}^{\pm}(E, \mathbf{r}) = Y_{L_1}(\hat{\mathbf{r}})h_{l_1}^{\pm}(ar) - \iiint_{|\mathbf{r}'|>r} K(E, \mathbf{r}, \mathbf{r}')v(\mathbf{r}')f_{L_1}^{\pm}(E, \mathbf{r}')d\mathbf{r}', \tag{6.79}$$

are also solutions of the differential equation. Assume there is a bounding sphere of radius A such that

$$v(\mathbf{r}) = 0 \quad \text{for} \quad |\mathbf{r}| > A. \tag{6.80}$$

This is the definition of a potential with a bounded domain. The domain does not have to be a sphere. When the potential satisfies to preceding condition, then

$$f_{L_1}^{\pm}(E, \mathbf{r}) = Y_{L_1}(\hat{\mathbf{r}})h_{l_1}^{\pm}(ar) \quad \text{for} \quad |\mathbf{r}| > A. \tag{6.81}$$

It is obvious that $f_{L_1}^{+}(E, \mathbf{r})$ and $f_{L_1}^{-}(E, \mathbf{r})$ are linearly independent, so it is possible to write

$$\phi_{L_0}(E, \mathbf{r}) = \frac{1}{2}\sum_{L_1}\left[f_{L_1}^{+}(E, \mathbf{r})a_{L_1L_0}(E) + f_{L_1}^{-}(E, \mathbf{r})b_{L_1L_0}(E)\right], \tag{6.82}$$

where $a_{L_1L_0}(E)$ and $b_{L_1L_0}(E)$ are independent of $\mathbf{r}$. By construction, the functions $\phi_{L_0}(E, \mathbf{r})$, $f_{L_1}^{+}(E, \mathbf{r})$, and $f_{L_1}^{-}(E, \mathbf{r})$ are all solutions of the differential equation for all $\mathbf{r}$. Another wave function that is a solution for all r is,

$$\psi_L(E, \mathbf{r}) = \frac{1}{2}\left\{f_L^{-}(E, \mathbf{r}) + \sum_{L_1,L_0}\left[f_{L_1}^{+}(E, \mathbf{r})a_{L_1L_0}(E)b_{L_0L}^{-1}(E)\right]\right\}. \tag{6.83}$$

Comparing this function in its asymptotic form for $r > A$ with equation (6.29), it can be seen that the s-matrix is related to the **a** and **b** matrices,

$$\mathbf{S}(E) = \mathbf{ab}^{-1}. \tag{6.84}$$

The matrices a and b can be calculated from the integrals

$$a_{L_1L_0} = \delta_{L_1L_0} - i\alpha\lim_{\delta\to 0}\iiint_{\mathbf{r}\subset V_\delta^A} \mathbf{h}_{L_1}^{-}(E, \mathbf{r})V(\mathbf{r})\phi_{L_0}(E, \mathbf{r})dv\delta_{L_1L_0}, \tag{6.85}$$

and,

$$b_{L_1L_0} = \delta_{L_1L_0} + i\alpha\lim_{\delta\to 0}\iiint_{\mathbf{r}\subset V_\delta^A} \mathbf{h}_{L_1}^{+}(E, \mathbf{r})V(\mathbf{r})\phi_{L_0}(E, \mathbf{r})dv\delta_{L_1L_0}, \tag{6.86}$$

where $\mathbf{h}_{L_1}^{\pm}(E, \mathbf{r}) = h_{l_1}^{\pm}(E, \mathbf{r})Y_{L_1}^{*}(\hat{\mathbf{r}})$. Alternative expressions are

$$a_{L_1L_0} = -i\alpha\iiint_{S_A} \left\{\mathbf{h}_{L_1}^{-}(E, \mathbf{r}), \phi_{L_0}(E, \mathbf{r})\right\}\cdot \mathbf{n}dS, \tag{6.87}$$

and,

$$b_{L_1 L_0} = i\alpha \iiint_{S_A} \left\{ \mathbf{h}_{L_1}^+(E, \mathbf{r}), \phi_{L_0}(E, \mathbf{r}) \right\} . \, \mathbf{n} dS, \tag{6.88}$$

where $\left\{ \mathbf{h}_{\bar{L}_1}(E, \mathbf{r}), \phi_{L_0}(E, \mathbf{r}) \right\}$ is the vector

$$\left\{ \mathbf{h}_{\bar{L}_1}(E, \mathbf{r}), \phi_{L_0}(E, \mathbf{r}) \right\} = \mathbf{h}_{\bar{L}_1}(E, \mathbf{r}) \nabla \phi_{L_0}(E, \mathbf{r}) - \phi_{L_0}(E, \mathbf{r}) \nabla \mathbf{h}_{\bar{L}_1}(E, \mathbf{r}). \tag{6.89}$$

Recalling that the standard s-matrix for spherically symmetric potentials is related to phase shifts by $S_l(E) = e^{i2\eta_l}$, it is reasonable to define sine and cosine matrices $\mathbf{s}$ and $\mathbf{c}$ by

$$\begin{aligned} \mathbf{a} &= \mathbf{c} + i\mathbf{s} \\ \mathbf{b} &= \mathbf{c} - i\mathbf{s} \end{aligned} . \tag{6.90}$$

From the above

$$\begin{aligned} c_{L_1 L_0} &= \delta_{L_1 L_0} - \alpha \lim_{\delta \to 0} \iiint_{\mathbf{r} \subset V_\delta^A} \mathbf{n}_{L_1}(E, \mathbf{r}) V(\mathbf{r}) \phi_{L_0}(E, \mathbf{r}) dv \\ &= -\alpha \iiint_{S_A} \left\{ \mathbf{n}_{L_1}(E, \mathbf{r}), \phi_{L_0}(E, \mathbf{r}) \right\} . \, \mathbf{n} dS \end{aligned} , \tag{6.91}$$

and

$$\begin{aligned} s_{L_1 L_0} &= -\alpha \lim_{\delta \to 0} \iiint_{\mathbf{r} \subset V_\delta^A} \mathbf{j}_{L_1}(E, \mathbf{r}) V(\mathbf{r}) \phi_{L_0}(E, \mathbf{r}) dv \\ &= -\alpha \iiint_{S_A} \left\{ \mathbf{j}_{L_1}(E, \mathbf{r}), \phi_{L_0}(E, \mathbf{r}) \right\} . \, \mathbf{n} dS \end{aligned} . \tag{6.92}$$

The t-matrix is

$$\begin{aligned} \mathbf{T} &= \frac{i}{2\alpha} [\mathbf{S} - \mathbf{I}] \\ \mathbf{T} &= -\frac{1}{\alpha} \mathbf{s} (\mathbf{c} - i\mathbf{s})^{-1} . \end{aligned} \tag{6.93}$$

Inserting the definition of the kernel into

$$\phi_{L_0}(E, \mathbf{r}) = Y_{L_0}(\hat{\mathbf{r}}) j_{l_0}(\alpha r) + \iiint_{|\mathbf{r}'| \leqslant r} K(E, \mathbf{r}, \mathbf{r}') v(\mathbf{r}') \phi_{L_0}(E, \mathbf{r}') d\mathbf{r}', \tag{6.94}$$

leads to

$$\phi_{L_0}(E, \mathbf{r}) = \sum Y_L(\hat{\mathbf{r}}) j_l(\alpha r) c_{LL_0}(r) - \sum Y_L(\hat{\mathbf{r}}) n_l(\alpha r) s_{LL_0}(r). \tag{6.95}$$

These $\mathbf{s}$ and $\mathbf{c}$ matrices are the generalizations of the matrices studied by Calogero [2] in his variable phase approach to scattering. He focused on the case where the potential is spherically symmetric. For this case, the scattering matrices can be identified with the famous Jost [1] functions, $a_l(E) = L_l^-(E)$ and $b_l(E) = L_l^+(E)$. Potentials that are non-spherical but have a finite cutoff radius as indicated in equation (6.80) are commonly used in multiple scattering calculations in condensed matter theory [3]. The integrals in equations (6.91) and (6.92) have been evaluated numerically for a wide range of potentials.

6.9 Spectral theory from scattering theory

To this point the goal has been to solve the Schrödinger equation in its entirety. That is, all of the wave functions, Green's functions, and energy eigenvalues are obtained. Many years ago mathematicians found that a great deal of information about the spectrum of an operator can be obtained without solving for all aspects of the problem. In the present context, the term spectrum refers to the distribution of the energy eigenvalues of the system under investigation.

Physicists have independently come upon relationships that fall in the category of spectral theory. One of the oldest of these is the Friedel sum

$$N(E) = \frac{2}{\pi} \sum_{l=0}^{l_{\max}} (2l + 1)\delta_l(E). \tag{6.96}$$

In this expression, $N(E)$ is the integrated density of states (IDOS) associated with a single atom embedded in a vacuum (or an impurity atom in an otherwise ordered solid). The $\delta_l(E)$ are the phase shifts that describe the scattering by the atom, and are defined in equation (6.38). The number of core states of this atom is given by Levinson's theorem

$$n_{core}(E) = \frac{2}{\pi} \sum_{l=0}^{l_{\max}} (2l + 1)\delta_l(0), \tag{6.97}$$

where

$$\delta_l(0) = \lim_{E \to 0} \delta_l(E). \tag{6.98}$$

The density of states for a system can be obtained from the Green function for the system as pointed out in equation (6.66). This can also be considered a spectral equation.

Although these formulae were arrived at in very different ways, a great deal of insight can be gained about them by relating them to a theorem proved by the Ukranian mathematician Mark Grigorievich Krein [4]. In the thirties, Krein created one of the strongest centers of functional analysis in the world at Odessa University. Many of the results of this period by him, as well as in collaborations with his friends, colleagues, and students are now characterized as classical and appear in textbooks on functional analysis.

6.10 Krein's theorem

Krein proved that, for any two self-adjoint operators whose difference is trace class, there exists a real function of real variables $\xi(E)$ such that

$$Tr[\phi(H) - \phi(H_0)] = \int_{-\infty}^{\infty} \frac{d\phi(x)}{dx} \xi(x) dx, \tag{6.99}$$

where $\phi(X)$ is any sufficiently smooth function. The notation $Tr[X]$ designates a process that is the abstract version of the trace operation. For applications to scattering theory, the self-adjoint operators are chosen to be the Hamiltonian operator for the system H and for a free particle H_0. The potential, $V = H - H_0$ is in trace class for most cases of physical interest. Also, the arbitrary smooth function is chosen to be

$$\phi(H) = (z - H)^{-1}, \qquad (6.100)$$

where z is a complex number. The variable of integration in equation (6.99) is chosen to be the energy E, which means that $\phi(E) = (z - E)^{-1}$. It follows that

$$Tr[(z - H)^{-1} - (z - H_0)^{-1}] = \int_{-\infty}^{\infty} \frac{\xi(E')}{(z - E')^2} dE'. \qquad (6.101)$$

Birman and Krein [5] proved that Krein's spectral shift function $\xi(E)$ is given by

$$e^{-i2\pi\xi(E)} = \det \mathbf{S}(E), \qquad (6.102)$$

where $\mathbf{S}(E)$ is the standard unitary S-matrix. The uniqueness of $\xi(E)$ is demonstrated in the papers quoted, and in the massive literature on the subject that followed. In simple terms, the definition includes the conditions that $\xi(E)$ is a continuous function of E and approaches zero as E approaches plus or minus infinity. The spectral shift function defined in this way is consistent with the known properties of the S-matrix. From this definition and Cauchy's integral theorem, it follows that

$$\lim_{z \downarrow E} \int_{-\infty}^{\infty} \frac{\xi(E')}{(z - E')^2} dE' = i\pi \frac{d\xi(E)}{dE}. \qquad (6.103)$$

As a simple example, Krein's theorem is applied to the case of a single potential embedded in free space. The potentials of interest are such that $v(\mathbf{r})$ is defined to be zero for $\mathbf{r}$ outside the finite domain Ω. It follows that the energy spectrum for a system with one such potential embedded in a vacuum is discrete for $E < 0$, being the set of bound state energies E_ν. The system has a continuous spectrum for $E \geqslant 0$.

The trace operation in equation (6.101) is carried out by taking the integral over all space, which leads to

$$Tr[(z - H)^{-1} - (z - H_0)^{-1}] = \int_{\infty} [G(E, \mathbf{r}, \mathbf{r}) - G_0(E, \mathbf{r}, \mathbf{r})] d\mathbf{r}, \qquad (6.104)$$

where $G(E, \mathbf{r}, \mathbf{r})$ is the diagonal element of Green's function defined in equation (6.62). The integral will give the sum of all the eigenvalues of the Hamiltonian H minus the eigenvalues of H_0

$$\left[2 \sum_\nu d_\nu \delta(E - E_\nu) + [n(E) - n_0(E)] \right] = -2 \frac{d\xi(E)}{dE}, \qquad (6.105)$$

where d_ν is the degeneracy of the bound state with energy E_ν, $n(E)$ is the density of states, and $n_0(E)$ is the free-electron density of states. From the properties of the potential, $v(\mathbf{r})$, $n(E)$ and $n_0(E)$ are zero for $E < 0$ and that all the bound state energies are negative. The 2 is included to account for the spin of the electrons. Integrating equation (6.105) from minus infinity to E, with $E \geqslant 0$, leads to

$$-2\xi(E) = N(E) - N_0(E) + n_c = N_K(E),$$ (6.106)

where n_c is the number of core electrons. The function $N_K(E)$ is called the Krein integrated density of states (IDOS).

Integrating equation (6.105) from minus infinity to zero leads to a generalization of Levinson's theorem

$$2\sum_{\nu} d_{\nu} = n_c = -2\xi(0) = N_K(0).$$ (6.107)

This equation, which holds for potentials that are not spherically symmetric, can be checked for consistency with the ordinary Levinson's theorem by applying it to potentials that are.

It is obvious that the advantage of Krein's formulation is its generality. The trace operation is independent of the orthonormal basis used, as long as it is complete. The versions familiar in quantum theory are spatial integrations, as in equation (6.65), or integrations over momenta. The requirement that the potential is trace class is essentially equivalent to the condition that the trace on the left side of equation (6.101) exists, with no shape approximation. As with the Green's function formulae in the preceding section, the ones based on Krein's theorem are made tractable by ignoring contributions from terms with $l > l_{max}$.

Krein's theorem, as stated in equation (6.101), holds for any energy. Calculations in the range $-\infty < E < 0$ are very difficult for the specific kind of potentials used here. For this reason, a numerical test of the generalized Levinson theorem, equation (6.107), for these potentials is carried out by calculating $\xi(E)$ for $0 \leqslant E < E_{max}$, where E_{max} is sufficiently large that $\xi(E)$ has become a constant. It can be shown that

$$\xi(E) \to 0 \quad \text{as} \quad E \to \infty,$$ (6.108)

so $\xi(E_{max})$ is set equal to zero, and integers are added to $\xi(E)$ over various energy intervals is such a way as to make it a continuous function. Following the function in this way from E_{max} to zero is a way of counting the bound states of the potential using equation (6.110). The result can be checked by finding the bound states with the numerical techniques normally used in atomic physics.

The S-matrix for the special case of a spherically symmetric potential is diagonal with elements

$$s_l(E) = e^{2i\delta_l(E)},$$ (6.109)

repeated $2l + 1$ times. Considering only the states that correspond to a given l, equation (6.107) is equivalent to

$$n_c^{(l)} = -2\xi_l(0) = \frac{2}{\pi}\delta_l(0),$$ (6.110)

the standard version of Levinson's theorem.

It is also clear from equation (6.109) that Krein's theorem for a spherical potential provides a mathematical justification for the Friedel sum, equation (6.96). The Friedel sum has been used for many years as a more or less ad hoc method for estimating the density of states (DOS) or the IDOS for an impurity imbedded in a free-electron solid. This connection with Krein's theorem clarifies the meaning of the Friedel sum beyond most of the derivations in the literature. To be specific, the impression is usually given that the change in the DOS caused be the potential $v(\mathbf{r})$ is restricted to the region of space within which $v(\mathbf{r}) \neq 0$. Actually, Green's function $G(E, \mathbf{r}, \mathbf{r})$ must be integrated over values of $\mathbf{r}$ that extend all the way to infinity in order to produce a DOS that is precisely equal to the one given by Krein's theorem. The interpretation of this, which becomes less surprising after reflection, is that a potential with a finite domain can influence states in all of space.

Problems

P6.1 Use Sokhotsky's formula

$$\underset{\varepsilon \to 0^+}{\text{Lim}} \frac{c}{x \pm i\varepsilon} = \text{Principal Value Integral} \left(\frac{c}{x}\right) \mp i\pi c \delta(x)$$

to prove equations (6.65), (6.66), and (6.67).
P6.2 Carry out the integration that leads from equations (6.13) to (6.14).
P6.3 From equations 6.28) and (6.35) show that $S_l = e^{i2\delta_l}$.
P6.4 Do the integral in equation (6.51) for the potential in equation (6.57).
P6.5 What happens to the equations in section 6.8 when $v(\mathbf{r})$ is spherically symmetric.
P6.6 The Schrödinger equation for a single delta-function scatterer is

$$-\frac{d^2\psi}{dx^2} - \frac{2P}{d}\delta(x)\psi = \alpha^2\psi$$

where $\alpha = \sqrt{E}$. In the neighborhood of the scatterer, the solutions can be written

$$\psi_I(x) = b_-e^{iax} + a_-e^{-iax} \quad x < 0$$
$$\psi_{II}(x) = a_+e^{iax} + b_+e^{-iax} \quad x > 0$$

as shown in the following drawing.

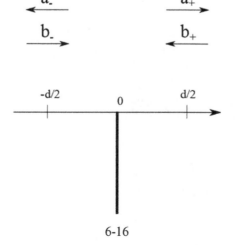

6-16

P6.6a Find the S-matrix that relates the coefficients

$$\begin{pmatrix} a_+ \\ a_- \end{pmatrix} = \mathbf{S} \begin{pmatrix} b_+ \\ b_- \end{pmatrix}.$$

P6.6b Show that $\mathbf{S}$ is unitary.

P6.6c Find the eigenvalues of $\mathbf{S}$.

P6.6d Find the determinant of $\mathbf{S}$.

P6.6e Find the Krein spectral displacement function for this potential $-2\xi(E)$.

P6.6f Choose units so that P/d is one. Plot the spectral displacement as a fiction of energy for $-4 \leqslant E \leqslant 12$.

References

[1] Res Jost and Yafaev D R 1992 *Mathematical Scattering Theory: General Theory* (Providence, RI: American Mathematical Society)

[2] Calogero F 1967 *Variable Phase Approach to Potential Scattering* (New York: Academic)

[3] Faulkner J S, Stocks G M and Wang Y 2019 *Multiple Scattering Theory: Electronic Structure of Solids* (Bristol: IOP Publishing Ltd)

[4] Krein M G 1953 *Matem. Sborn.* **33** 597

[5] Birman M L and Krein M G 1962 *Sov. Math.-Dokl.* **3** 740

IOP Publishing

Modern Quantum Mechanics and Quantum Information

J S Faulkner

Chapter 7

A practical tool

7.1 The exact equations

Most of the theories discussed to this point deal with the quantum mechanics of one electron moving in a potential. People who apply quantum theory to explain the materials and devices that are so important in our modern high tech economy have little interest in such systems. Their calculations are on models that contain N electrons, where N is usually of the order of Avogadro's number, 6.022×10^{23}. This chapter describes methods for treating atoms and small molecules for which N is of the order of 100. A much more general and complete discussion of many particle systems is given in chapter 12.

The Hamiltonian is a combination of kinetic energy and a potential energy, $H = T + V$. The kinetic energy for an N-electron system is

$$T = \sum_{i=1}^{N} \frac{p_i^2}{2m_i} = -\sum_{i=1}^{N} \frac{\hbar^2}{2m_i} \nabla_i^2, \tag{7.1}$$

and the potential energy is

$$V = V(\vec{r}_1, \vec{r}_2, \vec{r}_3, \dots, \vec{r}_N). \tag{7.2}$$

The wave function is

$$\Psi = \Psi(\vec{r}_1, \sigma_1, \vec{r}_2, \sigma_2, \vec{r}_3, \sigma_3, \dots, \vec{r}_N, \sigma_N, t), \tag{7.3}$$

where $\vec{r}_i$ is the position of the ith electron and σ_i is its spin. The Schrödinger equation

$$H\Psi = i\hbar \frac{\partial \Psi}{\partial t}, \tag{7.4}$$

is a partial differential equation in $3N + 1$ variables, and is very difficult to solve exactly.

doi:10.1088/978-0-7503-2167-9ch7

There is a case that falls within the above definition that can be solved, and that is the case where the particles move independently of each other. The potential for non-interacting particles is

$$V(\vec{r}_1, \vec{r}_2, \vec{r}_3, \dots, \vec{r}_N) = \sum_{i=1}^{N} V_i(\vec{r}_i), \tag{7.5}$$

which means that the Hamiltonian can be written by a sum of independent Hamiltonians

$$H = \sum_{i=1}^{N} H_i, \tag{7.6}$$

where

$$H_i = -\frac{\hbar^2}{2m}\nabla_i^2 + V_i. \tag{7.7}$$

It can be found by substitution that the wave function for non-interacting particles can be written as a product of one-particle wave functions

$$\Psi_E(\vec{r}_1, \sigma_1, \vec{r}_2, \sigma_2, \dots, \vec{r}_N, \sigma_N) = \psi_{\varepsilon_1}(\vec{r}_1, \sigma_1)\psi_{\varepsilon_2}(\vec{r}_2, \sigma_2)\dots\psi_{\varepsilon_N}(\vec{r}_N, \sigma_N), \tag{7.8}$$

which are solutions of

$$H_i\psi_{\varepsilon_i}(\vec{r}_i, \sigma_i) = \varepsilon_i\psi_{\varepsilon_i}(\vec{r}_i, \sigma_i). \tag{7.9}$$

The first investigations in quantum theory focused on obtaining the correct description of the hydrogen atom. When that problem was solved in 1926, it was obvious that the next step would be to apply the theory to larger atoms. In 1927, Douglas Hartree proposed a method for doing this. He assumed that the wave function for a polyelectron atom could be written as a product of single electron wave functions, even though the electrons must interact with each other.

The Hamiltonian for an atom is

$$H = \sum_{i=1}^{N}\left[-\frac{\hbar^2}{2m}\nabla_i^2 - \frac{Ze^2}{|\mathbf{r}_i - \mathbf{R}|} + \frac{1}{2}\sum_{\substack{j=1 \\ j\neq i}}^{N}\frac{e^2}{|\mathbf{r}_i - \mathbf{r}_j|}\right]. \tag{7.10}$$

Hartree's approximation leads to a set of one-electron equations

$$[-\nabla^2 + V_H^a(\vec{r})]\psi_a(\vec{r}, \sigma) = \varepsilon_a\psi_a(\vec{r}, \sigma), \tag{7.11}$$

where $V_H^a(\vec{r})$ is the Hartree effective potential

$$V_H^a(\vec{r}) = -\frac{2Z}{|\vec{r} - \vec{R}|} + \sum_{\substack{b=1 \\ (b\neq a)}}^{N} \int \psi_b^*(\vec{r}', \sigma')\frac{2}{|\vec{r} - \vec{r}'|}\psi_b(\vec{r}', \sigma')dv'. \tag{7.12}$$

The effective potential $V_H^a(\vec{r})$ has a superscript because it differs for each electron a. In words, it is the Coulomb potential for Z positive charges on the nucleus screened by the Coulomb potential arising from the charge density of all the electrons in the atom except for electron a. Hartree arrived at this potential by logical deduction, but a more rigorous derivation is in chapter 12.

The Hartree theory contains a feature that is very important in all treatments of N-electron systems. This feature is 'self-consistency'. Notice that the effective potential that is used to find $\psi_a(\vec{r}, \sigma)$ contains all the other eigenfunctions $\psi_b(\vec{r}, \sigma)$, which are also not known. The way around this quandary is to make a 'good guess' at the $\psi_a(\vec{r}, \sigma)$ and use them in the effective potentials to find a new iteration at these functions. The second iteration at the $\psi_a(\vec{r}, \sigma)$ is then used to calculate new effective potentials and hence another iteration of $\psi_a(\vec{r}, \sigma)$. This iteration process is continued until self-consistency is achieved, which means the last iteration differs from the preceding one by what is considered to be an acceptable error. An easy way to do this is to focus on the energies ε_a. When the ε_a for an iteration are the same, within an acceptable error, as the ones for a previous iteration, self-consistency is achieved.

Since the Hamiltonian doesn't contain spin, the spin-orbitals can be written, $\psi_a(\vec{r}, \sigma) = \phi_a(\vec{r})\chi_a(\sigma)$, where $\chi_a(\sigma) = |+\rangle$ for spin up, or $\chi_a(\sigma) = |-\rangle$ for spin down. The 'integration over spin' is a formal device to describe the inner product between spin vectors, $\langle+|+\rangle = \langle-|-\rangle = 1$ and $\langle+|-\rangle = 0$. The part of the electronic density that comes from the bth orbital is $\rho_b(\vec{r}) = \int \psi_b^*(\vec{r}, \sigma)\psi_b(\vec{r}, \sigma)d\sigma = \phi_b^*(\vec{r})\phi_b(\vec{r})$, so the effective potential can be rewritten,

$$V_H^a(\vec{r}) = -\frac{2Z}{|\vec{r} - \vec{R}|} + \sum_{\substack{b=1 \\ (b \neq a)}}^{N} \int \frac{2\rho_b(\vec{r}')}{|\vec{r} - \vec{r}'|} d\vec{r}'. \tag{7.13}$$

Equation (7.11) with this form for the effective potential is the eigenvalue equation for an electron moving in the electric field of the nuclei and all the other electrons. This is the actual equation used by Hartree. The spin never entered into his derivation.

The Hartree wave function for the atom is the product

$$\Psi_H(\vec{r}_1, \vec{r}_2, \dots \vec{r}_N) = \phi_1(\vec{r}_1)\phi_2(\vec{r}_2)\dots\phi_N(\vec{r}_N), \tag{7.14}$$

where 1,2,… denote the occupied states of the atom. There is nothing in Hartree's theory to prevent all the electrons from being put in the lowest energy state, but it was known that this is not the correct result. Bohr had earlier used the idea that the one-electron states were some generalization of the states of the hydrogen atom. He proposed that the first two electrons are in the lowest energy state (hydrogen and helium), but the next one must go into a higher energy state (lithium). By a close analysis of the chemistry of the periodic table, he was able to get a reasonable picture of the occupations of the electrons which he called the aufbauprinzip.

7.2 Pauli exclusion principle

Bohr's aufbauprinzip introduced the idea that, unlike any known classical system, the ground state of the atom would not be the state in which all electrons were in their lowest energy states. Wolfgang Pauli went the next step that no two electrons could have the same quantum numbers. In order for his scheme to work, he had to introduce an additional quantum number that could have two values. This quantum number was later identified as the spin of the electron. Thus the first electron would go into the state $n = 1, l = 0, m = 0, +$, and the second would have the quantum numbers $n = 1$, $l = 0, m = 0, -$. The next two electrons would have the quantum numbers $n = 2, l = 0$, $m = 0$ with $+$ and $-$. The next six electrons will have $n = 2$ and $l = 1$. They have $m = -1, 0, 1$ with $+$ and $-$. This describes the first ten atoms on the periodic table, hydrogen through neon.

Several authors noted that, if the N-electron wave functions have arguments that contain both positions and spin $a_i = \{x_i, y_i, z_i, \sigma_i,\}$, then the requirement that this function is antisymmetric under the interchange of two arguments

$$\Psi(\ldots, a_j, \ldots, a_i, \ldots) = -\Psi(\ldots, a_i, \ldots, a_j, \ldots), \tag{7.15}$$

leads to the Pauli exclusion principle. It was also pointed out by John Slater among others that, if the wave function was written as a product of one-electron wave functions, the easiest way to assure that it is antisymmetric is to write it as a determinant

$$\Psi_E(a_1, a_2, a_3, \ldots, a_N) = \frac{1}{\sqrt{N!}} \begin{vmatrix} \psi_{\varepsilon_1}(a_1) & \psi_{\varepsilon_1}(a_2) & \psi_{\varepsilon_1}(a_3) & \cdots & \psi_{\varepsilon_1}(a_N) \\ \psi_{\varepsilon_2}(a_1) & \psi_{\varepsilon_2}(a_2) & \psi_{\varepsilon_2}(a_3) & \cdots & \psi_{\varepsilon_2}(a_N) \\ \psi_{\varepsilon_3}(a_1) & \psi_{\varepsilon_3}(a_2) & \psi_{\varepsilon_3}(a_3) & \cdots & \psi_{\varepsilon_3}(a_N) \\ \cdots & \cdots & \cdots & \cdots & \cdots \\ \psi_{\varepsilon_N}(a_1) & \psi_{\varepsilon_N}(a_2) & \psi_{\varepsilon_N}(a_3) & \cdots & \psi_{\varepsilon_N}(a_N) \end{vmatrix}, \tag{7.16}$$

where the ε_i describe the ith quantum state. The interchange of arguments in Ψ_E is equivalent to interchanging columns in the determinant. It is well known that interchanging columns in a determinant changes the sign. If the quantum state ε_i is the same as ε_j then two rows of the determinant are the same. A determinant with two rows equal to each other must be zero, which is the same as the exclusion principle. Physicists call these Slater determinants because he championed their use from an early stage.

The Hartree–Fock theory is designed to find the best spin-orbitals that can be used in a Slater determinant like equation (7.16) because the Pauli exclusion principle is built in to this wave function. It is shown in chapter 12 that this leads to the Hartree–Fock equation

$$\left[-\nabla^2 - \frac{Ze^2}{|r_i - R|} \right] \psi_a(\mathbf{r}, \sigma) + \left\{ \sum_{b \neq a = 1}^{N} \int \psi_b^*(\mathbf{r}', \sigma') \frac{2}{|\mathbf{r} - \mathbf{r}'|} \psi_b(\mathbf{r}', \sigma') dv' \right\} \psi_a(\mathbf{r}, \sigma)$$

$$- \left\{ \sum_{b \neq a = 1}^{N} \int \psi_b^*(\mathbf{r}', \sigma') \frac{2}{|\mathbf{r} - \mathbf{r}'|} \psi_a(\mathbf{r}', \sigma') dv' \right\} \psi_b(\mathbf{r}, \sigma) = \varepsilon_a \psi_a(\mathbf{r}, \sigma)$$

$$\tag{7.17}$$

This equation differs from the Hartree equation by the introduction of the second integral on the left side. That term causes the Hartree–Fock equation to be an integro-differential equation, but it will be seen to be one of the most important contributions of quantum mechanics to the understanding of physical phenomena. It is called the exchange integral.

Since the Hamiltonian doesn't contain spin, the spin-orbitals can be written $\psi_a(\mathbf{r}, \sigma) = \phi_a(\mathbf{r})\chi_a(\sigma)$. The part of the electronic density that comes from the bth orbital is

$$\rho_b(\mathbf{r}) = \int \psi_b^*(\mathbf{r}, \sigma)\psi_b(\mathbf{r}, \sigma)d\sigma = \phi_b^*(\mathbf{r})\phi_b(\mathbf{r}), \tag{7.18}$$

so the above can be rewritten,

$$\left[-\nabla^2 - \frac{Ze^2}{|\mathbf{r}_i - \mathbf{R}|}\right]\phi_a(\mathbf{r}) + \left\{\sum_{b=1}^{N}\int\frac{2\rho_b(\mathbf{r}')}{|\mathbf{r} - \mathbf{r}'|}d\mathbf{r}'\right\}\phi_a(\mathbf{r})$$
$$-\left\{\sum_{b=1}^{N}\int\phi_b^*(\mathbf{r}')\frac{2}{|\mathbf{r} - \mathbf{r}'|}\phi_a(\mathbf{r}')d\mathbf{r}'\delta_{\sigma_b\sigma_a}\right\}\phi_b(\mathbf{r}) = \varepsilon_a\phi_a(\mathbf{r}) \tag{7.19}$$

This is the eigenvalue equation for an electron moving in the electric field of the nucleus and all the other electrons, but with the additional exchange term that only appears when the spins of the a and b electrons point in the same direction. The Hartree–Fock equations will be derived rigorously in chapter 12.

7.3 Atomic structure

Package programs are available at no charge for anyone who wants to do Hartree–Fock (H-F) atomic calculations. The ones based on Schrödinger's equations, as described above, are very accurate for light atoms. For heavier atoms, a simple change in the input file switches to relativistic H-F atomic calculations based on Dirac's equation. The NIST Multiconfiguration Hartree–Fock and Multiconfiguration Dirac-Hartree–Fock Database was developed by Charlotte Froese Fischer. She worked with Douglas Hartree, programming the first Electronic Digital Stored program Automatic Computer (EDSAC) for atomic structure calculations.

It is impossible to understand the periodic table of the elements, figure 7.1, without accurate atomic structure calculations. Starting from a simple picture based on the hydrogenic energy levels, it would be expected that the atoms with 3d outer shells, Sc, Ti, V, Cr, Mn, Fe, Co, Ni, Cu, and Zn would appear before the atoms with 4s orbitals K and Ca. It would also be expected that the 4f Lanthanides would appear before the 5s Rb and Sr as well as the 5p and 6s atoms. All of these arrangements are predicted by relativistic atomic structure calculations.

Highly accurate H-F calculations are the first step in perturbation theory atomic calculations that go beyond the H-F level of approximation. There are calculations that go beyond quantum theory. Feynman's quantum electrodynamic (QED) corrections cannot be compared with experiment without highly accurate H-F

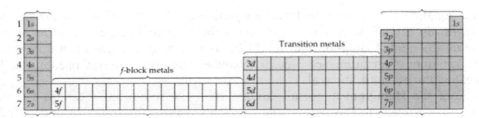

Figure 7.1. A schematic, but correct, picture of the periodic table of the elements.

calculations. Calculations on many atom systems like molecules and solids start from good atomic calculations. These calculations have revolutionized the fields of chemistry, condensed matter physics, and materials science. Modern technological devices like semiconductor chips also rely on H-F atomic calculations as a starting point.

7.4 The hydrogen molecule

Shortly after the publications of Heisenberg and Schrödinger in 1926, Linus Pauling used the new quantum mechanics to carry out the back of the envelope calculation on the hydrogen molecule shown below. He understood that his result explains a great mystery in chemistry. That is, why would two identical atoms decide to bond together and become a diatomic molecule? Not only that, but the bonding is strong, it requires considerable energy to break up a hydrogen molecule into two free atoms. Chemists call this kind of bond a covalent bond, and were unable to find a semiclassical explanation for it. Pauling was given the Nobel prize for his explanation of the covalent bond in 1954 [1].

A diagram that illustrates the structure of the hydrogen molecule is shown in figure 7.2.

In principle, the Hartree–Fock theory can be used to treat a system with two electrons moving in the field of two protons. It is a rather complicated calculation. However, it will be seen that the Rayleigh–Ritz (R-R) variational method can be used to find the lowest energy levels and wave functions with enough accuracy that the essential physics of the hydrogen molecule is clarified.

The wave functions for individual hydrogen atoms centered at the positions $\vec{R}_a$ and $\vec{R}_b$ will be the trial functions in the R-R equations. The ground state wave function corresponding to $n = 1$ and $l = 0$ is

$$R_0(r) = N_0 e^{-r/a_0}, \tag{7.20}$$

where a_0 is called the Bohr radius

$$a_0 = \frac{\hbar^2}{me^2} = 5.29 \times 10^{-11} \text{meters}. \tag{7.21}$$

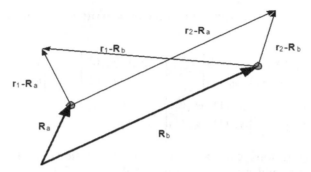

Figure 7.2. The two electrons in the hydrogen molecule are labeled 1 and 2. The two protons that are the hydrogen nuclei are labeled a and b.

The Hamiltonian for the hydrogen molecule in dimensionless units is

$$H = \sum_{i=1}^{2}\left[-\nabla_i^2 - \frac{2}{|\vec{r}_i - \vec{R}_a|} - \frac{2}{|\vec{r}_i - \vec{R}_b|} + \frac{1}{|\vec{r}_i - \vec{r}_i|} + \frac{1}{|\vec{r}_i - \vec{r}_2|} \right]. \qquad (7.22)$$

Associating electron 1 with proton a leads to

$$H = H_{1a} + H_{2b} + H_{ab}$$

$$H_{1a} = -\nabla_1^2 - \frac{2}{|\vec{r}_1 - \vec{R}_a|} \quad H_{2b} = -\nabla_2^2 - \frac{2}{|\vec{r}_2 - \vec{R}_b|}, \qquad (7.23)$$

$$H_{ab} = -\frac{2}{|\vec{r}_2 - \vec{R}_a|} - \frac{2}{|\vec{r}_1 - \vec{R}_b|} + \frac{2}{|\vec{r}_1 - \vec{r}_2|}$$

while associating electron 2 with proton a leads to

$$H = H_{2a} + H_{1b} + H_{ba}$$

$$H_{2a} = -\nabla_2^2 - \frac{2}{|\vec{r}_2 - \vec{R}_a|} \quad H_{1b} = -\nabla_1^2 - \frac{2}{|\vec{r}_1 - \vec{R}_b|}. \qquad (7.24)$$

$$H_{ba} = -\frac{2}{|\vec{r}_1 - \vec{R}_a|} - \frac{2}{|\vec{r}_2 - \vec{R}_b|} + \frac{2}{|\vec{r}_1 - \vec{r}_2|}$$

A simple approximation for the ground state of this molecule will be obtained using the **R-R** variational method with the ground state orbitals of the isolated hydrogen atoms as basis functions and writing them

$$a(\vec{r}) = Ne^{-|\vec{r}_i - \vec{R}_a|} \quad b(\vec{r}) = Ne^{-|\vec{r}_i - \vec{R}_b|}. \qquad (7.25)$$

The constant N is chosen to normalize these functions. In chapter 3 it was made clear that an electron has both a position wave function and a spin. Attaching spin functions to the orbitals defined above leads to four spin-orbitals

$$\psi_{a+}(\mathbf{r}_i, \sigma_i) = a(\mathbf{r}_i)|+\rangle \quad \psi_{b+}(\mathbf{r}_i, \sigma_i) = b(\mathbf{r}_i)|+\rangle$$

$$\psi_{a-}(\mathbf{r}_i, \sigma_i) = a(\mathbf{r}_i)|-\rangle \quad \psi_{b-}(\mathbf{r}_i, \sigma_i) = b(\mathbf{r}_i)|-\rangle. \qquad (7.26)$$

A general antisymmetric wave function can be written as a linear combination of determinants

$$\Psi(\mathbf{r}_1, \sigma_1, \mathbf{r}_2, \sigma_2) = C_1 \begin{vmatrix} \psi_{a+}(1) & \psi_{a+}(2) \\ \psi_{a-}(1) & \psi_{a-}(2) \end{vmatrix} + C_2 \begin{vmatrix} \psi_{b+}(1) & \psi_{b+}(2) \\ \psi_{b-}(1) & \psi_{b-}(2) \end{vmatrix} + C_3 \begin{vmatrix} \psi_{a+}(1) & \psi_{a+}(2) \\ \psi_{b+}(1) & \psi_{b+}(2) \end{vmatrix} +$$

$$+ C_4 \begin{vmatrix} \psi_{a-}(1) & \psi_{a-}(2) \\ \psi_{b-}(1) & \psi_{b-}(2) \end{vmatrix} + C_5 \begin{vmatrix} \psi_{a+}(1) & \psi_{a+}(2) \\ \psi_{b-}(1) & \psi_{b-}(2) \end{vmatrix} + C_6 \begin{vmatrix} \psi_{a-}(1) & \psi_{a-}(2) \\ \psi_{b+}(1) & \psi_{b+}(2) \end{vmatrix}. \tag{7.27}$$

These six configurations can be used as trial functions in a R-R variational calculation. That would give the energies of the ground state and some low level excited states, and the corresponding coefficients C_i.

As it turns out, the problem solves itself. The Hamiltonian doesn't contain spin, so it commutes with the total spin $\mathbf{S}^2 = \mathbf{S}_1^2 + \mathbf{S}_2^2 + 2S_{1x}S_{2x} + 2S_{1y}S_{2y} + 2S_{1z}S_{2z}$, and its z component $S_z = S_{1z} + S_{2z}$. The eigenfunctions of these operators are defined by $\chi_{l,m}$ where $S^2\chi_{l,m} = l(l+1)\hbar^2\chi_{l,m}$ and $S_z\chi_{l,m} = m\hbar\chi_{l,m}$. As was shown in chapter 1,

$$\chi_{00} = \frac{1}{\sqrt{2}}[|+\rangle\,|-\rangle - |-\rangle\,|+\rangle]$$

$$\chi_{1-1} = |-\rangle\,|-\rangle \quad \chi_{10} = \frac{1}{\sqrt{2}}[|+\rangle\,|-\rangle + |-\rangle\,|+\rangle] \quad \chi_{11} = |+\rangle\,|+\rangle \tag{7.28}$$

In the function $\Psi(\mathbf{r}_1, \sigma_1, \mathbf{r}_2, \sigma_2)$ defined above, ignore the first two configurations because they put both electrons on one atom, which will be a very high energy state. The third configuration is

$$\psi_3 = \psi_{11} = [a(\mathbf{r}_1)b(\mathbf{r}_2) - b(\mathbf{r}_1)a(\mathbf{r}_2)]\chi_{11}, \tag{7.29}$$

and the fourth is

$$\psi_4 = \psi_{1-1} = [a(\mathbf{r}_1)b(\mathbf{r}_2) - b(\mathbf{r}_1)a(\mathbf{r}_2)]\chi_{1-1}. \tag{7.30}$$

Operating the Hamiltonian in the two forms written above on the position part of these functions gives

$$H\big[a(\mathbf{r}_1)b(\mathbf{r}_2) - b(\mathbf{r}_1)a(\mathbf{r}_2)\big] = 2\varepsilon_0\big[a(\mathbf{r}_1)b(\mathbf{r}_2) - b(\mathbf{r}_1)a(\mathbf{r}_2)\big]$$
$$+ \big[H_{ab}a(\mathbf{r}_1)b(\mathbf{r}_2) - H_{ba}b(\mathbf{r}_1)a(\mathbf{r}_2)\big]. \tag{7.31}$$

This is almost an energy eigenstate while χ_{11} and χ_{1-1} are exact spin eigenstates. Thus ψ_{11} and ψ_{1-1} are already optimized trial functions.

The two configurations

$$\psi_5 = [a(\mathbf{r}_1)b(\mathbf{r}_2)|+\rangle\,|-\rangle - b(\mathbf{r}_1)a(\mathbf{r}_2)|-\rangle\,|+\rangle], \tag{7.32}$$

and

$$\psi_6 = [a(\mathbf{r}_1)b(\mathbf{r}_2)|-\rangle\,|+\rangle - b(\mathbf{r}_1)a(\mathbf{r}_2)|+\rangle\,|-\rangle], \tag{7.33}$$

are neither approximate energy nor spin eigenfunctions, so they are not satisfactory trial functions. However, linear combinations of these two configurations are

$$\frac{1}{\sqrt{2}}(\psi_5 + \psi_6) = \psi_{10} = [a(r_1)b(r_2) - b(r_1)a(r_2)]\chi_{10}, \qquad (7.34)$$

and

$$\frac{1}{\sqrt{2}}(\psi_5 - \psi_6) = \psi_{00} = [a(r_1)b(r_2) + b(r_1)a(r_2)]\chi_{00}, \qquad (7.35)$$

which are satisfactory.

Because of the spin functions, the R-R Hamiltonian matrix is diagonal by construction

$$\mathbf{H} = \begin{pmatrix} H_{00,\,00} & 0 & 0 & 0 \\ 0 & H_{11,\,11} & 0 & 0 \\ 0 & 0 & H_{10,\,10} & 0 \\ 0 & 0 & 0 & H_{1-1,\,1-1} \end{pmatrix}, \qquad (7.36)$$

as is the overlap matrix

$$\mathbf{\Delta} = \begin{pmatrix} \Delta_{00,\,00} & 0 & 0 & 0 \\ 0 & \Delta_{11,\,11} & 0 & 0 \\ 0 & 0 & \Delta_{10,\,10} & 0 \\ 0 & 0 & 0 & \Delta_{1-1,\,1-1} \end{pmatrix}. \qquad (7.37)$$

It follows that the step of diagonalizing these matrices is unnecessary.

The matrix elements of the Hamiltonian between the states with spin one are

$$H_{1m,1m'} = \int \int \psi_{1m}^* H \psi_{1m} dv_1 dv_2 \delta_{mm'}. \qquad (7.38)$$

The application of Schur's lemma from chapter 4 leads to the result that the three diagonal elements are the same, but that can be seen by simple inspection

$$H_{1m,1m} = \int \int [a(\vec{r}_1)b(\vec{r}_2) - b(\vec{r}_1)a(\vec{r}_2)]^* H [a(\vec{r}_1)b(\vec{r}_2) - b(\vec{r}_1)a(\vec{r}_2)] d\vec{r}_1 d\vec{r}_2. \qquad (7.39)$$

Using equation (7.31), this integral is

$$H_{1m,1m} = + \int \int [a(\vec{r}_1)b(\vec{r}_2) - b(\vec{r}_1)a(\vec{r}_2)]^* [H_{ab} a(\vec{r}_1)b(\vec{r}_2) - H_{ba} b(\vec{r}_1)a(\vec{r}_2)] d\vec{r}_1 d\vec{r}_2. \qquad (7.40)$$

Using the fact that this molecule is invariant under a reflection through a plane bisecting the line connecting the two nuclei, and hence the interchange of a and b

$$H_{1m,1m} = + 2 \int \int a(\vec{r}_1)b(\vec{r}_2) H_{ab} a(\vec{r}_1)b(\vec{r}_2) d\vec{r}_1 d\vec{r}_2 - 2 \int \int a(\vec{r}_1)b(\vec{r}_2) H_{ba} b(\vec{r}_1)a(\vec{r}_2) d\vec{r}_1 d\vec{r}_2, \qquad (7.41)$$

and

$$\Delta_{1m,1m} = \int\int \left[a(\vec{r}_1)b(\vec{r}_2) - b(\vec{r}_1)a(\vec{r}_2) \right]^* \left[a(\vec{r}_1)b(\vec{r}_2) - b(\vec{r}_1)a(\vec{r}_2) \right] d\vec{r}_1 d\vec{r}_2. \tag{7.42}$$

The matrix element

$$H_{00,\,00} = \int\int \psi_{00}^* H \psi_{00} dv_1 dv_2, \tag{7.43}$$

is

$$H_{00,\,00} = + \int\int \left[a(\vec{r}_1)b(\vec{r}_2) + b(\vec{r}_1)a(\vec{r}_2) \right]^* \left[H_{ab}a(\vec{r}_1)b(\vec{r}_2) + H_{ba}b(\vec{r}_1)a(\vec{r}_2) \right] d\vec{r}_1 d\vec{r}_2, \tag{7.44}$$

or

$$H_{00,\,00} = + 2\int\int a(\vec{r}_1)b(\vec{r}_2)H_{ab}a(\vec{r}_1)b(\vec{r}_2)d\vec{r}_1 d\vec{r}_2 + 2\int\int a(\vec{r}_1)b(\vec{r}_2)H_{ba}b(\vec{r}_1)a(\vec{r}_2)d\vec{r}_1 d\vec{r}_2, \tag{7.45}$$

and

$$\Delta_{00,\,00} = \int\int \left[a(\vec{r}_1)b(\vec{r}_2) + b(\vec{r}_1)a(\vec{r}_2) \right]^* \left[a(\vec{r}_1)b(\vec{r}_2) + b(\vec{r}_1)a(\vec{r}_2) \right] d\vec{r}_1 d\vec{r}_2. \tag{7.46}$$

Using the fact that the functions $a(\vec{r})$ and $b(\vec{r})$ are normalized, the above expressions can be simplified somewhat

$$\Delta_{1m,1m} = 2 - 2S^2, \tag{7.47}$$

and

$$\Delta_{00,\,00} = 2 + 2S^2, \tag{7.48}$$

where

$$S = \int a(\vec{r})b(\vec{r})d\vec{r}, \tag{7.49}$$

is called the overlap integral. It is a measure of the overlap of a hydrogen wave function on one site with that on the other site. The first integral in equations (7.41) and (7.45) can be written

$$\begin{aligned} 2J &= \int\int \rho_a(\vec{r}_1)\left[-\frac{2}{|\vec{r}_2 - \vec{R}_a|} - \frac{2}{|\vec{r}_1 - \vec{R}_b|} + \frac{2}{|\vec{r}_1 - \vec{r}_2|} \right]\rho_b(\vec{r}_2)d\vec{r}_1 d\vec{r}_2 \\ &= -2\int\int \frac{\rho_a(\vec{r}_1)\rho_b(\vec{r}_2)}{|\vec{r}_2 - \vec{R}_a|}d\vec{r}_1 d\vec{r}_2 - 2\int\int \frac{\rho_a(\vec{r}_1)\rho_b(\vec{r}_2)}{|\vec{r}_1 - \vec{R}_a|}d\vec{r}_1 d\vec{r}_2 + 2\int\int \frac{\rho_a(\vec{r}_1)\rho_b(\vec{r}_2)}{|\vec{r}_1 - \vec{r}_2|}d\vec{r}_1 d\vec{r}_2, \end{aligned} \tag{7.50}$$

and is obviously the part of the Coulomb energy that is not included in the one-particle Hamiltonians. The second integral in equations (7.41) and (7.45) can be written

$$2K = \int\int a(\vec{r}_1)b(\vec{r}_1)\left[-\frac{2}{|\vec{r}_2 - \vec{R}_a|} - \frac{2}{|\vec{r}_1 - \vec{R}_b|} + \frac{2}{|\vec{r}_1 - \vec{r}_2|} \right]a(\vec{r}_2)b(\vec{r}_2)d\vec{r}_1 d\vec{r}_2, \tag{7.51}$$

which cannot be put into any form that has a classical analog. This is the exchange energy, like the one discussed in the Hartree–Fock theory.

From the $1m$ part of the equations

$$H_{1m,1m} - \varepsilon_{1m}\Delta_{1m,1m} = 0 \tag{7.52}$$

it is found that the three states with parallel spins have the same energy

$$\varepsilon_{1m} = 2\varepsilon_0 + \frac{J - K}{1 - S^2}. \tag{.7.53}$$

From the 00 part of the equations

$$H_{00,\,00} - \varepsilon_{00}\Delta_{00,\,00} = 0, \tag{7.54}$$

and

$$\varepsilon_{00} = 2\varepsilon_0 + \frac{J + K}{1 + S^2}. \tag{7.55}$$

When the integral in equation (7.51) is carried out, it is found that the exchange term is negative, $K < 0$. From this it is found that the state corresponding to antiparallel electrons has the lowest energy

$$\varepsilon_{00} < \varepsilon_{1m}. \tag{7.56}$$

The fact that the Hamiltonian for the hydrogen molecule is symmetric under a reflection through a plane bisecting the line connecting the two protons means that it is invariant under the dihedral group D1. As explained in chapter 4, it follows from this that the eigenfunctions must be symmetric or antisymmetric under interchange of arguments. The eigenfunctions $\psi_{00}(\mathbf{r}_1, \sigma_1, \mathbf{r}_2, \sigma_2)$ and $\psi_{1m}(\mathbf{r}_1, \sigma_1, \mathbf{r}_2, \sigma_2)$ both antisymmetric when the space and spin components are interchanged together, as they must be to satisfy the antisymmetry principle. The eigenfunction $\psi_{00}(\mathbf{r}_1, \sigma_1, \mathbf{r}_2, \sigma_2)$ is antisymmetric under spin interchange, but symmetric under interchange of the spatial components, as seen in figure 7.3.

This is called the bonding state. The eigenfunction $\psi_{1m}(\mathbf{r}_1, \sigma_1, \mathbf{r}_2, \sigma_2)$ is symmetric under spin interchange, but antisymmetric under interchange of the spatial components (figure 7.4).

This is called the antibonding state.

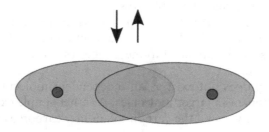

Figure 7.3. The ground state of the hydrogen molecule.

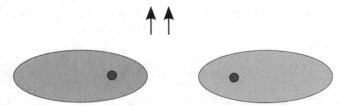

Figure 7.4. The excited state of the hydrogen molecule.

From the preceding algebra, the difference in energy between the bonding and antibonding state of the hydrogen molecule is

$$\varepsilon_{1m} - \varepsilon_{00} = \frac{-2K}{1 - S^2},$$ (7.57)

so the bonding of the hydrogen molecule is due to the exchange interaction. The exchange interaction has no analog in the classical world.

7.5 Covalent bonding

All of the field of organic chemistry is based on the covalent bond. After Pauling's work, various rules of thumb were developed that made it possible to predict the structure and behavior of organic molecules. One of these is the concept of spin pairing, although the above calculation shows that the Hamiltonian doesn't contain a spin term. Since humans are made out of complex organic molecules, they should find covalent bonds interesting. Today, quantum chemists use large computers to calculate the structure and behavior of organic molecules from first principles.

The simplest organic molecules are made up of carbon combined with one or more hydrogen atoms. The electronic structure of carbon is $1s^2 2s^2 2p^2$. The two 1s electrons have very low energies and play no role in the bonding process. The 2s and 2p orbitals have almost the same energy, and often mix.

A single 2s and 2p orbital can combine to make two sp orbitals as shown in figures 7.5 and 7.6.

Each lobe of the carbon atom can then form a covalent bond with a hydrogen atom according to the calculation in the preceding section as in figure 7.7.

The tratomic molecule CH_2 is called acetylene. It is highly combustible with the evolution of great heat, and is used in welding. Silver acetylide and copper acetylide are powerful and very dangerous explosives.

One 2s orbital can combine with two 2p orbitals to create three sp^2 orbitals. The carbon atom then has three lobes lying in a plane and pointing 120° from each other. Three carbon atoms can be attached to these lobes to form a methyl group CH_3 as illustrated in figure 7.8.

The methyl group does not exist as an independent molecule, but it frequently occurs as as a component in larger molecules. Another simple example is to remove one of the hydrogens and bond the carbon to another carbon as shown in figure 7.9 to make the extremely useful compound ethylene C_2H_4.

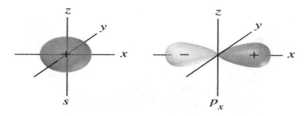

Figure 7.5. The 2s and 2p orbitals.

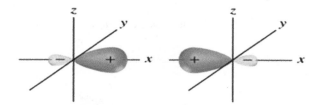

Figure 7.6. The two sp orbitals.

Figure 7.7. The structure of acetylene.

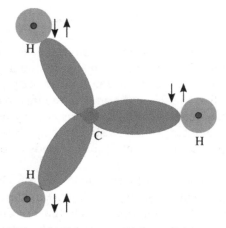

Figure 7.8. The structure of the methyl group.

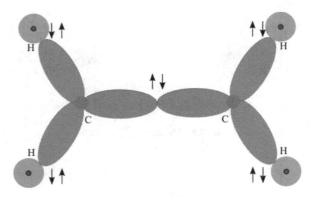

Figure 7.9. The structure of ethylene.

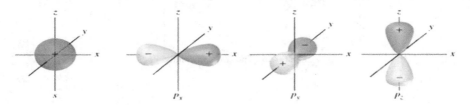

Figure 7.10. The 2s and 2p orbitals.

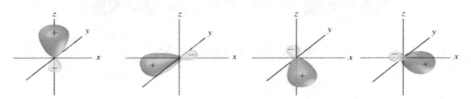

Figure 7.11. The sp^3 orbitals.

Because it contains a carbon–carbon double bond, ethylene is classified as an unsaturated hydrocarbon. Carbon–carbon bonds allow larger and larger organic molecules to be made. The molecules in the human body can contain millions of atoms, including atoms like sodium and iron. Combining an ethylene molecule with a water molecule produces ethyl alcohol C_2H_5OH, widely used in beverages.

An electron can be promoted from a 2s state to a 2p state. One 2s electrons then combines with three 2p electrons to get four sp^3 bonds as demonstrated in figures 7.10 and 7.11.

In figure 7.11 it is shown how four lobes radiate out from the carbon atom in three dimensions, and their geometry is such that they are as far apart as possible.

Bonding a hydrogen atom to each lobe leads to the molecule CH_4, which is called methane. Methane is the major component of natural gas, which is widely used in

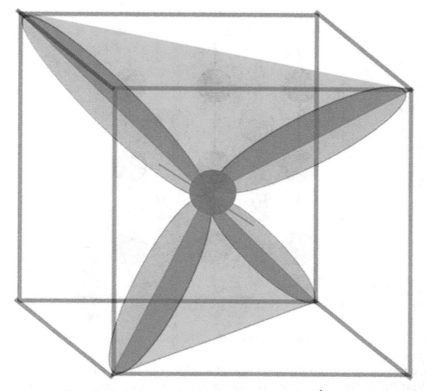

Figure 7.12. A diagram showing the directions of the sp^3 orbitals.

steam plants to produce electricity. Many large organic molecules have components that are sp^3 bonded.

Pure carbon exists as a solid in two different forms. In graphite all of the atoms are in the sp^2 state. They form carbon–carbon double bonds to create layers with hexagonal symmetry. The bonding within the layer is very strong, but that between layers is not. In diamond, all of the atoms are in the sp^3 state. The units look like the picture in figure 7.12, and the diamond structure has cubic symmetry. The strength of the carbon–carbon covalent bond is illustrated by the fact that diamond is one of the hardest materials known. Silicon and germanium have the same crystal structure as diamond, and are held together by covalent bonds. Crystals of these materials are the underlying structures for semiconductor chips. Modern technology is based on them.

7.6 Ionic bonding

Many common crystals, such as table salt NaCl, are held together by ionic bonds. The NaCl structure is two interpenetrating cubic lattices, with positive sodium ions on the lattice sites of one and negative chlorine ions on the sites of the other. A two dimensional slice of this structure is shown in figure 7.13.

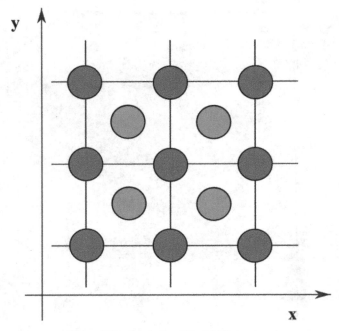

Figure 7.13. One layer of the NaCl structure.

Since the sodium atom transfers one of its electrons to the chlorine atom, it would seem that the bonding of the crystal is just a matter of electrostatics. However, why does this electron transfer take place? Chemists have come up with rules of thumb, ascribing a property to atoms called electronegativity. That simply describes the question but does not answer it. The bonding of NaCl can be explained quite easily using the basic tool of condensed matter physics, energy band theory.

Numerous books have been written on the electronic structure of solids, experimental methods for studying them, and advanced techniques for carrying out extremely accurate calculations of the electronic structure known as band-structure calculations. The oldest and simplest method for calculating energy bands is the tight-binding approximation, and that will be accurate enough to explain the ionic bond in NaCl.

Felix Bloch derived a theorem that states that if the wave function is constructed by placing one atomic orbital $a(\mathbf{r})$ on every lattice site, the eigenfunction of the crystal Hamiltonian will have the form

$$\psi_{\mathbf{k}}(\mathbf{r}) = \frac{1}{\sqrt{N}} \sum_{\mu=1}^{N} e^{i\mathbf{k}\cdot\mathbf{R}_{\mu}} a(\mathbf{r} - \mathbf{R}_{\mu}), \qquad (7.58)$$

where the real vector $\mathbf{k}$ indexes the crystal eigenfunctions. The electronic states are found by calculating the eigenvalues $E(\mathbf{k})$ that correspond to the $\psi_{\mathbf{k}}(\mathbf{r})$. Two electrons will be put in each state, starting with the lowest energy. Bloch's theorem has been derived several ways, but it arises from group theory. As pointed out in chapter 4, when a Hamiltonian is invariant under the operations of a group, the eigenfunctions transform like the basis functions for the irreducible representations of the group. The Hamiltonian is invariant under the group of crystal translations. The irreducible representations are one dimensional and correspond to the $\mathbf{k}$ vectors.

By defining a ket $|\mu\rangle$ so that $\langle \mathbf{r}|\mu\rangle = a(\mathbf{r} - \mathbf{R}_\mu)$, the eigenvector can be written

$$|\mathbf{k}\rangle = \frac{1}{\sqrt{N}} \sum_{\mu=1}^{N} e^{i\mathbf{k}\cdot\mathbf{R}_\mu} |\mu\rangle, \qquad (7.59)$$

and the tight-binding Hamiltonian is written

$$H = \varepsilon \sum_{\mu} |\mu\rangle\langle\mu| + \sum_{\mu\nu} W_{\mu\nu} |\mu\rangle\langle\nu|, \qquad (7.60)$$

where ε is the one-atom energy and $W_{\mu\nu}$ is the interaction potential between sites μ and ν. Using the fact that

$$\frac{1}{N} \sum_{\mu} e^{i(\mathbf{k}-\mathbf{k}')\cdot R_\mu} = \delta_{\mathbf{k}\mathbf{k}'}, \qquad (7.61)$$

the ket $|\mu\rangle$ is equal to

$$|\mu\rangle = \frac{1}{\sqrt{N}} \sum_{\mathbf{k}} e^{-i\mathbf{k}\cdot\mathbf{R}_\mu} |\mathbf{k}\rangle. \qquad (7.62)$$

This is called a lattice Fourier transform. In k-space, the diagonal elements of H are

$$\langle\mathbf{k}|H|\mathbf{k}\rangle = \left[\varepsilon + \sum_{\alpha} e^{i\mathbf{k}\cdot\mathbf{R}_\alpha} W_{\alpha 0} \right], \qquad (7.63)$$

and the off-diagonal elements are zero. Using this one dimensional matrix in the R-R variational method, the eigenvalue associated with $|\mathbf{k}\rangle$

$$E(\mathbf{k}) = \varepsilon + \sum_{\alpha} e^{i\mathbf{k}\cdot\mathbf{R}_\alpha} W_{\alpha 0}. \qquad (7.64)$$

The crystal that is being described by this simplest of tight-binding models is taken to be cubic as illustrated in figure 7.14.

It is conventional to plot $E(\mathbf{k})$ along one direction in k-space, for example as shown in figure 7.15 the x direction. Assuming that the interatomic potential only links nearest-neighbor atoms

$$E(k_x) = \varepsilon - 2U \cos k_x d, \qquad (7.65)$$

where d is the lattice spacing and $U = -W$.

It can be seen from equations (7.64) and (7.65) that $E(\mathbf{k})$ is periodic in k-space.

This simple tight-binding model can be extended to treat a crystal with two atoms per unit cell like NaCl. A picture of the lattice is in figure 7.13. The positions of the two atoms relative to the center of the unit cell are $\mathbf{r}_1$ and $\mathbf{r}_2$, and the atomic energies are ε^1 and ε^2. The tight-binding Hamiltonian for this case is

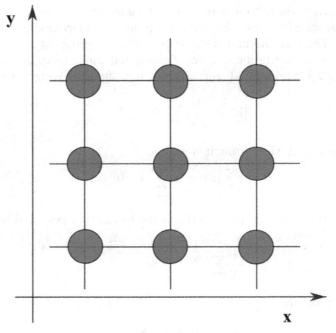

Figure 7.14. One layer of the simple cubic lattice.

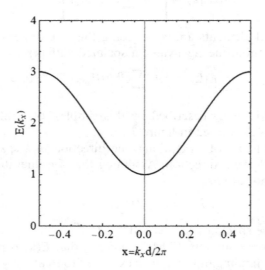

Figure 7.15. The energy as a function of k in the x direction. The abscissa is k_x in units of $2\pi/d$. The atomic energy ε is 2.0 and U is 0.5.

$$H = \varepsilon^1 \sum |1, \mu\rangle\langle 1, \mu | + \varepsilon^2 \sum |2, \mu\rangle\langle 2, \mu| + \sum_{ij\mu\nu} W_{\mu\nu}^{ij} |i, \mu\rangle\langle j, \nu|. \qquad (7.66)$$

The ket $|i, \mu\rangle$ is such that $\langle \mathbf{r}|i, \mu\rangle = a(\mathbf{r} - \mathbf{r}_i - \mathbf{R}_\mu)$ where, for simplicity, it is assumed that atoms 1 and 2 have the same orbitals. The ket $|i, \mu\rangle$ can be written

$$|i, \mu\rangle = \frac{1}{\sqrt{N}} \sum_{\mathbf{k}} e^{-ik\cdot(\mathbf{r}_i+\mathbf{R}_\mu)} |i, \mathbf{k}\rangle, \qquad (7.67)$$

where

$$|i, \mathbf{k}\rangle = \frac{1}{\sqrt{N}} \sum \mu e^{ik\cdot(\mathbf{r}_i+\mathbf{R}_\mu)} |i, \mu\rangle. \qquad (7.68)$$

There are four matrix elements of H for each $\mathbf{k}$. On the diagonal they are

$$\langle 1, \mathbf{k}|H |1, \mathbf{k}\rangle = \left[\varepsilon^1 + \sum_\alpha e^{ik\cdot\mathbf{R}_\alpha} W_{\alpha 0}^{11} \right]$$
$$\langle 2, \mathbf{k}|H|2, \mathbf{k}\rangle = \left[\varepsilon^2 + \sum_\alpha e^{ik\cdot\mathbf{R}_\alpha} W_{\alpha 0}^{22} \right]. \qquad (7.69)$$

The off-diagonal elements are

$$\langle 1, \mathbf{k}| H |2, \mathbf{k}\rangle = e^{ik\cdot(\mathbf{r}_1-\mathbf{r}_2)} \sum_\mu e^{ik\cdot\mathbf{R}_\mu} W_{\mu 0}^{12}$$
$$\langle 2, \mathbf{k}|H |1, \mathbf{k}\rangle = e^{ik\cdot(\mathbf{r}_2-\mathbf{r}_1)} \sum_\mu e^{ik\cdot\mathbf{R}_\mu} W_{\mu 0}^{21}. \qquad (7.70)$$

The solution of the R-R equations is the equivalent of diagonalizing this Hamiltonian matrix. The energies that are obtained are approximately

$$E_1(\mathbf{k}) = \varepsilon^1 + \sum_\alpha e^{ik\cdot\mathbf{R}_\alpha} W_{\alpha 0}^{11} + \Delta$$
$$E_2(\mathbf{k}) = \varepsilon^2 + \sum_\alpha e^{ik\cdot\mathbf{R}_\alpha} W_{\alpha 0}^{22} - \Delta, \qquad (7.71)$$

where

$$\Delta = \frac{\sum_\mu e^{ik\cdot\mathbf{R}_\mu} W_{\mu 0}^{12} \sum_{\mu'} e^{ik\cdot\mathbf{R}_{\mu'}} W_{\mu'0}^{21}}{\varepsilon^2 - \varepsilon^1}. \qquad (7.72)$$

The primary difference between the two energy bands is that $E_1(\mathbf{k})$ is centered about ε^1 and $E_2(\mathbf{k})$ is centered about ε^2. The function Δ is small and has little k dependence, so it will be ignored. The energy bands in the x direction in k-space are plotted using

$$E_1(k_x) \approx 2.0 - \cos k_x d$$
$$E_2(k_x) \approx -2.0 - \cos k_x d'$$

(7.73)

in which $\varepsilon^1 = 2.0$, $\varepsilon^2 = -2.0$, and $U = 0.5$ (figure 7.16).

The next point about band theory is that the number of **k** states in a filled energy band is N, the number of atoms in the crystal. Taking spin into account, these states can accommodate $2N$ electrons or two electrons per atom. The outer orbital of sodium contains one electron, so the initial state when the atoms are put into a crystal is such that the upper band and the lower band is half filled. In the equilibrium state, the electrons in the upper band drain into the lower band, filling the lower band and leaving an empty upper band as can be seen from figure 7.16. This process gives all of the chlorine atoms an electron creating negative ions. The sodium atoms that lose an electron become positive ions.

Accurate band-structure calculations on ionic crystals with complicated crystal structures produce very complicated band structures. The $E(\mathbf{k})$ curves are sometimes referred to as spaghetti. However, they all have the feature that is illustrated above. The $E(\mathbf{k})$ of the donor atoms are above those of the acceptor atoms, and the sets of bands are separated by a sizable gap. This not only explains the electron transfer, but also it explains the fact that ionic crystals are insulators. There are exchange terms in the band-structure calculations, but ionic bonding is not an exchange phenomena.

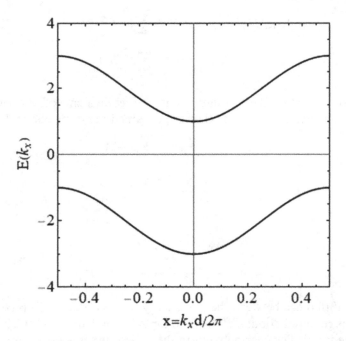

Figure 7.16. A crude approximation to the energy bands of NaCl. The lower band is associated with the chlorine atoms and the upper band with the sodium atoms.

7.7 Bonding in metals

Metals do not have covalent bonds. All of the atoms in a typical metal are the same, so there can be no ionic bonds. The energy bands of a metallic crystal with the atoms on the sites of a simple cubic lattice would look roughly like the band in figure 7.15. Simple metal like sodium has one electron outside of a inert gas core, in the case of sodium the inert gas is neon, so the energy band will be half full. The k states that are filled with two electrons are contained within a sphere in k-space. The radius of the sphere is the Fermi wave vector k_F and the surface of the sphere is called the Fermi surface. Electrons can easily be excited to energy states above the Fermi surface, and this explains why metals are conductors of electricity.

A hand waving argument for the bonding of metals is that, when an atom is in a metal, the outer electron has a wave function that stretches throughout the crystal. Such a wave function is smoother than the original atomic wave function, so the total kinetic energy of the electrons in a metal is reduced. The metallic bond is quite weak compared to the ionic or covalent bond, so it is difficult to explain it with a simple model.

Calculations of the electronic structure of simple metals is considered one of the easier problems in computational condensed matter. All-electron full-potential calculations including relativistic effects can be carried out in a small amount of time on a work station or high power laptop. All-electron means that even the core electrons are included in the calculations. Full-potential means that no simplifying approximations, such as assuming the potential is spherically symmetric, are made. The density functional theory, to be discussed in chapter 12, has been developed to the point that it is more accurate than H-F theory, even if that theory could be applied to crystals.

Calculating the binding energy of a material like sodium is so simple that it does not justify a stand alone publication. For that reason, a sketch of a hypothetical calculation that could be done is shown in figure 7.17.

It was pointed out in the early days of the development of band-structure calculations that the accuracy required to obtain the binding energy of a solid is comparable to calculating the weight of the captain of a cruise ship by weighing the ship with the captain on it and off it. The analogy is carried further by noting that the accuracy required to distinguish between different crystal structures is equivalent to finding the weight of the captains hat by weighing the ship with the captain wearing a hat and not wearing one. Such accuracies are routinely achieved in modern band-structure calculations.

Calculations on hypothetical sodium crystals with fcc, bcc, and hcp crystal structures are in figure 7.17. There are no metals with a simple cubic lattice structure. The calculations predict the correct bcc crystal structure for sodium. They give the correct lattice constant and binding energy to within a few percent.

The nature of the metallic bond can be studied in detail by analyzing the band-structure calculations that give the results in figure 7.17. Most of the bonding is due to the fact that the outer electrons are in band states that stretch throughout the crystal. Even though they are included, the changes in the core states contribute very

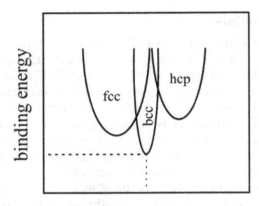

lattice constant

Figure 7.17. Hypothetical calculations of the binding energy of sodium with the atoms on the lattice sites of three different crystal lattices, the face-centered cubic (fcc), the body centered cubic (bcc), and the hexagonal closed packed (hcp). The calculations can be carried for a number of lattice constants for each structure, and the results fitted to a parabola. The dotted lines indicate that sodium is expected to have the bcc structure and they indicate the predicted binding energy and lattice constant.

little. However, it is simplistic to attribute all of the bonding to changes in kinetic energy. It that were the case, the calculations would show no difference between the binding energies of sodium with different crystal structures.

7.8 Conclusions

Quantum mechanical calculation on atoms, molecules, and solids has led to a fundamental understanding of nature that was not available before. With the help of faster computers, calculations on more complicated systems are now done. The richest source of areas for quantum theorists to study come from their colleagues in physics, chemistry, and materials science. They run across interesting new phenomena, either on purpose or through serendipity, and come to the theorist to do calculations that will help them understand their discovery. The calculations can provide a justification of their ideas, or point in a different direction.

Problems

P7.1 Show that the total energy of an atom calculated with the Hartree approximation is not just the sum over filled states of the ε_a.

P7.2 Prove that the states interchanged in equation (7.15) cannot have the same position and spin without assuming that Ψ is a determinant.

P7.3 How can a H-F atomic computer program be used to calculate the ionization energy of an atom?

P7.4 Use the symmetry of the hydrogen molecule and Schur's theorem to explain the shape of the eigenfunctions.

P7.5 Draw pictures to explain the structure and mechanical behavior of graphite.

P7.6 How can the band structure of NaCl be used to argue that it will not conduct electricity?

P7.7 Suppose atoms were rigid balls with a fixed radius a. What would be the density of a crystal made up of these atoms when the crystal structure is simple cubic, fcc, and bcc.

Reference

[1] Pauling L 1939 *The Nature of the Chemical Bond* (New York: Cornell University Press)

Chapter 8

An alternative reality

8.1 Gazing in wonder

Up to this point, the discussion of the quantum theory in this book has followed the standard pedagogical practice of laying things out in a calm and methodical fashion, implying that the discoveries of Einstein, Bohr, Heisenberg, Schrödinger, etc are part of a predictable evolution. Even so, some readers may have found aspects of quantum theory to be strange and counterintuitive or, in a word, weird. The position taken in this chapter is that the reader who finds quantum phenomena to be weird is wrong in the sense that quantum phenomena are much weirder than anything he or she could imagine. Some well known experiments are reanalyzed, emphasizing the aspects that are most difficult to understand. More recent experiments that challenge the conventional understanding of quantum theory are also described

8.2 The Einstein–Podolsky–Rosen experiment

The Einstein–Podolsky–Rosen (EPR) experiment was described in chapter 1, but here it is analyzed from a different angle. Suppose two observers, Alice and Bob are a very long distance from each other. In principle, they could be light years apart. Suppose Alice has two spin 1/2 particles that are bound into a spin 0 pair

$$|\psi\rangle = \frac{1}{\sqrt{2}}(|+\rangle\,|-\rangle - |-\rangle\,|+\rangle). \tag{8.1}$$

They are split apart. Alice keeps one particle and sends the other to Bob.

Stern–Gerlach devices were discussed in chapter 1. Assume Alice and Bob have such devices and both are oriented in the z-direction, (SG$_z$). The experiment is done N times. The outcomes of their measurements are written as in table 8.1.

Experimentally it will be found that $N_1 = N_2$, so the probabilities of these events, calculated quantum mechanically, are

Table 8.1. The N_i in column 1 are the number of times that the measurements described in columns 2 and 3 occur.

Number of occurrences	Alice's measurement	Bob's measurement
N_1	$(\hat{z}+)$	$(\hat{z}-)$
N_2	$(\hat{z}-)$	$(\hat{z}+)$

$$P_1 = \frac{N_1}{N} = 0.5 \quad P_2 = \frac{N_2}{N} = 0.5. \tag{8.2}$$

This experiment appears to show that Alice is communicating the sign of her spin to Bob faster than the speed of light, because the instant she looks at her data and sees a $(\hat{z}+)$, she knows Bob has a $(\hat{z}-)$, even if Bob is millions of miles away. The wave function has collapsed from the one in equation (8.1) to

$$|\psi\rangle = |+\rangle \, |-\rangle. \tag{8.3}$$

The literal interpretation of the wave function, that the particles are actually wave like, leads to the conclusion that the instantaneous collapse of the wave function when one observer makes a measurement means 'stuff' is moving faster than the speed of light. The more sophisticated Born picture of the wave function leads to other difficulties. Einstein believed that it would be manifestly wrong for the EPR experiment to work because it would violate what he called the locality principle. The result of an experiment cannot be influenced by an action a long way off. He proposed the EPR as a gedanken experiment to prove that quantum mechanics is wrong.

Einstein's argument has been refuted. The direct proof is that the EPR experiment can actually be done today and, when it is, it agrees exactly with the orthodox quantum mechanical prediction. How, then, can Einstein's arguments that the laws of locality and relativity are violated be refuted? The answer is that the EPR cannot be used to send information in the usual sense. In the course of the experiment, Bob simply has an apparently random list of $|+\rangle$ and $|-\rangle$ results. It is only after he and Alice compare their results, with a signal that is subluminal, that they understand the correlations.

Although a message cannot be sent with the EPR method, it does have practical applications. As will be discussed later, it can be used for quantum key distribution.

8.3 Hidden variables

Einstein conceded that his claim that the EPR experiment violates the tenets of relativity does not hold up, but he still did not like the conclusions. He went back to an idea that he had proposed when the quantum theory first began to take shape, called the hidden variables theory. The proposal is that quantum theory is like the early theories of thermodynamics and fluid dynamics. The mathematics developed in those areas is very sophisticated and highly predictive of the phenomena that are observed. However, it is now known that there is an underlying reality and that the materials being treated as continua are actually made up of atoms and molecules. The more recent theory of statistical mechanics and molecular dynamics calculations obtain the same results on the basis of atomic theory. Einstein's argument is that Born's statistical interpretation of the wave function is just a statement of our ignorance of the underlying physics, and that, later, all quantum phenomena can be explained on the basis of variables that are presently hidden from us.

The simplest hidden variable explanation of the EPR experiment is that the spin-zero wave function in equation (8.2) is a fiction. For a reason that is not understood now, Alice's electrons must be created in pairs with one having spin up and the other having spin down. If Alice sends an up electron to Bob, then classical reasoning says the one she kept is down. Since Alice can't see which way the electron spin points before she measures it, statistically the probability she will send Bob an up or down electron is 0.5.

Although the hidden variable theory described above gives the correct answer for the EPR experiment, it requires a complete restructuring of classical theory in order to predict the hidden variable state for the electron pairs. There are still some physicists who find quantum theory so logically unsatisfying that they have attempted this. When put into practice, hidden variable theories become extremely convoluted and most physicists ignore them.

The state described in equation (8.1) is now called an entangled state. In chapter 10 it will be seen that entangled states are fundamental components in the devices used in the field called quantum information (QI). They are so important that experimentalists have developed numerous practical methods for making entangled states with photons, electrons, and other things. QI devices are already being used in cryptography and computing, and a growing number of high tech companies are selling them. It might be said that entangled states are becoming standard engineering practice.

It would appear that hidden variable theory is finished, but it is necessary to understand both sides of the argument because there are some who believe that they offer something to quantum theory. Papers are still being written that revive some of the arguments.

8.4 Bell's inequalities

The physicist John Stewart Bell was a great admirer of Einstein, and found Einstein's arguments for hidden variables compelling. He developed theories for more advanced forms of the EPR experiment with the hope of supporting hidden variables. His best known attempt is the following.

Suppose Alice and Bob have three Stern–Gerlach devices oriented in three directions specified by the unit vectors $\hat{a}$, $\hat{b}$, $\hat{c}$. Alice and Bob can measure the spin direction of a particle with any one of these devices. The notation $(\hat{a} + , \hat{b} - , \hat{c}+)$ means that, if the observer uses the device with the orientation $\hat{a}$ they will see the particle has spin up, if they use their second device that has orientation $\hat{b}$ they will find it with spin down, or if they use their third device they

Table 8.2. The N_i in column 1 are the number of times that the measurements described in columns 2 and 3 occur.

Number of occurrences	Alice's measurement	Bob's measurement
N_1	$(\hat{a}+, \hat{b}+, \hat{c}+)$	$(\hat{a}-, \hat{b}-, \hat{c}-)$
N_2	$(\hat{a}+, \hat{b}+, \hat{c}-)$	$(\hat{a}-, \hat{b}-, \hat{c}+)$
N_3	$(\hat{a}+, \hat{b}-, \hat{c}+)$	$(\hat{a}-, \hat{b}+, \hat{c}-)$
N_4	$(\hat{a}+, \hat{b}-, \hat{c}-)$	$(\hat{a}-, \hat{b}+, \hat{c}+)$
N_5	$(\hat{a}-, \hat{b}+, \hat{c}+)$	$(\hat{a}+, \hat{b}-, \hat{c}-)$
N_6	$(\hat{a}-, \hat{b}+, \hat{c}-)$	$(\hat{a}+, \hat{b}-, \hat{c}+)$
N_7	$(\hat{a}-, \hat{b}-, \hat{c}+)$	$(\hat{a}+, \hat{b}+, \hat{c}-)$
N_8	$(\hat{a}-, \hat{b}-, \hat{c}-)$	$(\hat{a}+, \hat{b}+, \hat{c}+)$

will find that it has spin up. If the spins are entangled as in equation (8.1), if Alice has the possibilities $(\hat{a}+, \hat{b}-, \hat{c}+)$ Bob must have the possibilities $(\hat{a}-, \hat{b}+, \hat{c}-)$.

The possible observations that Bob and Alice can make and the number of such observations that will be made if the experiment is repeated many times are given in the Table 8.2. The number of occurrences for Alice having a given set of possibilities is hypothetical rather than experimental because it is not possible for her to know the results of all three experiments. It will be seen that the important properties of the N_i are only that they exist and are greater than zero. If hidden variables were assumed to exist, it would in principle be possible to calculate the N_i.

Suppose Alice chooses to use the device that measures in the $\hat{a}$ direction and Bob chooses to use the device that measures in the $\hat{b}$ direction. The probability that they will both see particles with spin up is $P(\hat{a}+; \hat{b}+)$. By scanning through the table it is seen that Alice sees $\hat{a}+$ at the same time Bob sees $\hat{b}+$ only for the occurrences in the third and fourth row, so

$$P(\hat{a}+; \hat{b}+) = \frac{N_3 + N_4}{N}, \tag{8.4}$$

where

$$N = N_1 + N_2 + N_3 + N_4 + N_5 + N_6 + N_7 + N_8. \tag{8.5}$$

Simply from the fact that $N_i > 0$ it is possible to write

$$\frac{N_3 + N_4}{N} \leqslant \frac{N_3 + N_4 + N_2 + N_7}{N} = \frac{N_4 + N_2}{N} + \frac{N_3 + N_7}{N}, \tag{8.6}$$

and again from the table the measurements that are common for Alice and Bob in rows two and four are $\hat{a}+$ and $\hat{c}+$. The common measurements in rows three and seven are $\hat{c}+$ and $\hat{b}+$. It follows that the manipulation in equation (8.6) leads to

$$P(\hat{a}+; \hat{b}+) \leqslant P(\hat{a}+; \hat{c}+) + P(\hat{c}+; \hat{b}+). \tag{8.7}$$

These probabilities can be calculated unambiguously by quantum mechanics because they are the absolute square of an inner product of two known states

$$P(\hat{\mathbf{a}}+;\hat{\mathbf{b}}+) = |\langle\hat{\mathbf{a}}+|\hat{\mathbf{b}}-\rangle|^2 \, P(\hat{\mathbf{a}}+;\hat{\mathbf{c}}+) = |\langle\hat{\mathbf{a}}+|\hat{\mathbf{c}}-\rangle|^2 \, P(\hat{\mathbf{c}}+;\hat{\mathbf{b}}+)$$
$$= |\langle\hat{\mathbf{c}}+|\hat{\mathbf{b}}-\rangle|^2.$$

(8.8)

The reason Bob's plus spin states are replaced by minus states in the inner products is that those states belong to Alice. In order for Bob to see a plus state, Alice must see a minus. According to the rules of quantum mechanics, Alice can create all of the required states and take the inner products. She has Stern–Gerlach devices pointing in the $\hat{\mathbf{a}}$, $\hat{\mathbf{b}}$, and $\hat{\mathbf{c}}$ directions, so it is only a matter of sending enough electrons through the devices to find one with the spin in the proper direction.

All of the equations derived so far are true for any choice of $\hat{\mathbf{a}}$, $\hat{\mathbf{b}}$, and $\hat{\mathbf{c}}$, and the same result would be obtained. However, there is a choice that makes the calculations easier. The choice is to have the vectors in one plane, and to choose the axes in that plane as shown in the following drawing (figure 8.1).

As can be seen from this figure, $|\hat{\mathbf{a}}+\rangle = |+\rangle$ and $|\hat{\mathbf{a}}-\rangle = |-\rangle$. It was shown in chapter 1 that the spin eigenfunction for an arbitrary direction $\hat{\mathbf{n}}$ in the x-z plane is

$$| \hat{\mathbf{n}}+\rangle = \cos\frac{\theta}{2} |+\rangle + \sin\frac{\theta}{2} |-\rangle,$$

(8.9)

and

$$| \hat{\mathbf{n}}-\rangle = -\sin\frac{\theta}{2} |+\rangle + \cos\frac{\theta}{2} |-\rangle.$$

(8.10)

The angle between $\hat{\mathbf{a}}$ and $\hat{\mathbf{b}}$ is called θ_{ab} so

$$\langle\hat{\mathbf{a}}+|\hat{\mathbf{b}}-\rangle = -\sin\frac{\theta_{ab}}{2}.$$

(8.11)

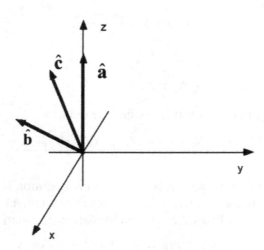

Figure 8.1. Bell's experiment.

It is also easy to see that

$$\langle \hat{\mathbf{a}}+|\hat{\mathbf{c}}-\rangle = -\sin \frac{\theta_{ac}}{2}. \tag{8.12}$$

Reorienting the z axis to be in the $\hat{\mathbf{c}}$ direction gives

$$\langle \hat{\mathbf{c}}+|\hat{\mathbf{b}}-\rangle = -\sin \frac{\theta_{cb}}{2}. \tag{8.13}$$

Equation (8.7) then becomes

$$1/2(\sin \theta_{ab}/2)^2 \leqslant 1/2(\sin \theta_{ac}/2)^2 + 1/2(\sin \theta_{cb}/2)^2 \tag{8.14}$$

where the 1/2 comes from the probability that Alice will measure a spin up in the first place. Choosing the simplest case $\theta_{ac} = \theta_{cb} = 2\varphi$ leads to

$$1/2(\sin 2\phi)^2 = 2(\sin \phi \cos \phi)^2 \leqslant (\sin \phi)^2, \tag{8.15}$$

or

$$\cos^2 \phi \leqslant \frac{1}{2}. \tag{8.16}$$

This is clearly not true for $0 < \phi < \frac{\pi}{4}$. Equation (8.15) is usually called Bell's inequality.

Bell's inequality is conceded to prove that the most common forms of the hidden variable picture are wrong. This caused him considerable discomfort, because it is opposite to what he was hoping for. He later wrote about the hidden variable explanation of a different experiment, the double slit interference experiment 'For me, it is so reasonable to assume that the photons in those experiments carry with them programs, which have been correlated in advance, telling them how to behave. This is so rational that I think that when Einstein saw that, and the others refused to see it, he was the rational man. The other people, although history has justified them, were burying their heads in the sand. I feel that Einstein's intellectual superiority over Bohr, in this instance, was enormous; a vast gulf between the man who saw clearly what was needed, and the obscurantist. So for me, it is a pity that Einstein's idea doesn't work. The reasonable thing just doesn't work'.

Although mortally wounded, this did not destroy the hidden variable program. There are those who simply added another twist to an already convoluted program in order to get around Bell's conclusions.

8.5 Double slit interference

Most physicists are familiar with Young's two slit interference experiment in optics. Feynman was fond of saying that all of quantum mechanics can be gleaned from carefully thinking through the implications of this experiment being applied to electrons. Feynman's statement becomes even more true when modern interference experiments are considered, as will be seen later.

Consider an incident plane wave of light generated, e.g. by a laser, passing through two slits. The ideal double slit interference pattern seen on the screen is $I_{ds} = C(1 + \cos(2\pi dy/\lambda L))$, where d is the distance between slits. The waves will interfere constructively at distances y from the center given by $y = n\frac{\lambda L}{d}$. An ideal interference pattern is shown in figure 8.2.

This is the aspect of the experiment that is of interest, but the experimental results do not look like the above because of the finite width of the slits. Light passing through a single slit produces a diffraction pattern on the screen that is described by

$$I_{ss} = I_0\left[\frac{\sin(2\pi ay/\lambda L)}{2\pi ay/\lambda L}\right]^2. \tag{8.17}$$

In this formula, a is width of slit, L is distance to screen, λ is the wave length, and y is the distance from the central line in the screen. The total interference and diffraction pattern is given by the product $I(y) = I_{ss}I_{ds}$ and shown in figure 8.3.

Young is famous for his experiment on light, and appeared to settle the debate as to whether light is a wave or a particle. It was later shown that Maxwell's equations can be manipulated to obtain a wave equation

$$\nabla^2 f(\mathbf{r}, t) = \frac{1}{c^2}\frac{\partial^2 f(\mathbf{r}, t)}{\partial t^2}, \tag{8.18}$$

that describes electromagnetic waves propagating with the speed of light c. This was interpreted to be the final proof of the wave nature of light, x-rays, radio waves, and all other electromagnet waves. Feynman's quantum electrodynamics (QED) reawakened this discussion, however, because the theory describes light as a collection of photons.

Schrödinger's equation predicts that electrons obey the wave equation

$$\nabla^2 \psi(\mathbf{r}, t) = i\hbar\frac{\partial \psi(\mathbf{r}, t)}{\partial t}. \tag{8.19}$$

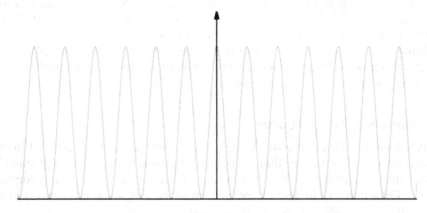

Figure 8.2. Ideal interference pattern.

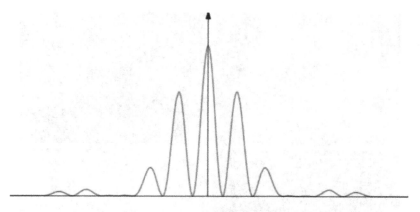

Figure 8.3. Experimental interference and diffraction pattern.

An obvious result of this is to try Young's experiment with electrons. The result is that they give the interference fringes expected for a wave with the de Broglie wave length $\lambda = h/p$. With modern equipment similar experiments can be done with atoms and even molecules.

These experiments seem to be definitive proof of the wave picture, but what happens if the intensity of the electron beam is made very low. A very sensitive experiment with one electron at a time was done at the Hitachi R & D Laboratory [1]. Using an advanced form of the kind of electron gun that was used in television sets at the time. A weak beam of electrons was allowed to pass through two slits. The sequence of spots that were found on the screen is shown in figure 8.4.

It is clear that electrons have a particle nature. All known detectors show the electron only when it interacts with matter, which is made of atoms. This means that they are localized, at least on the atomic scale. In photograph (a) the dots that indicate electrons appear to be random. They cannot be entirely random because, as more and more electrons hit the screen, the interference fringes develop.

There are areas of the screen where the probability for an electron spot to appear is very unlikely, and these regions don't change as the number of electrons increases or decreases. This is an illustration of Born's statistical interpretation of the wave function put forward in the early days of quantum mechanics. The wave function (or at least its absolute square) gives the probability for an event happening. This idea was carried further by J von Neumann who said the there is a state vector that evolves deterministically according to the Schrödinger equation, and that the process of measuring 'projects out' randomly one of the values that are allowed by the wave function. Each particle seems to strike at a random spot. It is only after a number of them have struck the screen that the diffraction pattern emerges.

Einstein noted in 1905 that the photoelectric effect, the ejection of electrons from a solid when light is shown on it, could only be explained by assuming that the light is made up of packets that have energy $h\upsilon$. These packets are the photons from QED. As with electrons, the photons are detected by their interaction with atoms in the screen, and are localized on the atomic scale. The results of double slit

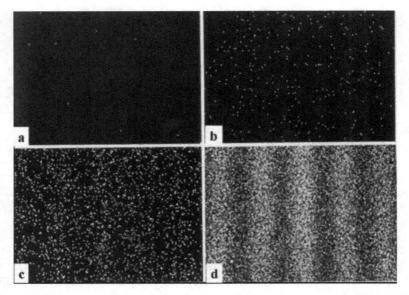

Figure 8.4. The number of electrons accumulated on the screen are; (a) 8 electrons; (b) 270 electrons; (c) 2000 electrons; (d) 160 000. The total exposure time from the beginning to the stage (d) is 20 min.

experiments with photons appear just like the electron results shown above if the beam intensity is low enough. The predictions of Huygens and Fresnel and the derivation from Maxwell's equations only hold in the limit of very many photons. This means that the questions arising in the interpretation of quantum mechanics are the same for photons and particles.

An aspect of quantum interference that can be seen from the low intensity measurements is that an electron or photon can only interfere with itself. If two electron guns are aimed at the slits, even though they fire the electrons simultaneously, there will be no interference. In addition, if an experiment is done to determine which slit the particle passes through, there is no interference. This will be made more clear in the quantum erasure experiments discussed later.

The concept of wave particle duality is introduced early in quantum mechanics courses. The idea is that sometimes an electron will act as a wave, and sometimes as a particle. The remarkable thing about the experiment illustrated in figure 8.4 is that it is showing wave and particle behavior simultaneously. As Bell pointed out, in order to explain this with hidden variables 'the photons in those experiments carry with them programs, which have been correlated in advance, telling them how to behave'. The electron that makes the first spot on the screen somehow has to know where all the other electrons are going to land. The quantum description is surprising, but the hidden variable description is unbelievable.

8.6 The adiabatic theorem

The next experiments to be discussed require a quantum theorem that was developed early. The standard reference for the adiabatic theorem is Volume II of

'*Quantum Mechanics*' by Albert Messiah [2]. He attributes a version derived for Heisenberg's matrix mechanics version of quantum theory to Ehrenfest. The theorem was first proved for Schrödinger's version of quantum mechanics by V Fock in 1928. T Kato and K O Friedrichs worked on a more mathematical form of the theorem, as have others. The derivation in Messiah is complicated and confusing because he talks about a Hamiltonian $H(t)$ that is time dependent without specifying how this time dependence comes about. As is well known, the Hamiltonian contains no explicit time dependence in the Schrödinger picture. It can only have a time dependence through the time dependence of parameters, such as the positions of the nuclei in a molecule in the Born–Oppenheimer approximation or external fields. The notation that makes this clear is $H(R(t))$.

In the Dirac notation, the time dependence of the state vector $|\psi(R(t), t)\rangle$ is given by

$$H(R(t)) |\psi(R(t), t)\rangle = i\hbar \frac{d \mid \psi(R(t), t)\rangle}{dt}, \tag{8.20}$$

or

$$|\psi(R(t), t)\rangle = U(t, t_0)|\psi(R(t), t_0)\rangle. \tag{8.21}$$

The eigenvalues of $H(R(t))$ are time dependent through the time dependence of the parameters

$$H(R(t))|\psi(R(t), t)\rangle = \varepsilon_j(R(t))|\psi(R(t), t)\rangle. \tag{8.22}$$

The adiabatic theorem is that

$$U(t, t_0) = e^{-\frac{i}{\hbar}\int_{t_0}^{t} \varepsilon_j(R(\tau))d\tau}. \tag{8.23}$$

The adiabatic theorem can be used to find the phase for an electron moving in a magnetic field. The Hamiltonian is

$$\frac{\left(\vec{p} - \frac{e}{c}\vec{A}\right)^2}{2m} + v(\vec{r}) = \frac{p^2}{2m} + v(\vec{r}) - \frac{e}{mc}\vec{A} \cdot \vec{p} + \frac{e^2}{2mc^2}A^2. \tag{8.24}$$

Ignoring the last term and using first order perturbation theory

$$\varepsilon = \varepsilon_0 - \frac{e}{mc}\vec{A} \cdot \langle \vec{p} \rangle \approx \varepsilon_0 - \frac{e}{c}\vec{A} \cdot \frac{d\langle \vec{r} \rangle}{dt}, \tag{8.25}$$

the phase in equation (8.23) is

$$-\frac{i}{\hbar}\int_{t_0}^{t} \varepsilon dt = -\frac{i}{\hbar}\varepsilon_0(t - t_0) + \frac{ie}{\hbar c}\int_{\vec{r}(t_0)}^{\vec{r}(t)} \vec{A} \cdot d\vec{r}. \tag{8.26}$$

Dirac [3] pointed out that if space is divided into cubes that are small on a macroscopic scale but large on the quantum scale, then $\vec{A}(\vec{r})$ is a constant in each cube so equation (8.25) holds. However, it changes from cube to cube so equation (8.26) leads to

$$|\psi(t)\rangle = e^{-\frac{i}{\hbar}\varepsilon_0(t-t_0) + \frac{ie}{\hbar c}\int_{\vec{r}(t_0)}^{\vec{r}(t)} \vec{A}(\vec{r})\cdot d\vec{r}} \, |\psi(t_0)\rangle. \tag{8.27}$$

8.7 The Bohm–Aharanov phase

The Bohm–Aharanov phase appears when a charged particle moves around a localized magnetic flux. The effect was analyzed theoretically by D Bohm and his student Y Aharanov [4], and was put forward as a gedanken experiment to demonstrate that orthodox quantum mechanics is wrong. A more modern analysis of the phase is given below.

If one imagines that the electron is in a wave-packet state that is highly localized, it is possible to talk about the position of the electron. Bohm and Aharonov noted that, if the electron follows a trajectory C that encircles a solenoid containing a magnetic field **B**, the wave function at the screen will take on the net phase that is the difference between the one from the upper path and the lower path, as illustrated in figures 8.1 and 8.5

$$\varphi(\vec{r}_0, \vec{r}_s) = \frac{e}{\hbar c}\int_{\vec{r}_0}^{\vec{r}_s} \vec{A}(upper) \cdot d\vec{r} - \frac{e}{\hbar c}\int_{\vec{r}_0}^{\vec{r}_s} \vec{A}(lower) \cdot d\vec{r} = \frac{e}{\hbar c}\int_C \vec{A} \cdot d\vec{r}. \tag{8.28}$$

This phase has been calculated with the adiabatic theorem, equation (8.27).

The magnetic field is constrained to be within a cylinder that is so small that it fits behind the part of the middle screen that separates the two slits. From classical electromagnetism it is known that the magnitude of the vector potential falls off like $1/r$ outside of the cylinder.

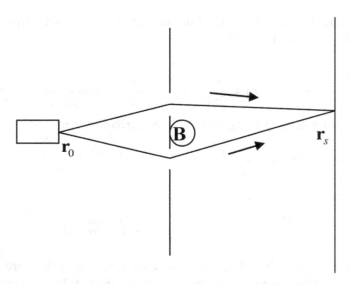

Figure 8.5. Trajectories of an electron in Bohm–Aharanov double slit experiment.

Using Stokes theorem and the fact that the magnetic field is the curl of $\vec{A}$, the phase can be manipulated

$$\varphi(\vec{r}_0, \vec{r}_s) = \frac{e}{\hbar c} \int_C \vec{A} \cdot d\vec{r} = \frac{e}{\hbar c} \int_S (\nabla \times \vec{A}) \cdot \hat{n} dS = \frac{e}{\hbar c} \int_S \vec{B} \cdot \hat{n} dS = \frac{e}{\hbar c} \Phi, \quad (8.29)$$

where Φ is the total magnetic flux in the cylinder. The electrons passing through the slits do not touch the cylinder, so every possible trajectory from $\vec{r}_0$ to $\vec{r}_s$ will have the same phase shift.

If there is no magnetic field, the wave function at the screen will manifest the ordinary double slit interference pattern as shown in figure 8.3. The addition of a magnetic field will cause a shift in the interference pattern as illustrated in figure 8.6.

Bohm and Aharanov's argument that this result cannot be correct is that the path of a particle cannot be modified by a potential alone. In classical electromagnetism, potentials are a mathematical convenience. Forces are caused by fields. For example, the force on an electron is

$$\vec{F} = e\vec{v} \times \vec{B}. \quad (8.30)$$

In this experiment, the electrons pass through a region in which $\vec{B} = 0$ but $\vec{A} \neq 0$. Quantum theory in both the Heisenberg and Schrödinger formulations makes use of Hamiltonians, and Hamiltonians necessarily contain potential functions.

Somewhat surprisingly this experiment can be carried out [5]. Bohm and Aharanov were visiting the University of Bristol when they proposed it. A professor at that university, R G Chambers, was aware of a newly discovered form of matter known as iron whiskers. These are single crystals that are long but only nanometers wide. They are even better at constraining a magnetic field than soft iron. They can be grown using a technique in which ferrous chloride is reduced in a hydrogen gas flow at high temperature. Their small diameter and high susceptibility make them the perfect material for constructing a Bohm–Aharanov device.

Chambers observed exactly the shift in interference fringes predicted by quantum theory. The argument that only fields can modify the behavior of particles is simply another illustration of the difference between the classical world and the quantum world.

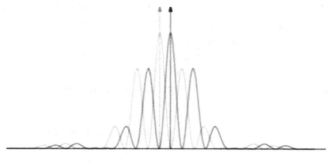

Figure 8.6. Shift in pattern caused by presence of magnetic flux.

8.8 The Berry phase

The Berry geometrical phase [6] appears when time-dependent parameters in a quantum Hamiltonian change around a closed path in parameter space. It is an addition to the standard dynamical phase $\varphi = \int_{t_0}^{t} \varepsilon(\tau)d\tau$ predicted by the adiabatic theorem described above. Traditionally, any observable in quantum mechanics is regarded as the eigenvalue of a Hermitean operator. Berry showed by construction that there are observables of a completely different nature. The Berry phase is a well defined gauge-invariant phase of the state vectors that can be measured experimentally, but it cannot be expressed as the eigenvalue of any operator because it depends on the particular path in parameter space that one chooses to traverse. This is the reason for the term 'geometrical phase'.

It will be seen that the Berry phase only exists when the Hamiltonian that is used to describe the system contains time-dependent parameters. Such a Hamiltonian does not describe an isolated system. The time-dependent parameters describe the rest of the Universe that is not included in the Hilbert space that is being considered, e.g., external fields that are not being treated quantum mechanically. In a truly isolated system, there will be no time-dependent parameters and hence no Berry phase. In this sense, the Berry phase is an unnecessary semiclassical concept. Its value stems from the fact that on many occasions a parameterized Hamiltonian is the most convenient way to treat a physical problem. On such occasions, a Berry phase will also be useful and unavoidable.

The Berry phase is based on the concepts of holonomy and anholonomy [7] in the abstract space of the parameters in the quantum Hamiltonian. Anholonomy is the failure of certain variables to return to their original values in a system that appears to be periodic. Anholonomy can be illustrated in a non quantum mechanical context by considering the parallel transport of a vector around a closed path in curved space. A sphere can be considered to be a curved two-dimensional space, for example, a globe that shows the map of the Earth. Put a pencil on the north pole of such a globe pointing along any longitude. Move the pencil in the direction of its point along the longitude until it reached the equator. Now move the pencil along the equator with the eraser end always pointing toward the north pole. When it reaches some other longitude, move it in the direction of the eraser back to the north pole. Although the pencil is kept parallel throughout this circuit in the sense that the eraser always points in the same direction, the curvature of the space leads to the result that the final position of the pencil makes an angle ϕ with the original position. This experiment is shown in figure 8.7.

In order to use this concept in a quantum mechanical context, consider a Hamiltonian that depends on the momentum operators for N particles $\vec{p}_i$, the position operators $\vec{r}_i$, the spins of the particles s_i, and some number d of time-dependent parameters $R_\alpha(t)$,

$$H(\vec{p}_1, \vec{p}_2, \ldots, \vec{p}_N, \vec{r}_1, \vec{r}_2, \ldots, \vec{r}_N, s_1, s_2, \ldots, s_N, R_1(t), R_2(t), \ldots, R_d(t)). \quad (8.31)$$

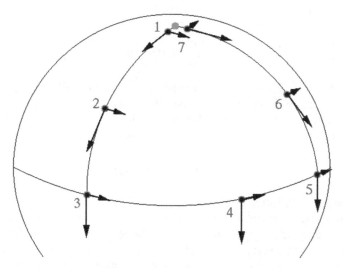

Figure 8.7. Sphere to illustrate anholonomy.

In the following equations, the parameters that are not of immediate interest will be suppressed, and the ones that appear are abbreviated. The wave function

$$\Psi(\mathbf{r}_1, \mathbf{r}_2, \mathbf{r}_3, \dots, \mathbf{r}_N, s_1, s_2, s_3, \dots, s_N, R_1(t), R_2(t), R_3(t), \dots, R_d(t)), \qquad (8.32)$$

is a solution of

$$H\Psi(R(t), t) = i\hbar\frac{\partial\Psi(R(t), t)}{\partial t}, \qquad (8.33)$$

and hence

$$\Psi(R(t), t) = U(t, t_0)\Psi(R(t_0), t_0). \qquad (8.34)$$

The Hamiltonian has eigenvalues and eigenvectors that depend on time through the variation of the $R_\alpha(t)$

$$H\psi_j(R(t), t) = \varepsilon_j(R(t))\psi_j(R(t), t). \qquad (8.35)$$

According to the adiabatic approximation

$$\psi_j(R(t), t) = U(t, t_0)\psi_j(R(t_0), t_0) = e^{-\frac{i}{\hbar}\int_{t_0}^{t}\varepsilon_j(R(t'))dt'}\psi_j(R(t), t_0). \qquad (8.36)$$

Suppose the system is cyclic in the sense that the time-dependent parameters all return to their original values $R_\alpha(t_1) = R_\alpha(t_0)$ at time t_1. Then the preceding formula becomes

$$\psi_j(R(t_0), t_1) = U(t_1, t_0)\psi_j(R(t_0), t_0) = e^{i\varphi}\psi_j(R(t_0), t_0), \qquad (8.37)$$

with

$$\varphi = -\frac{1}{\hbar}\int_{t_0}^{t_1} \varepsilon(R(t'))dt'. \tag{8.38}$$

The dynamical phase factor φ can be measured experimentally by interference if the cycled system is recombined with another that was separated from it at an earlier time and for which $R_\alpha(t) = R_\alpha(t_0)$ for all t. The dynamical phase factor for the system for which the $R_\alpha(t)$ are constant is obviously

$$\varphi_0 = -\frac{1}{\hbar}\varepsilon(R(t_0))(t_1 - t_0) \tag{8.39}$$

and the experiment will measure $\varphi - \varphi_0$.

Berry showed that, for certain physical systems, there will be a geometrical phase factor $\gamma(C)$ that depends on the circuit C that the $R_k(t)$ trace out in parameter space, which is in addition to the dynamical phase factor φ. It follows that the interference experiment will measure $\gamma(C) + \varphi - \varphi_0$. He presented a derivation of $\gamma(C)$, and studied the conditions that are necessary for it to be non-zero.

If Berry's postulate is correct, the standard adiabatic formula must be modified to read

$$\psi_j(R(t), t) = U(t, t_0)\psi_j(R(t_0), t_0) = e^{i\gamma_j - \frac{i}{\hbar}\int_{t_0}^{t} \varepsilon_j(R(t'))dt'}\psi_j(R(t), t_0). \tag{8.40}$$

Insert this wave function into the time-dependent Schrödinger equation

$$H\psi_j(R(t), t) = \varepsilon_j(R(t))\psi_j(R(t), t) = i\hbar\frac{\partial\psi_j(R(t), t)}{\partial t}, \tag{8.41}$$

leads to

$$i\hbar\frac{\partial\psi_j(R(t), t)}{\partial t} = \varepsilon_j e^{i\gamma_j - \frac{i}{\hbar}\int_{t_0}^{t} \varepsilon_j(R(t'))dt'}\psi_j(R(t), t_0)$$
$$+ i\hbar e^{i\gamma_j - \frac{i}{\hbar}\int_{t_0}^{t} \varepsilon_j(R(t'))dt'}\left\{i\frac{d\gamma_j}{dt}\psi_j(R(t), t_0) + \frac{\partial\psi_j(R(t), t_0)}{\partial t}\right\} \tag{8.42}$$

The Berry form for the wave function can only be consistent with the Schrödinger equation if

$$\frac{d\gamma_j}{dt}\psi_j(R(t), t_0) = i\frac{\partial\psi_j(R(t), t_0)}{\partial t}. \tag{8.43}$$

The normalization condition $\int\psi_j^*\psi_j dv = 1$ leads to

$$\frac{d\gamma_j}{dt} = i\int\psi_j^*(R(t), t_0)\frac{\partial\psi_j(R(t), t_0)}{\partial t}dv, \tag{8.44}$$

or, because the time dependence comes from the time-dependent parameters

$$\frac{d\gamma_j}{dt} = i \int \psi_j^*(R(t), t_0) \sum \frac{\partial \psi_j(R(t), t_0)}{\partial R_\alpha} \frac{dR_\alpha}{dt} dv. \tag{8.45}$$

Integrating the left side of this equation from t_0 to t_1 obviously gives

$$\int_{t_0}^{t_1} \frac{d\gamma_j}{dt} dt = \gamma_j(t_1) - \gamma_j(t_0). \tag{8.46}$$

The integral on the right is the same as a line integral in d-dimensional space. The line follows a path defined by the values the $R_\alpha(t)$ take as t increases from t_0 to t_1. The case for which all of the $R_\alpha(t_1)$ equal $R_\alpha(t_0)$ is focused on. That is, a closed contour C in R_α space. Thus,

$$\gamma_j(C) = i \oint_C \int \psi_j^*(R(t), t_0) \sum \frac{\partial \psi_j(R(t), t_0)}{\partial R_\alpha} dv dR_\alpha \tag{8.47}$$

where $\gamma_j(C) = \gamma_j(t_1) - \gamma_j(t_0)$.

At this point, the notation will be switched to a version that is similar to the vector notation used in the conventional vector analysis that appears in elementary physics texts. In the field of differential geometry it is shown that operations like gradients, curls, etc can be extended to multidimensional space. With this notation,

$$\gamma_j(C) = i \oint_C \int \psi_j^*(R(t), t_0) \nabla_\alpha \psi_j(R(t), t_0) dv \cdot d\vec{R} \tag{8.48}$$

where a generalization of the gradient concept to the d-dimensional space of the parameters R_α has been used. The subscript on the gradient operator is to remind us that the derivatives are with respect to the R_a. The gradient of the normalization condition can be written

$$\nabla_\alpha \int \psi_j^* \psi_j dv = 0 = \int \nabla_\alpha \psi_j^* \psi_j dv + \int \psi_j^* \nabla_\alpha \psi_j dv, \tag{8.49}$$

so it follows that $\int \psi_j^* \nabla_\alpha \psi_j dv$ is pure imaginary.

The equation for $\gamma_j(C)$ may be rewritten

$$\gamma_j(C) = \oint_C \vec{A}_j \cdot d\vec{R}, \tag{8.50}$$

where

$$\vec{A}_j = -\text{Im} \int \psi_j^*(R(t), t_0) \nabla_\alpha \psi_j(R(t), t_0) dv. \tag{8.51}$$

Stokes theorem in three dimensions is

$$\int_S (\nabla_\alpha \times \vec{A}) \cdot \hat{n} ds = \oint \vec{A} \cdot d\vec{R} = \oint \sum_\alpha A_\alpha dR_\alpha, \tag{8.52}$$

and this is another operation that generalizes to multiple dimensions. A vector quantity $\vec{B}_j$ can be defined

$$\vec{B}_j = \nabla_\alpha \times \vec{A}_j = -\text{Im} \int (\nabla_\alpha \psi_j^* \times \nabla_\alpha \psi_j) dv. \tag{8.53}$$

Using the preceding equations

$$\gamma_j(C) = \int_S \vec{B}_j \cdot \hat{n} ds. \tag{8.54}$$

The normalization integral can be written

$$\int \psi_j^* \psi_j dv = \langle j | j \rangle, \tag{8.55}$$

in the Dirac notation where $|j, R(t)\rangle$ is an abstract vector that depends on the R_α. With this notation and using the fact that the eigenvectors are a complete set

$$\sum |k\rangle\langle k| = I, \tag{8.56}$$

leads to

$$\vec{B}_j = -\text{Im} \sum_k \nabla_\alpha \langle j|k \rangle \times \langle k| \nabla_\alpha |j\rangle. \tag{8.57}$$

Taking the derivative of the eigenvalue equation leads to

$$\vec{B}_j = -\text{Im} \sum_k \nabla_\alpha \langle j|k \rangle \times \langle k| \nabla_\alpha |j\rangle. \tag{8.58}$$

Premultiplying by $\langle k|$ leads to

$$\langle k| \nabla_\alpha |j\rangle = -\frac{\langle k| \nabla_\alpha H |j\rangle}{\varepsilon_k - \varepsilon_j}, \tag{8.59}$$

if $k \neq j$, so

$$\vec{B}_j = -\text{Im} \sum_{k \neq j} \frac{\langle j| \nabla_\alpha H|k \rangle \times \langle k| \nabla_\alpha H |j\rangle}{(\varepsilon_k - \varepsilon_j)^2}. \tag{8.60}$$

The advantage to this equation is that it eliminates the need for differentiating all of the eigenvectors and replaces it with finding the derivatives of $H(R(t))$ with respect to the R_α. It can be shown that $\nabla \cdot \vec{B}_j = 0$. From this it follows that integrating over any surface S that has the trajectory C as an edge will lead to the same Berry phase.

If there are no singularities, the Berry phase will be zero. The typical way that the Berry phase will be non-zero is that a surface S with a boundary C passes through a point for which $\varepsilon_j = \varepsilon_k$ for some value of k. Of course, this cannot be the case for a point on the trajectory C because then the adiabatic theorem would not hold.

This sounds like Cauchy's theorem in complex variable theory that leads to the result that the value of a contour integral is given by the poles of the singularities included within the contour. This is an analogy, but Berry and others have pointed out that it should not be pushed too far. The underlying mathematics is not the same.

As an example, consider the case of a spin 1/2 particle in a time-dependent magnetic field. Calling the magnetic field $\vec{R}$, the Hamiltonian is

$$H = -\frac{\mu}{2}\vec{\sigma} \cdot \vec{R}(t), \tag{8.61}$$

where $\mu = \frac{e\hbar}{mc}$. If it is assumed that $\vec{R}$ is in the $\hat{z}$ direction, then

$$H = -\frac{\mu}{2}\sigma_z R_z = \begin{pmatrix} \varepsilon_- & 0 \\ 0 & \varepsilon_+ \end{pmatrix} \tag{8.62}$$

and the two eigenvalues of this system are

$$\varepsilon_\pm = \pm\frac{\mu}{2}R_z. \tag{8.63}$$

They are degenerate when $R_z = 0$. Obviously, the gradient of the Hamiltonian in equation (8.61) is

$$\nabla_a H = \frac{\mu}{2}\vec{\sigma} \tag{8.64}$$

where

$$\vec{\sigma} = \begin{pmatrix} 0 & 1 \\ 1 & 0 \end{pmatrix}\hat{x} + \begin{pmatrix} 0 & -i \\ i & 0 \end{pmatrix}\hat{y} + \begin{pmatrix} 1 & 0 \\ 0 & -1 \end{pmatrix}\hat{z}. \tag{8.65}$$

The eigenvectors for the eigenvalues $\varepsilon_\pm$ are $|+\rangle$ and $|-\rangle$. The matrix elements for equation (8.60) are

$$\langle +| \vec{\sigma} |-\rangle = \hat{x} - i\hat{y}$$
$$\langle -| \vec{\sigma} |+\rangle = \hat{x} + i\hat{y}. \tag{8.66}$$

If the state $|j\rangle$ is chosen to be $|+\rangle$, then the vector $\vec{B}_+$ is

$$\vec{B}_+ = -\text{Im}\frac{\frac{\mu^2}{4}\langle +| \vec{\sigma} |-\rangle \times \langle -| \vec{\sigma} |+\rangle}{(\mu R_z)^2} \tag{8.67}$$

or

$$\vec{B}_+ = -\frac{\hat{z}}{2R_z^2} = -\frac{\hat{R}}{2 |R|^2} \tag{8.68}$$

where, more generally, the magnetic field is chosen to point in the direction $\hat{R}$. Assume that the trajectory C lies on the surface of a sphere of radius R that is centered at the origin of parameter space. The easiest surface S to use in the integral

$$\gamma_+ = \int_S \vec{B}_+ \cdot \hat{n}\,ds, \tag{8.69}$$

is the portion of the surface of that sphere inside of the trajectory. The area of that surface is $R^2\Delta\Omega$, where $\Delta\Omega$ is the solid angle subtended by the surface. Then

$$\gamma_+ = \int_S \vec{B}_+ \cdot \hat{R} R^2 d\Omega = -\frac{1}{2}\Delta\Omega. \qquad (8.70)$$

Several conclusions can be drawn from this derivation. It provides a concrete example of a model for which the Berry phase is non-zero. It demonstrates that the singularity at $R = 0$ in parameter space where $\varepsilon_+ = \varepsilon_-$ is crucial to the existence of $\gamma_+(C)$. Finally, the physical example of the rotation of spins in a magnetic field can be realized in the laboratory.

There are two experiments in which the spins of neutrons under the influence of a rotating magnetic field are studied [8]. Neutrons are the particle of choice for this experiment because they have no charge but they have the same spin as an electron. Neutron diffraction devices are a ready source of a monoenergetic stream of neutrons that can be sent through a field. As a neutron with fixed velocity in the z-direction travels through a cylinder wrapped in a helical pattern with super-conducting wires carrying very large currents it sees a time-dependent magnetic field perpendicular to the z-direction $B_\perp(t)$.

There is also a fixed field in the z-direction B_z although the solenoid that creates that is not shown in figure 8.8. The easiest way to calculate the Berry phase for this experiment is to use equation (8.69), which leads to

$$\gamma_+ = \pi B_z B_\perp^2. \qquad (8.71)$$

When the neutrons were sent through the device illustrated in figure 8.8 they measured the phase shift predicted by Berry's theory. This differs from ordinary precession discussed previously which takes place in a magnetic field that is fixed in direction.

There have been hundreds of papers on the Berry phase, and many other experiments have been done that demonstrate its existence. The Bohm–Aharanov phase described above can be looked upon as a special case of the Berry phase. There is an effect known as the molecular Aharonov–Bohm effect that can best be treated as a Berry phase. In this effect, the Jahn–Teller distortion in a molecule

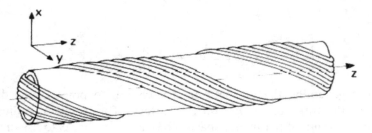

Figure 8.8. Sketch of a cylinder wrapped with wires in a helical pattern.

treated in the Born–Oppenheimer approximation causes a linkage between the electronic states and the vibrational states. The effect is observed in metallic trimmers such as Na_3 and Li_3. Many phenomena that were known before Berry's work have been reanalyzed using his theory. The modern theory of polarization in solids and ferroelectricity is based on the Berry phase. The effects of anholonomy in the scattering of polarized optical waves in a crystal was recognized by S Pancharatnam before Berry did his work.

John Hannay set out to answer a question posed by Berry, namely, what is the classical limit $\hbar \to 0$ of the Berry phase? Instead of accomplishing that, Hannay derived a different anholonomy effect that arises when the parameters in a classical Hamiltonian trace out a closed trajectory in parameter space adiabatically. The Hannay phase that appears in the angle variable in an action-angle analysis of an integrable classical system. There are limits on the systems for which it will occur that do not exist in the quantum case. Berry showed the semiclassical connection between the Hannay phase and the Berry phase. The precession of a Foucault pendulum is an example of the Hannay phase.

8.9 Quantum erasure

8.9.1 First experiment

This is the standard Young's double slit experiment. Light leaves the source one photon at a time (figure 8.9). The photons that pass through the upper slit are said to be on path one, and the ones that pass through the lower slit are on path two. The wave function at the screen for a photon seen by a counter at the position shown may be written as a superposition

$$|\psi(\text{ronscreen})\rangle = Ae^{i\delta_1}|\alpha_1\rangle + Ae^{i\delta_2}|\alpha_2\rangle, \tag{8.72}$$

where $\delta_1 = \frac{2\pi(l_1 - l_0)}{\lambda}$ and $\delta_2 = \frac{2\pi(l_2 - l_0)}{\lambda}$ and the kets are vectors that describe the polarization of the wave. The intensity at the screen is the absolute value of this wave function

$$I = \langle\psi|\psi\rangle = |A|^2(1 + 1 + \langle\alpha_1|\alpha_2\rangle e^{i\delta} + \langle\alpha_2|\alpha_1\rangle e^{-i\delta}), \tag{8.73}$$

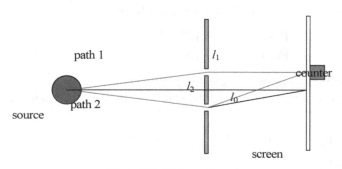

Figure 8.9. First experiment

where $\delta = \frac{2\pi(l_2 - l_1)}{\lambda}$. Normally, the polarization vector that applies to both paths is the same, the one that the photon had when it left the source. For this case, the intensity simplifies to

$$I = \langle \psi | \psi \rangle = 2\,|A|^2(1 + \cos \delta). \tag{8.74}$$

This is the standard equation for two slit interference, ignoring the diffraction due to the finite widths of the slits.

8.9.2 Quarter-wave plate

In order to understand the next experiments the reader must be familiar with an experimental tool called a quarter-wave plate. This device consists of a carefully adjusted thickness of a birefringent material such that the light associated with the larger index of refraction is retarded by 90° in phase (a quarter wave length) with respect to that associated with the smaller index. The material is cut so that the optic axis is parallel to the front and back sides of the plate. Any linearly polarized light which strikes the plate will be divided into two components with different indices of refraction. One of the useful applications of this device is to convert linearly polarized light to circularly polarized light and vice versa by adjusting the plane of the incident light so that it makes 45° angle with the optic axis. This gives equal amplitude o- and e-waves. When the o-wave is slower, as in calcite, the o-wave will fall behind by 90° in phase, producing circularly polarized light.

$$\begin{aligned} |R\rangle &= e^{i\delta}\,|x\rangle + e^{i(\delta - \pi/2)}\,|y\rangle = e^{i\delta}(|x\rangle - i\,|y\rangle) \\ |L\rangle &= e^{i\delta}\,|x\rangle + e^{i(\delta + \pi/2)}\,|y\rangle = e^{i\delta}(|x\rangle + i\,|y\rangle) \end{aligned} \tag{8.75}$$

8.9.3 Second experiment

This starts out as a standard interference experiment, but quarter-wave plates are placed behind the two slits (figure 8.10). They are oriented so that a photon with polarization in the x direction is converted to one with right circular polarization by the red plate and left polarized by the blue. A y-polarized photon is transformed in the opposite way, as illustrated in the figure above

$$\begin{aligned} \text{Path1:} \quad |x\rangle &\rightarrow |L\rangle \quad |y\rangle \rightarrow i\,|R\rangle \\ \text{Path2:} \quad |x\rangle &\rightarrow |R\rangle \quad |y\rangle \rightarrow -i\,|L\rangle \end{aligned} \tag{8.76}$$

These may be looked upon as experimental results.

If a linear polarizer is put behind the source so that only x-polarized photons pass through

$$\psi_x = Ae^{i\delta_1}\,|x\rangle + Ae^{i\delta_2}\,|x\rangle \rightarrow Ae^{i\delta_1}\,|L\rangle + Ae^{i\delta_2}\,|R\rangle, \tag{8.77}$$

and the intensity is

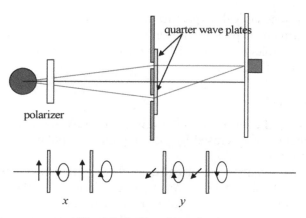

Figure 8.10. Second experiment.

$$I_x = \langle \psi_x | \psi_x \rangle = |A|^2 (1 + 1 + \langle L|R \rangle e^{i\delta} + \langle R|L \rangle e^{-i\delta}) = 2\,|A|^2. \tag{8.78}$$

If a y-polarizer is put behind the source,

$$\psi_y = Ae^{i\delta_1} |y\rangle + Ae^{i\delta_2} |y\rangle \to iAe^{i\delta_1} |R\rangle - iAe^{i\delta_2} |L\rangle \tag{8.79}$$

and the intensity is

$$I_y = \langle \psi_y | \psi_y \rangle = |A|^2 (1 + 1 + \langle R|L \rangle e^{i(\delta-\pi)} + \langle L|R \rangle e^{-i(\delta-\pi)}) = 2\,|A|^2. \tag{8.80}$$

Obviously, there is no interference pattern for these cases.

Suppose now the polarizer is oriented so that the polarization of the photon is at an angle α away from the x direction, which is the state

$$|\alpha+\rangle = \cos \alpha\, |x\rangle + \sin \alpha\, |y\rangle. \tag{8.81}$$

Then

$$\psi_{\alpha+} = Ae^{i\delta_1}(\cos \alpha\, |L\rangle + i \sin \alpha\, |R\rangle) + Ae^{i\delta_2}(\cos \alpha\, |R\rangle - i \sin \alpha\, |L\rangle), \tag{8.82}$$

and the intensity of this state is

$$I_{\alpha+} = 2\,|A|^2 (1 + \sin 2\alpha \sin \delta). \tag{8.83}$$

The polarization state vector orthogonal to $|\alpha+\rangle$ is

$$|\alpha-\rangle = -\sin \alpha\, |x\rangle + \cos \alpha\, |y\rangle, \tag{8.84}$$

and the intensity from this state is

$$I_{\alpha-} = 2\,|A|^2 (1 - \sin 2\alpha \sin \delta). \tag{8.85}$$

These intensities are consistent with the I_x and I_y derived before because $\sin 2\alpha$ is zero for $\alpha = 0$ or $\pi/2$.

The quantum mechanical explanation for these experiments is that in an interference experiment a photon can only interfere with itself. The quarter-wave plates make it possible to distinguish the path that the photon takes and this information destroys the interference.

8.9.4 Third experiment

In this experiment, the information given by using quarter-wave plates in the previous experiment is now erased by inserting another polarizer after the two slits as shown in figure 8.11.

Suppose the first polarizer is oriented in the x direction. This light is passed through the two slits where it is converted to circularly polarized light of two different kinds. As shown above, light in this condition will show no interference fringes. Passing the light through a linear polarizer after the two slits filters out the linearly polarized light. Experiments show that this light shows interference fringes when it reaches the screen

$$
\psi_x = A e^{i\delta_1} |x\rangle + A e^{i\delta_2} |x\rangle \rightarrow A e^{i\delta_1} |L\rangle + A e^{i\delta_2} |R\rangle
$$
$$
\rightarrow A e^{i\delta_1} |x\rangle + A e^{i\delta_2} |x\rangle. \tag{8.86}
$$

The effect of the quarter-wave plates has thus been erased.

Whoa! We claimed above that the conversion to circularly polarized light introduced information about the slit that the photon went through, and this information makes it impossible for the photon to interfere with itself. It is intuitively obvious that when interference is destroyed, it cannot be brought back. The quantum mechanical answer to this conundrum is that light, which was described above with simple equations, is made up of a huge number of photons. The quarter-wave plates put the photons into a statistical state, but, as seen in equation (8.75), circularly polarized light can be looked on as a superposition of light linearly polarized in two directions. Some photons were always in a state in which they could interfere with themselves. The last linear polarizer filters out these photons so that the interference pattern can be seen.

The relation of photons to light is analogous to the relation of molecules to liquid. If there is a high enough density of molecules, they behave according to the fluid

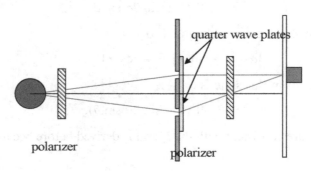

Figure 8.11. Third experiment.

dynamics laws like the Euler equations. A large enough number of photons will behave according the classical laws of optics.

8.9.5 Fourth experiment

A more interesting quantum erasure experiment using photons and a Young's double slit apparatus was done by S P Walborn *et al* from the Universidade Federal de Minas Gerais in Brazil [9].

In order to understand this experiment, the reader must familiarize themselves with two devices that have not been described above. Entangled particles are used in the EPR experiment, but this experiment requires entangled photons. Present research on QI has led to devices to produce them. Walborn *et al* used the following. The entangled photons are produced by a process called spontaneous parametric down conversion. This takes place in a special nonlinear crystal called beta-barium borate (BBO). A photon from an argon ion pump laser (351.1 nm) is converted to two longer wave length (702.2 nm) photons. The two photons go off in two different directions.

The other device is a coincidence counter. Coincidence counting has been used for years in experimental particle physics. The determination that two events occur at the same time is made electronically with a coincidence system. This unit operates on standardized pulses and determines whether events occur within a certain time interval, called the resolving time. The standard pulses from any single channel analyzer are used as input, with one input from each detector.

The ordinary laser is replaced by a BBO crystal which will emit entangled photons with wave length 2λ when it is excited by a photon with wave length λ. The photons are sent on two different paths. Path s leads from the source to Alice's friend Bob. The signal that Bob has counted a photon after processing it is sent to the coincidence counter. Photons on path p reach the coincidence counter at a point on the screen by passing through double slits with half-wave plates as in the second experiment above. This set of operations is shown in figure 8.12.

8.9.5.1 Double slit experiment

There is no polarizer in the path from Alice to Bob and there are no quarter-wave plates next to the slits. The entangled s and p photons are in the state

$$|\psi\rangle = 1/\sqrt{2}\left[\psi_s\,|x\rangle\psi_p\,|y\rangle + \psi_s\,|y\rangle\psi_p\,|x\rangle\right]. \tag{8.87}$$

The counting at the screen is done as follows. The counter is placed at the top of the screen and counts the number of pulses that it senses for x seconds. The number of counts is put in a bin. It is then moved down by an increment, and counts the pulses for another x seconds. Those counts are put in the next bin. This process is repeated until it reaches the bottom of the screen. It is assumed that the spatial increments are small so that plotting the number of counts in the bins gives an approximately continuous curve.

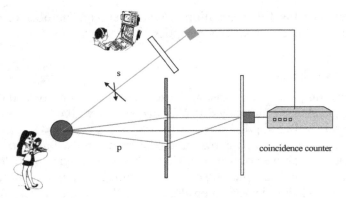

Figure 8.12. Fourth experiment. The red box is Bob's counter. The green box measures photons that go through path p and reach the screen.

If the coincidence counter is ignored, the counting described above gives a curve that shows no interference.

Leaving everything else the same, the counting is done differently. Any pulse that is not in coincidence with a pulse that reaches Bob's counter is ignored. This does not determine the polarization of the photon that left the source, but it has the effect of determining it. The resulting curve shows interference as in the first experiment.

8.9.5.2 Which-way experiment
There is still no polarizer in the path from Alice to Bob and there are quarter-wave plates next to the slits. There is no interference pattern. The only pulses counted are the ones that are coincident with the ones seen by Bob. This caused an interference pattern in the preceding case, but the presence of the quarter-wave plates destroys it. The reason for this was discussed in detail in connection with the second experiment described above.

8.9.5.3 Quantum erasure
There is a linear polarizer in the path from Alice to Bob and there are quarter-wave plates next to the slits. The direction of polarization makes no difference. There is no interference pattern in the counts at the screen but counting only the pulses that are coincident with the ones counted by Bob brings back the interference pattern. The reason for this is that the photons being counted at the screen are in the same orientation state as they were when they left the source. Using the coincidence counter has the same effect as the polarizer after the slits in the third experiment.

The explanation for this effect is the same as for the third experiment, but it is made more interesting because it has the extra wrinkle of entanglement. In the first experiment, all of the photons are allowed to interfere with themselves. Even with the quarter-wave plates, some will. The coincidences with photons that all have the same polarization filters out the ones that will interfere.

8.9.5.4 Delayed erasure

To this point the implication has been that Bob measures the pulses that hit his counter at exactly the same time as they hit the counter at the screen. This does not have to be the case. The time stamps on all of the electrons in a bin can be retained. The path from the source to Bob's counter can be lengthened so that his copy of the entangled photons is measured later than the copy that hit the screen. The time increment for the delay Δt is the same for all the photons, and it can be calculated from the experimental parameters. The photons from a given entangled pair can thus be found by combining Bob's counts with the ones on the screen even when they are not counted at the same time.

It is thus possible to reproduce the preceding experiments with delayed photons. Somehow the fact that interference is identified with information that did not exist when the photons struck the screen makes the erasure experiment seem even more mysterious.

8.10 Resume

It could be argued that the EPR effect and the Bohm–Aharanov effect are discoveries of interesting quantum phenomena that were found for the wrong reason. The predictions of quantum theory seemed so outlandish that very good scientists proposed experiments that they were convinced would not work. The experiments suggested by Berry and Walborn also seem to contradict intuition, but they expected quantum theory to be validated.

Problems

P8.1 Is the idea of a collapsing wave function a general feature of the measurement process?

P8.2 Draw a cartoon showing what Alice putting electrons in her box must look like according to hidden variable theory.

P8.3 Work out the Bell's inequalities for $P(\hat{\mathbf{b}} +; \hat{\mathbf{c}} +)$.

P8.4 Suppose there were four men and four women in a quantum mechanics class. They are standing in the hall outside the classroom. There is a row of eight chairs in the classroom. One at a time, they enter the classroom, pick a chair, write their name on a piece of tape, and place it on the bottom of the seat. When the process is finished, they go into the classroom and sit in the chair they picked. What is the chance that no two students picked the same chair and that the arrangement of the students would be alternating man, woman, man, woman, ...?

P8.5 Why is the Hamiltonian in quantum mechanics normally time independent?

P8.6 Is the Bohm–Aharanov effect gauge invariant?

P8.7 Hold a cat with his feet pointing toward the ceiling. Drop the cat. Hopefully the cat will land with his feet on the floor. Is angular momentum conserved? Is this an example of anholonomy?

P8.8 When entangled electrons are used in a two slit experiment, is the rule that an electron can only interfere with itself broken?

References

[1] http://hqrd.hitachi.co.jp/em/doubleslit.cfm
[2] Messiah A 1966 *Quantum Mechanics* vol II (New York: North-Holland)
[3] Dirac P A M 1931 *Proc. Roy. Soc.* A **133** 60
[4] Aharonov Y and Bohm D 1959 *Phys. Rev.* **115** 485
[5] Chambers R G 1960 *Phys. Rev. Lett.* **5** 3
[6] Berry M V 1984 *Proc. R. Soc. London* **392** 45
[7] Henderson D W 1997 *Differential Geometry: A Geometric Approach* (New York: Prentice Hall)
[8] Bitter T and Dubbers D 1987 *Phys. Rev. Lett.* **59** 251
    Richardson D J, Livingston A I, Green K and Lamoreaux S K 1988 *Phys. Rev. Lett.* **61** 2030
[9] Walborn S P, Terra Cunha M O, Padua S and Monken C H 2001 *Phys. Rev.* A **65** 033818

IOP Publishing

Modern Quantum Mechanics and Quantum Information
J S Faulkner

Chapter 9

What does it all mean?

9.1 What are we to make of quantum experiments?

From the earliest days of quantum mechanics, physicists found it difficult to understand its meaning. They had become experts on classical mechanics, studying that field in much more detail than today's students do. The switch from the classical to the quantum way of thinking was very difficult. Born's statistical interpretation of the wave function was considered a significant step toward understanding the Schrödinger equation by many, but it was an anathema to others.

The brilliant Hungarian physicist John von Neumann realized very early that the enigma at the center of quantum theory is the problem of measurements. He developed a very sophisticated mathematical theory of measurements, based on the equations of Schrödinger and Heisenberg. But it left questions.

Physicists were attracted to Schrödinger's formulation of quantum mechanics because basic theorems on differential equations developed in the preceding century, such as the Sturm–Liouville (S-L) theory, could be put to use. Applying Born's interpretation to the eigenfunctions and eigenvalues of S-L theory led to the simple result that a measurement of a property to a system in an eigenstate would definitely give a result that is the eigenvalue. What if the quantum system is in a superposition of such states? Born's rule gives the probability of obtaining a result from a measurement. The wave function that exists before a measurement collapses to an eigenfunction in the measurement process. The group of researchers who were addressing the problem were unhappy with the philosophical implications of that conclusion. Do observables have a specific value in the mixed state that is revealed by a measurement, or do the specific values only occur because of the measurement process? It will be seen that recent research using the consistent histories theory answers many of these questions, but in 1926 these questions led to a great deal of angst among the group.

doi:10.1088/978-0-7503-2167-9ch9

9.2 The Orthodox Copenhagen interpretation (Bohr)

Heisenberg at first took the view that measurement alters the state of the system that existed before measurement. This implies that the system was in a definite state before measurement, and that the quantum mechanical formalism gives an incomplete description of the way that measurement changes it.

N Bohr clearly felt that the combination of Heisenberg's equation, the Schrödinger equation, and Born's interpretation embodied a complete theory. It was his job to answer the philosophical objections to this view. Einstein took the opposite view, and the other members of the group were not sure.

Bohr's position was that it only makes sense to attribute values to the observable quantities of a physical system when the system is being measured in a particular way. This is Bohr's solution to the puzzling wave–particle duality exhibited by entities such as photons and electrons. The 'wave' and 'particle' aspects of these entities are 'complementary' in the sense that it is physically impossible to construct a measuring device that will measure both aspects simultaneously. Bohr concluded that from a physical standpoint it only makes sense to speak about the wave or particle aspects of quantum entities as existing relative to particular measurement procedures.

Bohr's view was that one cannot even ask what the state of a physical system is between measurements, since an answer to this question would imply knowledge of the state without a measurement. Heisenberg's position was that the observables always have definite values between measurement, but we can never know what those values are since they can only be determined by measurement, which indeterministically disturbs the system. Another view mistakenly attributed to Bohr is that the values of the system's observable quantities before measurement are 'smeared out' between the particular values that the observable quantity could have upon measurement.

In Bohr's view, the world is divided into two realms of existence, that of quantum systems, which behave according to the formalism of quantum mechanics and do not have definite observable values outside the context of measurement, and of classical systems, which always have definite values but are not described within quantum mechanics. The line between the two realms is arbitrary. A subsystem can be a part of the quantum system or a measuring device that behaves classically though indeterministically.

There are several difficulties with Bohr's view. To begin with it gives no reason why physics should not be able to give a complete description of the measurement process. After all, a measuring device is a physical system, and in performing a measurement it simply interacts with another physical system such as a photon or an electron. Bohr's view offers no precise characterization that would mark off those physical interactions that are measurements, and hence classical, from those physical interactions that are not. This hardly seems satisfactory from a physical standpoint.

Schrödinger pointed out that the orthodox interpretation of Bohr allows for inconsistent descriptions of the state of macroscopic systems. For example, suppose that you place a cat in an enclosed box along with a device that will release a poison gas if a counter measures that a certain radium atom has decayed. According to the quantum mechanical formalism, the radium atom is in a superposition of decaying and not decaying. Since there is a correlation between the state of the radium atom and the Geiger counter, and between the state of the Geiger counter and the state of the cat, the cat should also be in a superposition of being alive and dead. That is to say, according to the orthodox interpretation, if the cat is not a measuring device then it is both alive and dead. On the other hand, if the cat is a measuring device, then according to the orthodox interpretation the cat will either be definitely alive or definitely dead. Since these are contradictory states, Bohr's claim that the definition of a subsystem as a measuring device is a matter of choice cannot be correct.

Wigner also argued against Bohr's view that the distinction between a subsystem and a measuring device can be made arbitrarily. What if he put one of his friends in the box with the cat. The 'measurement' you make at a given time is to ask Wigner's friend if the cat is dead or alive. If the friend is considered to be a subsystem, he would be in a superposition of believing the cat is dead or alive, but people simply do not exist in superposed belief-states. Wigner's solution was that the presence of a conscious observer is what distinguishes between a subsystem and a measuring device.

Most of the scientists who originally used Planck's constant (Planck, de Broglie, and Einstein) disliked the shape of the quantum mechanics that they had helped introduce. They particularly did not like the indeterminacy that seems to be inherent in the theory. They carried these prejudices throughout their lives.

Bohr's ideas won the debates at the Fifth Solvay International Conference on Electrons and Photons in 1927, where the world's most notable physicists met to discuss the newly formulated quantum theory. Einstein, unconvinced, remarked 'God does not play dice'. Bohr replied, 'Einstein, stop telling God what to do'.

Bohr's orthodox interpretation appears in many textbooks on quantum mechanics. As quantum theory was successfully used to an ever greater extent as a practical tool to explain the behavior of materials and high-tech devices, interest in interpretations of it waned. However, it has become a field of active research in recent years because questions that arise in such new fields as quantum cosmology and quantum information cannot be answered within the orthodox interpretation.

9.3 Bohm's interpretation

In 1952, physicist David Bohm formulated a complete alternative to standard quantum mechanics. Bohm's theory was put into a relatively simple and elegant mathematical form by John Bell in 1982. Bohm's theory arises from the general class of interpretations known as hidden variable theories, which were favored by Einstein and de Broglie. The way that Bohm does this is by postulating that there is a quantum force that moves the particles around so that they behave exactly as standard quantum mechanics predicts. His theory is deterministic. If you knew the initial configuration of every particle in the Universe, applying Bohm's theory would allow you to predict with certainty every subsequent position of every particle in the Universe. However, the Universe is set up so that it is a physical, rather than a merely practical, impossibility to know the configuration of particles in the Universe. This means that the world behaves just as if it's indeterministic even though the apparent indeterminism is simply a matter of ignorance. In addition, although the wave function governing the particle's motion never collapses, the particle moves around so that it looks as if measurement causes it to collapse.

Bohm's theory has not been generally accepted by physicists for two reasons. It can't be used for anything, since it's impossible to know the configuration of all particles in the Universe at any given time. Because of our ignorance of this configuration, the physical theory that has to be used to generate predictions is standard quantum mechanics. Why then bother with Bohm's theory at all? What is more important, in Bohm's theory all the particles in the Universe are intimately connected so that every particle instantaneously affects the quantum force governing the motions of the other particles. This is a throwback to the type of 'action at a distance' expunged by relativity theory. Moreover, Bohm's theory assumes that simultaneity is absolute and that makes it inconsistent with relativity theory, because it assumes that information about the positions can be sent faster than light.

Bohm was inspired by Einstein to work on the theory of hidden variables. However, Einstein did not like the product of his research for the reasons explained above.

9.4 The many-worlds interpretation

A view first developed in 1957 by H Everett in his PhD thesis is called the many-worlds interpretation (MWI). It was later developed further by B de Witt. Everett was a student of Prof. J A Wheeler at Princeton. This view states that there is no collapse when a measurement occurs; instead, at each such point where a measurement occurs the Universe 'branches' into separate, complete worlds, a separate world for every possible outcome that a measurement could have. In each branch, it looks like the measuring devices indeterministically take on a specific value, and the empirical frequencies that occur upon repeated measurement in every branch converge to the probabilities predicted by the Born rule and von Neumann's projection postulate. The deterministically evolving wave function correctly describes the quantum mechanical state of the Universe as a whole, including all of its branches. Wheeler was interested in developing a theory like this because he wanted to apply quantum mechanics to cosmology. That could not be done if the theory required an external observer as in Bohr's theory.

The MWI has several advantages. First, it can be developed with mathematical precision. Second, it postulates that the Universe as a whole evolves deterministically according to Schrödinger's equation, so that you only need one type of evolution, not Schrödinger's equation plus collapse during a measurement. Third, the notion of a measurement (which is needed to specify where the branching occurs) can be spelled out precisely and non-arbitrarily as a certain type of physical interaction. Fourth, it makes sense to speak of the quantum state of the Universe as a whole, as needed for cosmology.

On the other hand, the MWI has several serious shortcomings. First, it's simply weird conceptually. Second, it does not account for the fact that we never experience anything like a branching of the world. How is it that my experience follows one particular path in the branching universe and not others? Third, as noted above, the MWI predicts that there will be worlds where the observed empirical frequencies of repeated measurements will not fit the predictions of the quantum mechanical formalism. In other words, if the theory is true, there will be worlds at which it looks as if quantum theory is false! Finally, the theory does not seem to give a clear sense to locutions such as 'the probability of this electron going up after passing through this Stern–Gerlach device is 1/2.'. What exactly does that number 1/2 mean, if the Universe simply branches into two worlds, in one of which the electron goes up and in the other of which it goes down?

Another objection is related to the strategy of a believer in many worlds who is given an offer to play a Russian roulette game with a million dollar prize. The argument is that the believer should agree to play the roulette any number of times. At the end, there will be one world in which he is a multi-millionaire and all other worlds in which he is not alive. Since he is here talking with you, he must be in the Universe in which he won all the bets and is a rich and happy guy.

9.5 The Ghirardi–Rimini–Weber (GRW) interpretation

In 1986, G C Ghirardi, A Rimini and T Weber proposed a way of accounting for the fact that macroscopic objects are never observed in superpositions, whereas microscopic systems (such as photons and electrons) are. Their views were developed further in 1987 by John Bell. According to the Ghirardi–Rimini–Weber interpretation (GRWI), there is a very small probability (one in a trillion) that the wave functions for the positions of isolated, individual particles will collapse spontaneously at any given moment without a measurement. This is achieved by adding terms to the Schrödinger equation that localize quantum particles in the same manner as a Gaussian wave packet. When particles couple together to form a macroscopic object, the tiny probabilities of spontaneous collapse quickly add up so that the system as a whole is in a collapsed and hence well defined state. Thus, the GRWI can explain why we never observe macroscopic objects in superpositions, but often observe interference effects due to superposition when we're looking at isolated microscopic particles. Moreover, they give a precise and non-arbitrary characterization of the measurement process as a type of physical interaction.

There are two problems with the GRWI. First, though microscopic particles are localized in wave packets, they never have precise positions. Instead, particles have positions that fit inside a Gaussian curve. Gaussian functions have tails that extend to infinity. Macroscopic objects that are composed of subatomic particles have this same property. In other words, according to the GRWI you are mostly in this room, but there's a vanishingly small part of you at every other point in the Universe, no matter how distant! Second, the GRWI predicts that energy is not conserved. While the total violation of conservation of energy predicted by the GRWI is too small to be observed, even over the lifetime of the Universe, it does discard a feature that many physicists consider to be essential.

9.6 Consistent (decoherent) histories interpretation

The consistent histories approach was first proposed by Robert Griffiths in 1984 [1] and further developed by Roland Omnès in 1988 [2]. The same theory was put forward in 1990 by Murray Gell-Mann and James Hartle, who used the term decoherent histories [3].

A consistent history combines wave functions and probabilities in a way which does not rely upon the use of measurements. Probabilities can be assigned to histories provided certain *consistency conditions* are satisfied. Histories can be used to describe how a particle interacts with a measuring apparatus, and how the outcome of a measurement, e.g., the position of a pointer, is related to some property of the particle before the measurement took place. However, they can also be employed for a single particle, or any number of particles, in the absence of any measurement. For example, by using consistent histories it is possible to assign a probability for the time at which a radioactive atom will decay, even if it is out in interstellar space far from any measuring device.

Consistent histories can be used to analyze various quantum paradoxes, such as the interference produced by a particle passing through a double slit, or the correlated pair of particles considered by Einstein, Podolksy and Rosen. This allows the paradox to be understood in quantum terms, without any need to invoke peculiar long-range influences or other ghostly effects. The consistent histories approach has been employed to analyze problems in quantum computation and quantum cryptography.

A mathematically correct definition of consistent histories is very complex. Whole articles and book chapters have been written on it. There can be numerous histories for the same experiment. Every history is consistent with the Schrödinger equation. No two histories can be true at the same time.

What does consistent histories offer? It shows that quantum mechanics can give an unambiguous answer to a range of questions. Statements that appear to be paradoxical are only that if they ask the probability for something that is not in a consistent history. In the early days of quantum theory famous scientists almost came to blows in arguments about what particles were doing before an experiment was completed. The answer from consistent histories is that you can look at it in many ways and hold several conclusions at the same time. A set of consistent histories for the simple Stern–Gerlach experiment shown in figure 9.1 are shown in table 9.1. The table illustrates that if you choose to believe that the particle went into a particular state at some time before the final measurement, fine, there is a probability for that.

The Stern–Gerlach experiment is arranged so that at $t = t_0$, the particle is in the state $|S_x^+\rangle = (|\uparrow\rangle + |\downarrow\rangle)/\sqrt{2}$. The electron has a fixed velocity in a direction so that it passes through the inhomogeneous magnetic field created by the shaped magnets. The mathematical analysis of the experiment is given in chapter 1. Two electron counters are placed on the other side of the magnet. If the electron reaches the upper counter, this is taken as proof that it is in the spin up state, and if it reaches the lower counter it must be in a spin down state. The device that creates the initial state of the electron, the magnets, and the counters are all macroscopic, so this is the kind of experiment envisioned by Bohr. The purpose of the consistent histories theory is not to challenge the results of the experiment, but to clarify the meaning of it. There is not one but many consistent histories to describe this experiment. They are listed in the table 9.1.

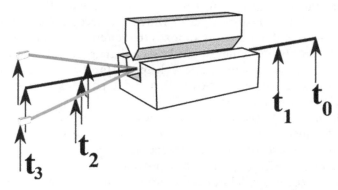

Figure 9.1. A Stern–Gerlach device for consistent histories. Credit: Wikimedia: Master of the Universe 322 (CC-BY SA 4.0).

Table 9.1. Table of consistent histories for Stern–Gerlach experiment.

	t_0	t_1	t_2	t_3
History 1a	$|S_x^+\rangle$	$|S_x^+\rangle$	$|S_x^+\rangle$	$|\uparrow\rangle$
History 1b	$|S_x^+\rangle$	$|S_x^+\rangle$	$|S_x^+\rangle$	$|\downarrow\rangle$
History 2a	$|S_x^+\rangle$	$|S_x^+\rangle$	$|\uparrow\rangle$	$|\uparrow\rangle$
History 2b	$|S_x^+\rangle$	$|S_x^+\rangle$	$|\downarrow\rangle$	$|\downarrow\rangle$
history 3a	$|S_x^+\rangle$	$|\uparrow\rangle$	$|\uparrow\rangle$	$|\uparrow\rangle$
History 3b	$|S_x^+\rangle$	$|\downarrow\rangle$	$|\downarrow\rangle$	$|\downarrow\rangle$
History 4	$|S_x^+\rangle$	$|S_x^+\rangle$	Both	Both

Histories 1a and 1b are equivalent to the standard description of the Stern–Gerlach experiment. At $t = t_0$, the particle is in the state $|S_x^+\rangle$. It is in the same state at $t = t_1$ and at $t = t_2$. This means we know nothing about the state of the particle until the last step at $t = t_3$ when it reaches the counter. Since the particle is essentially in the state $|S_x^+\rangle$ when it reaches the counters, we know the probability of being in $|\uparrow\rangle$ or $|\downarrow\rangle$ is 1/2.

In histories 2a and 2b, we know the particle goes into states $|\uparrow\rangle$ or $|\downarrow\rangle$ at $t = t_2$. The probability at that point becomes 1/2. It remains 1/2 at $t = t_3$ because the only consistent processes are for a particle in path $|\uparrow\rangle$ to arrive at counter $|\uparrow\rangle$ and one in $|\downarrow\rangle$ must end in $|\downarrow\rangle$. This means that the experimental results predicted in the second set of histories is the same as the ones predicted by the first set, the standard description.

In histories 3a and 3b the particle goes into states $|\uparrow\rangle$ or $|\downarrow\rangle$ at $t = t_2$ with probability 1/2. The probabilities don't change after that because there is only one path that they can follow consistently. Again, the final prediction is the same as with the two earlier histories.

The final history, number 4 is consistent but trivial in that it doesn't tell us anything. It is assumed that no effort is made to observe anything.

The existence of these histories, all of which are acceptable, demonstrates that many of the debates and apparent paradoxes that circulate around quantum theory are simply irrelevant. In particular, it can be seen that arguments about the state of a particle before a measurement are not helpful. In the books and articles on consistent histories it has been shown that most paradoxes arise because they assume a history that is not consistent.

An example of an inconsistent history is shown in table 9.2. The steps from $t = t_2$ to $t = t_3$ are obviously physically inconsistent, and it can be shown that they are impossible under the laws of quantum mechanics.

The consistent histories approach has proved to be helpful for analyzing the devices that are used in the field of quantum information. There are many papers being written on the subject, and these continue to clarify some of the subtle points of quantum theory that are being discovered. There are some who feel that the consistent (decoherent) histories interpretation is the same thing as the many-worlds interpretation. Wheeler stated that he sometimes regretted his description in terms of alternate universes because it leads to arguments that are not focused on the point he was trying to make (Russian roulette). It is likely that he would prefer the consistent histories language because it does not come with such baggage.

Table 9.2. Inconsistent histories.

	t_0	t_1	t_2	t_3
History 5a	$|S_x^+\rangle$	$|S_x^+\rangle$	$|\downarrow\rangle$	$|\uparrow\rangle$
History 5b	$|S_x^+\rangle$	$|S_x^+\rangle$	$|\uparrow\rangle$	$|\downarrow\rangle$

9.7 Most widely held interpretation

SHUT UP AND CALCULATE!!!!!

9.8 Decoherence

If we get on an airplane and fly from our home city to another city or even another continent, we expect to find a physical environment that is not changed in a fundamental way. Perhaps the most frequently used language is different and the traffic laws may change, but devices and constructions will work in the way that we expect. This obvious statement is made in order to emphasize the point that, as much as we have come to accept quantum theory and could not function without its technological applications, there is a classical world that can be relied on.

In order to discuss theoretically the connection between the probabilistic world of quantum mechanics and the reliable classical world, it is first necessary to have a mathematical definition of the difference between those worlds. Since the quantum world must be considered the most fundamental, it is necessary to find a mechanism for transforming from the quantum to the classical worlds. That mechanism is called decoherence. Decoherence theory is the mathematical formulation of the process whereby a quantum system transforms into one that is described by classical laws.

9.9 Density matrices

The mathematical foundation for decoherence theory are the density matrices that were used by von Neumann in his research on measurements and quantum mechanics [4]. Consider a Hilbert space that is spanned by N vectors, $|a_1\rangle, |a_2\rangle, |a_3\rangle, \ldots, |a_N\rangle$, where N can be finite or infinite. Any state vector can be expanded in terms of the basis vectors,

$$|\psi\rangle = c_1|a_1\rangle + c_2|a_2\rangle + c_3|a_3\rangle + \ldots + c_N|a_N\rangle. \tag{9.1}$$

The expectation value of an operator A is

$$\langle A\rangle_\psi = \langle \psi | A | \psi\rangle = \sum_{i,j=1}^{N} c_i^* A_{ij} c_j, \tag{9.2}$$

which can be manipulated into another form

$$\langle A\rangle_\psi = \sum_{i,j=1}^{N} c_i^* A_{ij} c_j = \sum_{i,j=1}^{N} c_j c_i^* A_{ij} = \sum_{i,j=1}^{N} \rho_{ji} A_{ij} = \mathrm{Tr}(\rho A), \tag{9.3}$$

where

$$\rho_{ji} = c_j c_i^* \quad A_{ij} = \langle a_i|A|a_j\rangle. \tag{9.4}$$

Defining an operator

$$\rho = \sum_{i=1}^{N}\sum_{j=i}^{N} |a_i\rangle \rho_{ij}\langle a_j|, \tag{9.5}$$

leads to

$$\langle A \rangle_\psi = \text{Tr}(\rho A). \tag{9.6}$$

The trace operation on operators used in this equation

$$\text{Tr}(X) = \sum_{i=1}^{N} \langle a_i | X | a_i \rangle. \tag{9.7}$$

As in matrix theory, it can be shown that the trace does not depend on the particular set of vectors used to span the Hilbert space.

So far, nothing more has been done than to rewrite well-known formulas in a different way. However, von Neumann noted that the density matrix can be used to describe a quantum mechanical state that can be described in no other way. Consider the discussion of the Stern–Gerlach experiment with the magnets rotated so that they polarize the electrons along a direction defined by the unit vector $\hat{n}$.

After the particle has passed the magnets, it is known that its spin state vector is one of the states

$$|S_{\hat{n}}, +\rangle = \cos\frac{\theta}{2}\,|+\rangle + \sin\frac{\theta}{2}e^{i\phi}\,|-\rangle$$
$$|S_{\hat{n}}, -\rangle = -\sin\frac{\theta}{2}\,|+\rangle + \cos\frac{\theta}{2}e^{i\phi}\,|-\rangle \tag{9.8}$$

The expectation value of S_z when the electron is in the state $|S_{\hat{n}}, +\rangle$

$$\langle S_z \rangle_{\hat{n}+} = \langle S_{\hat{n}}, +|S_z|S_{\hat{n}}, +\rangle = \text{Tr}(\rho_{\hat{n}+}S_z) = \frac{\hbar}{2}\cos\theta, \tag{9.9}$$

where

$$\rho_{\hat{n}+} = \begin{pmatrix} \cos^2\dfrac{\theta}{2} & \sin\dfrac{\theta}{2}\cos\dfrac{\theta}{2}e^{-i\phi} \\ \sin\dfrac{\theta}{2}\cos\dfrac{\theta}{2}e^{i\phi} & \sin^2\dfrac{\theta}{2} \end{pmatrix}, \tag{9.10}$$

and

$$S_z = \frac{\hbar}{2}\begin{pmatrix} 1 & 0 \\ 0 & -1 \end{pmatrix}. \tag{9.11}$$

All of this is straightforward, but how can the state of the particles that have come out of the furnace but have not gone through the first Stern–Gerlach filter be described? There is no vector that can describe that state because there is a statistical indeterminacy in addition to the quantum mechanical indeterminacy. It was pointed out by von Neumann that the spin state of the particle can be described by a superposition of density matrices

$$\rho_{\text{mixed}} = \frac{1}{2}\rho_{\hat{n}+} + \frac{1}{2}\rho_{\hat{n}-}. \tag{9.12}$$

The expectation value of an operator has a meaning in this description

$$\langle S_z \rangle_{\text{mixed}} = \text{Tr}(\rho_{\text{mixed}} S_z) = \frac{1}{2}\text{Tr}(\rho_{\hat{n}+} S_z) + \frac{1}{2}\text{Tr}(\rho_{\hat{n}-} S_z) = 0. \tag{9.13}$$

A density matrix can be defined using an arbitrary set of eigenstates $|\psi_i\rangle$ of some Hamiltonian H using statistical weightings p_i

$$\rho = \sum_i p_i \rho_{\psi_i}. \tag{9.14}$$

This density matrix can be considered as an operator in its own right. It was shown by von Neumann that the commutator

$$[H, \rho] = H\rho - \rho H, \tag{9.15}$$

satisfies the equation

$$[H, \rho] = i\hbar \frac{\partial \rho}{\partial t}. \tag{9.16}$$

He also showed that the entropy of a mixed state can be defined

$$S = -Tr(\rho \ln \rho). \tag{9.17}$$

The importance of the general density matrix will become clear below.

9.10 Defining decoherence

The picture of the Stern–Gerlach filter given thus far is not complete. The description is that an incoming beam made up of a linear combination of $|S_{\hat{n}}, +\rangle$ and $|S_{\hat{n}}, -\rangle$ is split into beams in the pure spin states. There is, however, nothing in the notation that indicates how this is done. A more complete description of the wave function would be

$$|S_{\hat{n}}, +\rangle |d\uparrow\rangle$$
$$|S_{\hat{n}}, -\rangle |d\downarrow\rangle \tag{9.18}$$

where $|d\uparrow\rangle$ and $|d\downarrow\rangle$ describe the spatial trajectories of the particles. The density matrix (in operator form) of the incoming state is

$$\rho_{\hat{n}+} = \cos^2 \frac{\theta}{2} |d\uparrow\rangle |+\rangle\langle +| \langle d\uparrow| + \sin \frac{\theta}{2} \cos \frac{\theta}{2} e^{-i\phi} |d\uparrow\rangle |+\rangle\langle -| \langle d\downarrow| +$$
$$\sin \frac{\theta}{2} \cos \frac{\theta}{2} e^{i\phi} |d\downarrow\rangle |-\rangle\langle +| \langle d\uparrow| + \sin^2 \frac{\theta}{2} |d\downarrow\rangle |-\rangle\langle -| \langle d\downarrow|. \tag{9.19}$$

The density matrices of the outgoing waves are one of these

$$\rho_+ = |d\uparrow\rangle |+\rangle\langle +| \langle d\uparrow| \quad \rho_- = |d\downarrow\rangle |-\rangle\langle -| \langle d\downarrow|. \tag{9.20}$$

The probability of the particle being in the $|+\rangle|d\uparrow\rangle$ state is

$$|\langle d\uparrow| \langle +||S_{\hat{n}}, +\rangle|^2 = \text{Trace}\langle d\uparrow| \langle +| \rho_{\hat{n}+} |+\rangle |d\uparrow\rangle = \cos^2\frac{\theta}{2}, \qquad (9.21)$$

and the probability of it being in the $|-\rangle|d\uparrow\rangle$ state is $\sin^2\frac{\theta}{2}$.

At this point, the results do not seem different from a classical system. It can be assumed that the state is really a mixture of $|+\rangle$ and $|-\rangle$ particles, and the off-diagonal elements do not seem to play a necessary roll. When the particles are sent through the measuring device, they appear with the expected probabilities. However, the measuring device can be reoriented so that the z axis points in another direction $\hat{z}$. The results of the new experiment show that the particles in state $|S_n, +\rangle$ can equally well be considered to be a mixture of $|\hat{+}\rangle$ and $|\hat{-}\rangle$ particles, the new notation meaning that they are simply spin up and down relative to the new z direction. The probabilities of finding a particle in these states are $\cos^2\frac{\hat{\theta}}{2}$ and $\sin^2\frac{\hat{\theta}}{2}$, where the direction of the unit vector $\hat{n}$ is now measured relative to the new z direction. The off-diagonal elements in the density matrix are now seen to be necessary. If they were not there, it would be impossible to transform ρ into a new coordinate system. Thus, in a quantum system, it is impossible to say if the particles are really in the states $|+\rangle$ and $|-\rangle$ or $|\hat{+}\rangle$ and $|\hat{-}\rangle$.

It was pointed out by von Neumann that the transformation from the quantum density matrix to a classical one consists in setting the off-diagonal elements of the density matrix equal to zero. For simplicity, focus on a Hilbert space that is spanned by two basis vectors, $|1\rangle$ and $|2\rangle$ so any state vector can be written as $\psi = \alpha_1|1\rangle + \alpha_2|2\rangle$. The pure state density matrix is

$$\rho_\psi = \begin{pmatrix} |\alpha_1|^2 & \alpha_1\alpha_2^* \\ \alpha_2\alpha_1^* & |\alpha_2|^2 \end{pmatrix}. \qquad (9.22)$$

Consider the simple transformation

$$|1\rangle = \frac{1}{\sqrt{2}}\left(|\hat{1}\rangle + |\hat{2}\rangle\right)$$
$$|2\rangle = \frac{1}{\sqrt{2}}\left(|\hat{1}\rangle - |\hat{2}\rangle\right) \qquad (9.23)$$

In the new basis, the density matrix becomes

$$\rho_\psi = \begin{pmatrix} |\hat{\alpha}_1|^2 & \hat{\alpha}_1\hat{\alpha}_2^* \\ \hat{\alpha}_2\hat{\alpha}_1^* & |\hat{\alpha}_2|^2 \end{pmatrix}, \qquad (9.24)$$

where

$$\hat{\alpha}_1 = \frac{1}{\sqrt{2}}(\alpha_1 + \alpha_2)$$
$$\hat{\alpha}_2 = \frac{1}{\sqrt{2}}(\alpha_1 - \alpha_2) \qquad (9.25)$$

The system can equally well be considered to be in the states $|1\rangle$ and $|2\rangle$ or $|\hat{1}\rangle$ and $|\hat{2}\rangle$.

The density matrix that has been made classical by setting the off-diagonal elements equal to zero

$$\rho_\psi^c = \begin{pmatrix} |\alpha_1|^2 & 0 \\ 0 & |\alpha_2|^2 \end{pmatrix}, \tag{9.26}$$

does not have this property. Inserting the transformation in equation (9.25) into it gives

$$\rho_\psi^r = \begin{pmatrix} \frac{1}{2}(|\alpha_1|^2 + |\alpha_2|^2) & \frac{1}{2}(|\alpha_1|^2 - |\alpha_2|^2) \\ \frac{1}{2}(|\alpha_1|^2 - |\alpha_2|^2) & \frac{1}{2}(|\alpha_1|^2 + |\alpha_2|^2) \end{pmatrix} \tag{9.27}$$

which shows the form of ρ_ψ^c is not invariant when the states are changed. Therefore measurements on a system described by ρ_ψ^c will yield one of the states $|1\rangle$ or $|2\rangle$, and nothing else. Von Neumann pointed out that this is the kind of density matrix that appears in classical mechanics. It can be written as a statistical mixture of two pure states,

$$\rho_\psi^r = |\alpha_1|^2 \rho_1 + |\alpha_2|^2 \rho_2 \tag{9.28}$$

where $\rho_1 = |1\rangle\langle1|$ and $\rho_2 = |2\rangle\langle2|$.

At this point decoherence can be defined. It is the process whereby the off-diagonal elements in the quantum density matrix ρ_ψ are reduced to zero, leading to the classical density matrix ρ_ψ^c. It is argued by Zeh [5] and Zurek [6] that the interactions with the environment in which the system embedded can do this under certain circumstances.

The picture in figure 9.2 is an effort to illustrate decoherence. It is an analog, not a depiction because quantum effects cannot be drawn. The image in the upper right contains all of the unknown quantum phase information. The next one on the right contains less. The process continues, moving to the right and then wrapping down. The final image in the lower right corner is the classical image.

9.11 Simple example of decoherence

Consider the state vector $|\psi\rangle = \alpha_1|1\rangle + \alpha_2|2\rangle$. The pure state density matrix is

$$\rho_\psi = \begin{pmatrix} |\alpha_1|^2 & \alpha_1\alpha_2^* \\ \alpha_2\alpha_1^* & |\alpha_2|^2 \end{pmatrix}. \tag{9.29}$$

The simplest way to give the state vector a 'kick' is to use a rotation operator

$$R(\theta)\,|\psi\rangle = \alpha_1 e^{-i\frac{\theta}{2}}\,|1\rangle + \alpha_2 e^{i\frac{\theta}{2}}\,|2\rangle. \tag{9.30}$$

Figure 9.2. Depiction of decoherence.

The transformed density matrix is

$$\rho(\theta) = \begin{pmatrix} |\alpha_1|^2 & \alpha_1 \alpha_2^* e^{-i\theta} \\ \alpha_2 \alpha_1^* e^{i\theta} & |\alpha_2|^2 \end{pmatrix}. \tag{9.31}$$

The system is given a lot of kicks

$$\rho = \sum p(\theta)\rho(\theta) \rightarrow \int_{-\infty}^{\infty} \rho(\theta) p(\theta) d\theta, \tag{9.32}$$

with

$$p(\theta) \rightarrow \frac{1}{\sqrt{4\pi\lambda T}} e^{-\frac{\theta^2}{4\lambda T}}. \tag{9.33}$$

That is, it is given an infinity of random kicks with probabilities described by a Gaussian. The width of the Gaussian is proportional to the square root of the number of kicks N. That number is proportional to the time T that the system is exposed to the environment, which is the source of the kicks

$$N \propto \lambda T. \tag{9.34}$$

The number that multiplies the off-diagonal element of the density matrix is obtained using the well-known Fourier transform

$$\frac{1}{\sqrt{4\pi\lambda T}} \int_{-\infty}^{\infty} e^{-i\theta} e^{-\frac{\theta^2}{4\lambda T}} d\theta = e^{-\lambda T}. \tag{9.35}$$

It follows that after being exposed to the environment for a time T, the density matrix is

$$\rho = \begin{pmatrix} |\alpha_1|^2 & \alpha_1\alpha_2^* e^{-\lambda T} \\ \alpha_2\alpha_1^* e^{-\lambda T} & |\alpha_2|^2 \end{pmatrix}. \tag{9.36}$$

This simple model shows decoherence because the off-diagonal elements of the density matrix go to zero for large exposure time T.

9.12 Back to Schrödinger's cat

It is difficult to work out the mathematics of decoherence except for simple models, but it is easy to picture how it works and what the outcome must be. It is easy to imagine that the quantum phases associated with an automobile will disappear quickly, but there are some large objects that remain quantum mechanical for long periods of time. The LIGO gravity-wave detector contains a cryogenic version of the Weber bar which must be treated as a quantum harmonic oscillator even though it weighs a ton. Another such object is a superconductor. The electrons remain in a quantum condensed state as long as the temperature remains lower than a critical value which can be as high as 70K, even when the superconducting wire is the wrapping for a magnet in a particle accelerator or a neighborhood MRI facility.

The two forms of ammonia, with the nitrogen above or below the plane of hydrogens, are enantiomers, molecules that are mirror images of each other. It is known that the maser and the ammonia clock will not work unless the molecule is in a quantum state in which it is equally probable to find the nitrogen in each of its two possible positions. As molecules get larger, they no longer appear as superpositions. The two forms of sugar, dextrose and levulose, are also enantiomers. They body reacts in completely different ways to these sugars because humans evolved in a world that doesn't contain levulose naturally. Somewhere between ammonia and dextrose, decoherence goes from being infinitesimal to being almost instantaneous.

It can be seen from these examples, that size alone does not define decoherence. It is necessary to study the way that the object interacts with its environment.

What, then, about Schrödinger's cat. The most obvious observation on this old debate that comes from the previous discussion is that, for a cat, decoherence would be so rapid that all efforts to treat it as a quantum object would fail. This, however, does not affect the argument that Schrödinger was making. In quantum information theory, large systems that do exist as quantum superpositions are called Schrödinger cat (SC) states. They are very useful in realizations of quantum computers.

In a sense, decoherence turns the question raised at the Solway Conference on its head. Instead of addressing the question 'How do quantum states come about?' it gives an answer to 'How do classical states come about?'

Problems

P9.1 How many women were invited to the fifty Solvay conference on physics?

P9.2 State as succinctly as possible the basic assumptions of the pilot-wave theory.

P9.3 How does von Neumann's concept of wave function collapse connect with the many-worlds interpretation.

P9.4 Why is the GRW called a spontaneous collapse theory?

P9.5 What is the difference between the consistent histories and the many-worlds interpretations?

P9.6 How does a quantum system know what classical system to decohere to?

References

[1] Griffiths R B 1984 *J. Stat. Phys.* **36** 219

[2] Omnès R 1988 *J. Stat. Phys.* **53** 893

[3] Gell-Mann M and Hartle J B 1990 Quantum mechanics in the light of quantum cosmology *Complexity, Entropy, and the Physics of Information* ed W Zurek (Reading: Addison Wesley)

[4] von Neumann J 1952 *Mathematische Grundlagen der Quantenmechanik* (Berlin: Springer) English translation by Beyer R T 1955 (Princeton, NJ: Princeton University)

[5] Zeh H D 1970 *Found. Phys.* **1** 69

[6] Zurek W H 2003 *Rev. Mod. Phys.* **75** 715

Chapter 10

Quantum information

10.1 Information science

The goals and nomenclature of quantum information science are built to a large extent on the concepts that today are studied in the fields of information science and cryptography. The field of information science was initiated by Claude E Shannon. Many of these concepts were originated by mathematicians and physicists, particularly by Alan Turing and John von Neumann. Von Neumann has already been quoted extensively in this book for his work on quantum theory.

It would be difficult to function in toady's society without interacting extensively with computers. However, few who use these devices give much thought to how they work or how they came to be. Most users of computers have confidence that they can do their banking and purchasing on a computer without worrying about identity theft. They occasionally read the small print on the license for an application which mentions something about the encryption of the information that you make available to it, but what is encryption and how reliable is it.

In this chapter, the history of information theory and an introduction to quantum information theory is given.

10.2 Turing machine

A Turing machine is an abstract representation of a computing device. It consists of a read/write head that scans a possibly infinite one-dimensional bi-directional tape divided into squares, each of which is inscribed with a 0 or 1. Computation begins with the machine, in a given 'state', scanning a square. It erases what it finds there, prints a 0 or 1, moves to an adjacent square, and goes into a new state. This behavior is completely determined by three parameters: (1) the state the machine is in, (2) the number on the square it is scanning, and (3) a table of instructions. The table of instructions specifies, for each state and binary input, what the machine should write, which direction it should move in, and which state it should go into. (E.g., 'If in state 1 scanning a 0: print 1, move left, and go into state 3'.) The table can list only a finite number of states, each of which becomes implicitly defined by the role it plays in the table of instructions. These states are often referred to as the 'functional states' of the machine.

A Turing machine, therefore, is more like a computer program (software) than a computer (hardware). It can be realized or implemented on any computing device. Computer scientists and logicians have shown that Turing machines—given enough time and tape—can compute any function that digital computers can compute.

Example: A Turing machine can be visualized as a two way tape all of whose cells are blank except for the four contiguous cells containing the symbols '0110'. The read/write head is positioned 'on' the first (left-most) non-blank cell containing a '0'. The machine is 'in' (internal) state 'A'. This description is illustrated in figure 10.1.

The operation of the Turing machine is simple. The current (internal) state and the current symbol read by the read/write head, determine the next (internal) state, the next symbol which overwrite the current symbol and whether the read/write head moves left or right. Turing machine instructions can be represented as quintuples (current state, current symbol, next symbol, next state, move left or right)

For example the quintuple (A, 0, 1, B, R) will cause the above Turing machine configuration to change to figure 10.2, where the current state = 'A' and current symbol = '0' causes the Turing machine to overwrite the '0' with a '1', change the internal state to 'B' and move the read/write head to the Right.

Turing machines were first proposed by Alan Turing, in an attempt to give a mathematically precise definition of 'algorithm' or 'mechanical procedure'. Early work by Turing and Alonzo Church spawned the branch of mathematical logic now known as recursive function theory.

Turing, on the basis of Gödel's theorem, proved that there exist certain uncomputable functions. Gödel's theorem states that, in any formal system rich enough to describe arithmetic, there are theorems that are true but can never be

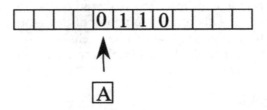

Figure 10.1. Turing machine 1.

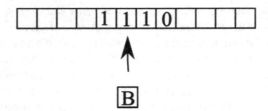

Figure 10.2. Turing machine 2.

proved. Turing's proof of the existence of uncomputable functions is one of the most profound results of this century. It had an enormous effect on all of mathematics and related fields. The unproven yet so far not disproved important result called the Church–Turing thesis states; any 'computable function' can be computed using the Turing machine.

As an example of an uncomputable function, consider the following computer algorithm.

```
x = 8 = 5 + 3
while (x = sum of two prime numbers)
x = x + 2
else {print x; halt}
```

Nobody knows whether this process will halt (Goldbach's conjecture).

This leads to the question: is there a general way to establish whether algorithms will halt? This is equivalent to asking: is there a general way to establish whether a given proposition can be proved or disproved? If it cannot decided whether an algorithm will halt or not, it is deemed uncomputable.

10.3 Bits and bytes and Boolean gates

The Turing machine works with 0s and 1s. No matter how complex the computer becomes, it still uses only those two numbers. Binary arithmetic is the basis of all calculations. A single 0 or 1 is called a bit. A collection of 8 bits is called a byte. Information transfer is also based on bits and bytes.

The computer can be visualized as a network of logic gates, with bits carried on wires. The network model can be used to describe the Turing machine, i.e., the models are equivalent. The network model is closer to the way a real computer is designed and constructed, and is easier to work with. The gates are a physical realization of the operations defined in the logical algebra of the nineteenth century mathematician George Boole, and hence are called Boolean gates. The translation of logical statements to bits should be obvious. It will be seen that quantum computers took over the idea of using Boolean gates.

NOT gate (inverter)

The output Q is true when the input A is NOT true, the output is the inverse of the input: Q = NOT A. A NOT gate can only have one input. In an actual computer, there would be one wire going in and one coming out. The icon that represents a not gate is designed to illustrate this. A NOT gate is also called an inverter and is illustrted in figure 10.3.

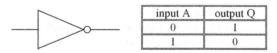

input A	output Q
0	1
1	0

Figure 10.3. Icon and truth table for the NOT gate.

AND gate

The output Q is true if input A AND input B are both true: Q = A AND B.

An AND gate can have two or more inputs, its output is true if all inputs are true. In a computer there are two wires going in and one coming out. If the inputs are bits, the output is the product $Q = A \times B$ as shown in figure 10.4.

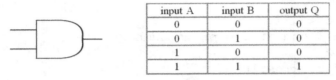

input A	input B	output Q
0	0	0
0	1	0
1	0	0
1	1	1

Figure 10.4. Icon and truth table for the AND gate.

NAND gate (NAND = Not AND)

This is an AND gate with the output inverted, as shown by the 'o' on the output. The output is true if input A AND input B are NOT both true: Q = NOT (A AND B) A NAND gate can have two or more inputs, its output is true if NOT all inputs are true as shown in figure 10.5.

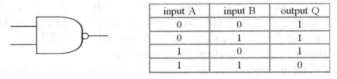

input A	input B	output Q
0	0	1
0	1	1
1	0	1
1	1	0

Figure 10.5. Icon and truth table for the NAND gate.

OR gate

The output Q is true if input A OR input B is true (or both of them are true): Q = A OR B. An OR gate can have two or more inputs, its output is true if at least one input is true, see figure 10.6.

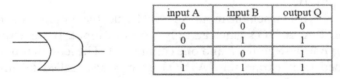

input A	input B	output Q
0	0	0
0	1	1
1	0	1
1	1	1

Figure 10.6. Icon and truth table for the OR gate.

NOR gate (NOR = Not OR)

This is an OR gate with the output inverted, as shown by the 'o' on the output. The output Q is true if NOT inputs A OR B are true: Q = NOT (A OR B). A NOR gate can have two or more inputs, its output is true if no inputs are true as shown in figure 10.7.

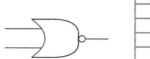

input A	input B	output Q
0	0	1
0	1	0
1	0	0
1	1	0

Figure 10.7. Icon and truth table for the NOR gate.

EX-OR (EXclusive-OR) gate

The output Q is true if either input A is true OR input B is true, but not when both of them are true: Q = (A AND NOT B) OR (B AND NOT A).

This is like an OR gate but excluding both inputs being true.

The output is true if inputs A and B are DIFFERENT.

EX-OR gates can only have 2 inputs. In the language of bits, the output is the sum of A and B modulo 2, $Q = A \oplus B$ as illustrated in figure 10.8.

input A	input B	output Q
0	0	0
0	1	1
1	0	1
1	1	0

Figure 10.8. Icon and truth table for the EX-OR gate.

EX-NOR (EXclusive-NOR) gate

This is an EX-OR gate with the output inverted, as shown by the 'o' on the output. The output Q is true if inputs A and B are the **SAME** (both true or both false): **Q = (A AND B) OR (NOT A AND NOT B).** EX-NOR gates can only have 2 inputs, see figure 10.9.

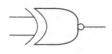

input A	input B	output Q
0	0	1
0	1	0
1	0	0
1	1	1

Figure 10.9. Icon and truth table for the EX-NOR gate.

10.4 Universality

It can be shown that all logic gates can be simulated with a combination of NAND gates and an operation called fanout.

The fanout takes a bit 'a' and copies it on two wires. This operation is also called cloning. The drawings in figures 10.10 and 10.11 show how some of the logic gates are constructed from fanout and NAND gates.

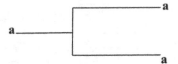

Figure 10.10. Operation fanout.

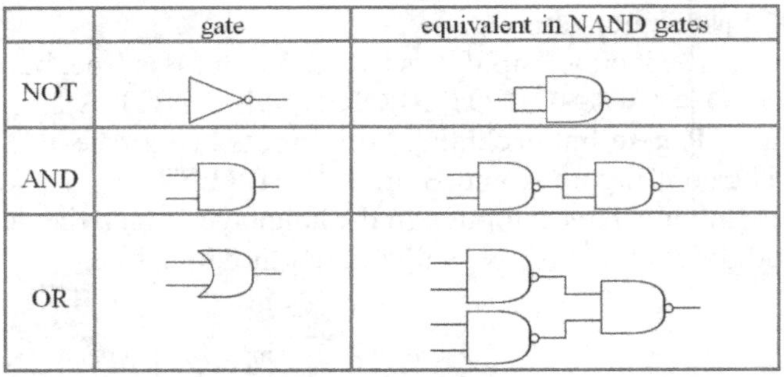

	gate	equivalent in NAND gates
NOT		
AND		
OR		

Figure 10.11. Logic gates made with fanouts and NAND gates.

10.5 Measuring information

In 1948 Claude E Shannon published two papers on '*A Mathematical Theory of Communication*' that are the foundation for the field of information science. He was born in 1916, so he was only thirty two when he published them. At the age of twenty two Shannon published a masters thesis in which he explained how to design computers using bits and bytes and Boolean gates as described in the preceding section. He later derived a fundamental theorem in cryptography that will be discussed later.

Shannon pointed out that the thing that will be communicated is information, so he had to create a mathematical definition of information. In order to do this, it is necessary to separate the concept from the ordinary everyday use of the term. A speaker may produce a large number of words, but, if the listener doesn't know any more after listening to them, no information has been transmitted. To make the idea more precise, suppose the thing to be communicated is the outcome of a Stern–Gerlach experiment. It is sufficient to send a 1 or a 0 to tell if the spin is up or down. The same bits can give the result of a coin flip. It is reasonable to believe that it will take more effort to communicate the results of an experiment that can have three outcomes.

Shannon defined the measure of information to be

$$H = -\sum_i p_i \log p_i,$$ (10.1)

where p_i is the reciprocal of the number of equally likely outcomes and the logarithm uses the base two. The number p_i is also the probability that a person would get the right answer with one guess. The information content increases with the possibility of making a mistake.

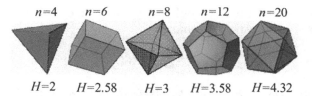

$$n=4 \quad n=6 \quad n=8 \quad n=12 \quad n=20$$

$$H=2 \quad H=2.58 \quad H=3 \quad H=3.58 \quad H=4.32$$

Figure 10.12. Dice made from Platonic solids. The number of faces appear above, and the Shannon entropy below.

Examples: The information measure for a coin flip is

$$H = -\sum_{i=1}^{2} 1/2 \log 1/2 = 1. \tag{10.2}$$

If the faces of a Platonic solid are numbered, the information content of giving the face that is on the table if the object is thrown randomly is

$$H = -\sum_{i=1}^{n} 1/n \log 1/n = \log n, \tag{10.3}$$

where n is the number of faces. The platonic solids and shannon entropies are shown above.

In all of these simple examples, the probabilities p_i are all the same, but that is by no means necessary.

Another advantage to Shannon's definition of information in equation (10.1) is that it treats joint events correctly. If the probability that event A will happen is p_a and the probability of event B is p_b, then the probability A and B will happen is $p_a p_b$. The information required to give the results of A and B is

$$H = -\sum_{a=1}^{n_a}\sum_{b=1}^{n_b} p_a p_b \log p_a p_b$$
$$= -\sum_a p_a \log p_a - \sum_b p_b \log p_b = H_a + H_b, \tag{10.4}$$

where the fact that

$$H = \sum_{a=1}^{n_a} p_a = \sum_{b=1}^{n_b} p_b = 1, \tag{10.5}$$

has been used. This is exactly the result that would be expected from a proper mathematical definition of information. For example, the information required to describe the outcome throwing 5 dice

$$H = 5 \log 6 = 12.9248. \tag{10.6}$$

The formula for information looks very much like Boltzmann's formula for entropy in books on statistical mechanics

$$H = -k\sum_i p_i \log p_i, \tag{10.7}$$

where k is Boltzmann's constant. People frequently refer to this as information entropy. An increase in entropy is coupled with an increase in the disorder of a system. It can be argued that the more disordered the system, the more information is required to describe it. However, it is dangerous to push the analogy too far.

After defining information, Shannon analyzed the ability to send information through a communications channel. He showed that a channel with a given bandwidth has a certain maximum transmission rate that cannot be exceeded. He went on to demonstrate mathematically that even in a noisy channel with a low bandwidth, essentially perfect, error-free communication could be achieved by keeping the transmission rate within the channel's maximum and by using error-correcting schemes. Error correction requires the transmission of additional bits that enable the information to be extracted from the noise-ridden signal.

10.6 Landauer's theory of the energy required for calculations

In thermodynamics, a reversible process changes the state of a system in such a way that the net change in the combined entropy of the system and its surroundings is zero. Reversible processes define the boundaries on the efficiency of ideal heat engines. A reversible process is one where no heat is lost from the system as 'waste' and the machine is thus as efficient as it can possibly be. The Carnot cycle describes the most efficient heat engine that can be made because it is reversible. No energy is wasted in unretrieved entropy production.

The relation between the entropy of information and thermodynamic entropy led to the speculation that each floating point operation in a computer (flop) would require energy. It was a surprise when it was discovered that this is not the case.

Landauer's Principle, first argued in 1961 by Rolf Landauer of IBM labs, holds that any logically irreversible manipulation of information, such as the erasure of a bit or the merging of two computation paths, must be accompanied by a corresponding entropy increase in non-information bearing degrees of freedom of the information processing apparatus or its environment.

Specifically, each bit of lost information will lead to the release of an amount $kT\ln2$ of heat, where k is the Boltzmann constant and T is the absolute temperature of the circuit. On the other hand, if no information is erased, computation may in principle be achieved which is thermodynamically reversible and require no release of heat. This has led to considerable interest in the study of reversible computing.

Landauer's principle can be proved with the help of a simple mechanical model. Consider a computer storage system in which a particle is placed in one of two possible segments of a cylinder. If the particle is in the upper segment, the location has value 1. If it is in the lower segment, the value is 0. This description is illustrated in figure 10.13.

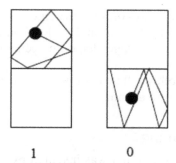

1   0

Figure 10.13. A representation of the bits 1 and 0 by a particle in a location.

The fact that no energy is expended in the process of changing the content of a storage location is demonstrated by imagining a set of pistons that can be used to move the particle from the 1 position to the 0 position. As shown in figure 10.14, if the space between the pistons is kept constant, no work is done in this process.

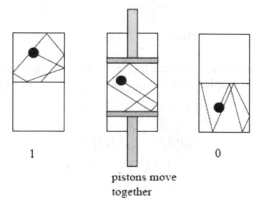

1   0

pistons move
together

Figure 10.14. Demonstration of change from 1 to zero without work.

On the other hand, erasing everything from a location is equivalent to removing the partition between the upper and lower segment as shown in figure 10.15.

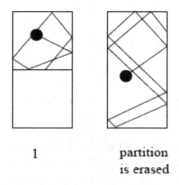

1   partition
is erased

Figure 10.15. Demonstration that erasing a bit changes the entropy.

It is proved in elementary classes on statistical mechanics that this process leads to an increase of entropy given by $\Delta S = k \ln(V/V_1) = k \ln 2$. From this it follows that erasing a number from a storage location generates an amount of heat $\Delta Q = T\Delta S = kT \ln 2$.

From this analysis, it follows that computations are reversible if they do not contain operations that require erasing data.

10.7 Reversible computing

Consider the gate in figure 10.16 that was invented by T Toffoli in 1980.

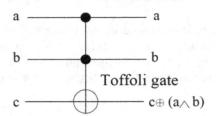

Figure 10.16. A Toffoli gate.

The symbol $\oplus$ means addition modulo 2 and the $\wedge$ means the multiplication of the bits. It follows that the truth table for the gate is the one in figure 10.17.

Input			Output		
0	0	0	0	0	0
0	0	1	0	0	1
0	1	0	0	1	0
0	1	1	0	1	1
1	0	0	1	0	0
1	0	1	1	0	1
1	1	0	1	1	1
1	1	1	1	1	0

Figure 10.17. Truth table for the Toffoli gate.

Figure 10.18 shows that the effect of the Toffoli gate can be written as a matrix equation. The input and outputs are written as column matrices. When this is done, the output matrix is related to the input matrix by matrix multiplication.

$$\begin{pmatrix} 0 & 0 & 0 \\ 0 & 0 & 1 \\ 0 & 1 & 0 \\ 0 & 1 & 1 \\ 1 & 0 & 0 \\ 1 & 0 & 1 \\ 1 & 1 & 1 \\ 1 & 1 & 0 \end{pmatrix} = \left(\begin{array}{cccc|cc|cc} 1 & 0 & 0 & 0 & & & & \\ 0 & 1 & 0 & 0 & & & 0 & \\ 0 & 0 & 1 & 0 & & & & \\ 0 & 0 & 0 & 1 & & & & \\ \hline & & & & 1 & 0 & & 0 \\ & & 0 & & 0 & 1 & & \\ \hline & & & & & 0 & 0 & 1 \\ & & & & 0 & & 1 & 0 \end{array} \right) \begin{pmatrix} 0 & 0 & 0 \\ 0 & 0 & 1 \\ 0 & 1 & 0 \\ 0 & 1 & 1 \\ 1 & 0 & 0 \\ 1 & 0 & 1 \\ 1 & 1 & 0 \\ 1 & 1 & 1 \end{pmatrix}$$

Figure 10.18. The matrix that relates the input matrix to the output matrix for the Toffoli gate.

From this representation it is clear that the Toffoli gate is reversible because

$$T^{-1} = T. \tag{10.8}$$

It can be shown that a universal reversible Boolean logic gate must have at least three bits.

10.8 Universality

The Toffoli gate is universal because if the input c is set equal to 1 it can be seen from figure 10.19 that it is equivalent to an NAND gate. The input is in the first two columns, and the output is the last column.

Input			Output		
0	0	1	0	0	1
0	1	1	0	1	1
1	0	1	1	0	1
1	1	1	1	1	0

Figure 10.19. Truth table of the Toffoli gate used as a NAND.

A fanout operation can also be obtained with a Toffoli gate by restricting a to be 1 and c to be 0. This is demonstrated in figure 10.20 where the input is the second

Input			Output		
1	0	0	1	0	0
1	1	0	1	1	1

Figure 10.20. Truth table of the Toffoli gate used as a fanout.

column, and the output is the last two columns. From the statement in the preceding section, any gate can be made with these two.

10.9 Zero power computing

In theory a computer made only of Toffoli gates would use no power at all. The price that must be paid for this is that, since no bits are erased, the number of bits carried in the circuit would increase without limit. In 1989 C H Bennett proposed a scheme whereby the extra bits could be modified and reused in the calculation without erasing them. It follows that zero power classical computing is possible. It is even more necessary that quantum computers must be run in the zero power mode.

10.10 Computational complexity

After it is determined that a problem is computable, the next question is if the computation is practical. The justification for working on quantum computers is based on the proposition that the computational complexity class for certain problems is lower for a quantum computer than for a classical computer.

The complexity of a problem is classified according to the number of steps $s = s(L)$ required by a computer to solve a problem specified by L amount of information. The number of digits in a binary number, n, is like the number of faces on a die. As pointed out in equation (10.3), the information required to give the outcome of throwing an n faced die is $L = \log_2 n$.

For example, the computation of some function $f(x)$ of an argument x that has n digits is desired. If the number of steps s required in the computation is polynomial in L

$$s = \sum_{i=1}^{max} c_i L^i \tag{10.9}$$

then the problem is tractable and placed in complexity class P. If s increases exponentially with L, i. e. max $\rightarrow \infty$, then the problem is hard.

Note that it is easier to verify a solution than to find one. Complexity class NP is the set of problems for which the solutions can be verified in polynomial time. Clearly, P $\in$ NP. In general, membership in a complexity class is independent of the model of the computer (whether Turing or network model, etc).

An important intractable (hard) problem is prime factorization when the number to be factored is large. The fundamental theorem of arithmetic states that every positive integer (except the number 1) can be represented in exactly one way apart from rearrangement as a product of one or more primes. A well-known method for solving this problem is the general number field sieve method. A theoretical and experimental formula for the number of steps required to a factor an n digit binary number with the general number field sieve is

$$s = e^{[2L^{1/3}(\log L)^{2/3}]}, \tag{10.10}$$

where $L = \log_2 n$. The first example has $n = 2^{300}$ binary digits, which seems big but in the standard base 10 arithmetic it only a 90 digit number. Inserting $L = 300$ into the above formula, the number of steps required is 5×10^{23}. Assuming each step is a floating point operation (flop) and that an exaflop computer is used (10^{18} flops per second), it will take 5.8 days to factor the number. However, if the number of digits is doubled and the problem is to factor a 180 digit number (base 10), on the same computer it will take 5.4 million years. It will be seen in the next chapter that the purpose for factoring these large numbers is to decode a message. It is safe to say that no one will be interested in the contents of the message after millions of years.

10.11 Quantum devices

The discussion to this point concerns the operation of devices that behave in a classical way. It is not necessary to understand quantum mechanics to grasp there function and use. In the late 1970s Feynman began considering the problem of calculating the properties of systems at the quantum level, and, in a paper published in 1982, *Simulating Physics with Computers*, he postulated that to simulate quantum systems you would need to build quantum computers. The computer that Feynman envisioned is like an analog computer. Devices have been built in which continuously variable physical quantities such as electrical potential, fluid pressure, or mechanical motion are represented in a way analogous to the corresponding quantities in the problem to be solved. The devices are set up according to initial conditions and then allowed to change freely. The evolution of the devices gives an insight into the physics of the problem, and hence they are called computers.

Analog computers have been replaced for the most part by digital computers that manipulate 1s and 0s with such blinding speed that they can do a wide range of calculations efficiently. Theorists have developed techniques for treating the problems that Feynman worried about with calculations on modern massively parallel supercomputers. With such machines, quantum mechanical calculations are routinely done on atoms, molecules, and condensed matter.

There are still some problems that cannot be dealt with using a digital computer, no matter how fast it is. These are the hard problems described in the preceding section. For this reason, there is interest in creating quantum devices including quantum computers. As with classical computers, the analog approach has been avoided in favor of quantum bits and quantum logic gates [1].

10.12 Quantum bits (qubits)

Instead of the binary bits 0 and 1 used in most classical computers, the quantum computer uses superposed states, called qubits

$$|\psi\rangle = \alpha\,|0\rangle + \beta\,|1\rangle \tag{10.11}$$

where $|\alpha|^2 + |\beta|^2 = 1$. The biggest difference between a bit and a qubit is that the qubit cannot be measured until and unless the programmer is prepared to destroy it. A measurement of $|\psi\rangle$ will project out the state $|0\rangle$ with probability $|\alpha|^2$ or $|1\rangle$ with probability $|\beta|^2$. No single measurement can distinguish between $|\psi_1\rangle = \alpha_1|0\rangle + \beta_1|1\rangle$ and $|\psi_2\rangle = \alpha_2|0\rangle + \beta_2|1\rangle$. A convenient way of picturing a qubit is as a point on a unit sphere determined by the angles θ and

$$|\psi\rangle = \cos\frac{\theta}{2}\,|0\rangle + e^{i\phi}\sin\frac{\theta}{2}\,|1\rangle. \tag{10.12}$$

10.13 Single qubit gates

As with classical gates, it is useful to picture quantum gates as physical objects with wires going in and wires coming out. A wire can carry at most one qubit. The input to all of the single qubit gates in this section is a generic qubit.

The **X** gate or **NOT** gate converts a qubit into the state $|\psi\rangle = \beta|0\rangle + \alpha|1\rangle$.

$$|\psi\rangle = \alpha|0\rangle + \beta|1\rangle \quad \boxed{X} \quad |\psi\rangle = \beta|0\rangle + \alpha|1\rangle$$

The **Hadamard** gate or square root of **NOT** gate converts a qubit into the state $|\psi\rangle = \alpha\frac{|0\rangle + |1\rangle}{\sqrt{2}} + \beta\frac{|0\rangle - |1\rangle}{\sqrt{2}}$.

$$|\psi\rangle = \alpha|0\rangle + \beta|1\rangle \quad \boxed{H} \quad |\psi\rangle = \alpha\frac{|0\rangle + |1\rangle}{\sqrt{2}} + \beta\frac{|0\rangle - |1\rangle}{\sqrt{2}}$$

The **Z** gate converts a qubit into the state $|\psi\rangle = \alpha|0\rangle - \beta|1\rangle$.

$$|\psi\rangle = \alpha|0\rangle + \beta|1\rangle \quad \boxed{Z} \quad |\psi\rangle = \alpha|0\rangle - \beta|1\rangle$$

These gates can be expressed as 2 by 2 matrices acting on a column vector $\begin{pmatrix} \alpha \\ \beta \end{pmatrix}$.

$$X = \begin{pmatrix} 0 & 1 \\ 1 & 0 \end{pmatrix}, \quad H = \frac{1}{\sqrt{2}}\begin{pmatrix} 1 & 1 \\ 1 & -1 \end{pmatrix}, \quad Z = \begin{pmatrix} 1 & 0 \\ 0 & -1 \end{pmatrix}. \tag{10.13}$$

The names of the X and Z gates can be understood by making explicit the parallel between the qubits and the wave functions for spin 1/2 particles, which is seen by writing $|0\rangle = |+\rangle$ and $|1\rangle = |-\rangle$. Then the matrix representative for the X gate is the Pauli matrix σ_x, and the matrix representative for the Z gate is the Pauli matrix σ_z.

All of these matrices have the feature that $A^2 = I$. Since they are real and symmetric, they are Hermitian, $A = A^\dagger$, so it follows that they are unitary $AA^\dagger = I$. Any unitary matrix can represent an acceptable gate in quantum mechanics. All quantum mechanical gates are reversible, which means that if you know the output you can infer the input. From Landauer's principle, this means that quantum computers will use no energy. A quantum Toffoli gate is universal, so universal reversible computing is the rule for quantum computers.

10.14 Random number generator

Even at this level, a quantum computer can do something a classical computer can never do. Consider the circuit shown in figure 10.21. That is, start with the pure cubit $|0\rangle$ and send it through the Hadamard gate to obtain the superposition $|\psi_2\rangle = \frac{|0\rangle + |1\rangle}{\sqrt{2}}$.

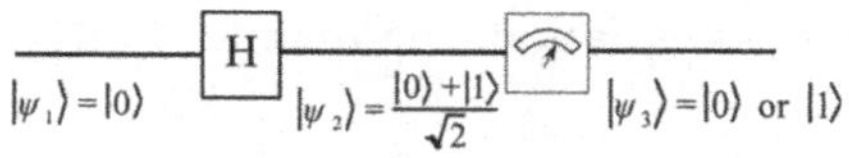

$|\psi_1\rangle = |0\rangle$ — H — $|\psi_2\rangle = \dfrac{|0\rangle + |1\rangle}{\sqrt{2}}$ — $|\psi_3\rangle = |0\rangle$ or $|1\rangle$

Figure 10.21. A quantum random number generator.

The dial icon represents a measurement. The result is one of two states with a probability that is absolutely random.

Classical computers use random number generator programs to achieve this result, but such programs do not produce truly random numbers. They are properly called pseudo-random number generators. The sequences of numbers that they produce are actually periodic, although the period is very long for a good generator program and they are usable in calculations that require random numbers.

10.15 A two qubit gate

As mentioned above, it is useful to think of gates as physical objects with wires going in and wires coming out. A wire can only carry one cubit. If a gate has two input wires, the top one carrying $|\psi_1\rangle = \alpha_1|0\rangle + \beta_1|1\rangle$ and the bottom one carrying $|\psi_2\rangle = \alpha_2|0\rangle + \beta_2|1\rangle$, then it is frequently useful to consider the input to the gate to be

$$|\psi\rangle = \alpha_1\alpha_2\,|00\rangle + \alpha_1\beta_2\,|01\rangle + \beta_1\alpha_2\,|10\rangle + \beta_1\beta_2\,|11\rangle = \sum_{i=0}^{3}a_i\,|i\rangle. \qquad (10.14)$$

The convention being used is that a ket written $|ab\rangle$ has a in the upper wire and b in the lower wire. The numbers in the kets in the sum $|i\rangle$ make sense if they are binary numbers.

The basic quantum mechanical two cubit gate is the CNOT. The qubit in the top wire passes through. The outgoing qubit on the bottom wire depends on the incoming qubit and also the qubit that is in the upper wire. Suppose the top wire carries the qubit $|0\rangle$, then the qubit on the bottom wire is unchanged. If the top wire carries the qubit $|1\rangle$, then an input of $|0\rangle$ on the bottom wire is converted to a $|1\rangle$ and an input of $|1\rangle$ becomes a $|0\rangle$. To put this another way, assume the ket $|a\rangle$ with length a is on the upper wire and the ket $|b\rangle$ with length b is the input on the lower wire, then the ket $|c\rangle$ with length c will be the output on the lower wire. The rule for calculating $|c\rangle$ is that c is the sum of a and b mod 2 as shown in figure 10.22.

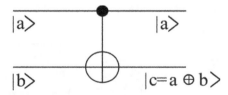

Figure 10.22. A CNOT gate.

As shown in the following equation, the quantum CNOT can be written as the unitary matrix.

$$\begin{pmatrix} |00\rangle \\ |01\rangle \\ |11\rangle \\ |10\rangle \end{pmatrix} = \begin{pmatrix} 1 & 0 & 0 & 0 \\ 0 & 1 & 0 & 0 \\ 0 & 0 & 0 & 1 \\ 0 & 0 & 1 & 0 \end{pmatrix} \begin{pmatrix} |00\rangle \\ |01\rangle \\ |10\rangle \\ |11\rangle \end{pmatrix}, \qquad (10.15)$$

where the column matrix on the right is the input and that on the left is the output. The truth table is given in table 10.1.

Table 10.1. The table showing the input and output of the CNOT gate.

IN	OUT
$|00\rangle$	$|00\rangle$
$|01\rangle$	$|01\rangle$
$|10\rangle$	$|11\rangle$
$|11\rangle$	$|10\rangle$

10.16 No cloning theorem

The CNOT is one of the most important gates in quantum circuits. It is possible to imagine a classical gate that is analogous to the quantum one. Figure 10.23 shows that it is possible to create such a gate from the well-known classical XOR gate. Figure 10.24 shows that a classical CNOT gate can be used to clone a bit. The bit is put on the upper wire, and the bit 0 is on the lower one. The operation of the gate clones the bit on the upper wire.

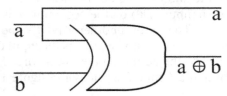

Figure 10.23. Method to construct a classical CNOT gate.

Figure 10.24. Cloning the bit a with a hypothetical classical CNOT gate.

Figure 10.25 shows an effort to clone a cubit with a quantum CNOT circuit. Putting the generic qubit $|\psi_{in}\rangle = \alpha|0\rangle + \beta|1\rangle$ on the upper wire and $|0\rangle$ on the lower one leads to the qubit

$$|\psi_f\rangle = \alpha|00\rangle + \beta|11\rangle, \tag{10.16}$$

as an outcome. If the gate had cloned the incoming bit the result would be

$$|\psi_f\rangle = \alpha\alpha|00\rangle + \alpha\beta|01\rangle + \beta\alpha|10\rangle + \beta\beta|11\rangle, \tag{10.17}$$

which is only the same as equation (10.16) for particular values of α and β. The above discussion is not a proof of the no cloning theorem. It shows that the most obvious method does not work. To date, no other method has produced cloning.

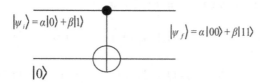

Figure 10.25. An effort to clone a qubit with a CNOT gate.

10.17 Bell or EPR states

The four basis vectors used so far for the two qubit state, $|00\rangle, |01\rangle, |10\rangle, |11\rangle$, form an orthonormal set, but other orthonormal sets can be constructed. In quantum information (QI), a particularly common choice is

$$\begin{aligned}
|\beta_{00}\rangle &= \frac{|00\rangle + |11\rangle}{\sqrt{2}} & |\beta_{01}\rangle &= \frac{|01\rangle + |10\rangle}{\sqrt{2}} \\
|\beta_{10}\rangle &= \frac{|00\rangle - |11\rangle}{\sqrt{2}} & |\beta_{11}\rangle &= \frac{|01\rangle - |10\rangle}{\sqrt{2}}
\end{aligned} \tag{10.18}$$

These states are said to be entangled. The physics of entangled states was first discussed in connection with the Einstein–Podulsky–Rosen (EPR) effect and Bell's inequality theorem, which explains the nicknames. All of these states are equally entangled, and the most commonly used one in QI is $|\beta_{00}\rangle$. These states are created by an entanglement circuit.

10.18 Entanglement and disentanglement

The entanglement circuit is shown in figure 10.26. The input can be written $|\psi_{in}\rangle = |ab\rangle$. When the cubit $|a\rangle$ passes through the Hadamard gate, the outcome can be written

$$H|a\rangle = \frac{1}{\sqrt{2}}\sum_{c=0}^{1}(-1)^{ac}|c\rangle. \tag{10.19}$$

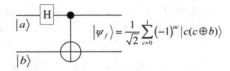

Figure 10.26. The entanglement circuit.

Using the rules of the CNOT gate with this qubit in the upper wire and $|b\rangle$ in the lower one, the final output of the circuit is

$$|\psi_f\rangle = \frac{1}{\sqrt{2}} \sum_{c=0}^{1} (-1)^{ac} |c(c \oplus b)\rangle. \tag{10.20}$$

Applying this formula to various inputs leads to

$$\begin{aligned} \text{E}\,|00\rangle &= \frac{1}{\sqrt{2}}(|00\rangle + |11\rangle)) & \text{E}\,|01\rangle &= \frac{1}{\sqrt{2}}(|01\rangle + |10\rangle)) \\ \text{E}\,|10\rangle &= \frac{1}{\sqrt{2}}(|00\rangle - |11\rangle)) & \text{E}\,|11\rangle &= \frac{1}{\sqrt{2}}(|01\rangle - |10\rangle)) \end{aligned} \tag{10.21}$$

This exercise demonstrates the reason for calling the circuit in figure 10.26 an entanglement circuit. The notation $\text{E}|ab\rangle$ indicates that the cubit $|ab\rangle$ has been sent through the entanglement circuit. Since $\text{E}|ab\rangle = |\beta_{ab}\rangle$, the notation in equation (10.18) anticipates the present results.

The circuit in figure 10.27 can be used to disentangle an entangled state. This can be seen by passing $|\psi_f\rangle$ through a CNOT gate and then putting the result through the Hadamard gate. Entanglement and disentanglement are used frequently either implicitly or explicitly in QI circuits.

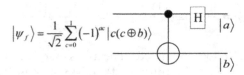

Figure 10.27. A disentanglement circuit.

10.19 Quantum teleportation

A use for entangled qubits is teleportation, which sounds like something from a television show. QI scientists are very excited about it because it obviates an inherent problem in quantum circuits. However, members of the general public are usually disappointed when they find it has nothing to do with the brain and the object teleported is a qubit. The circuit shown below will transport a qubit from Alice to Bob with no connection between them except by a classical communication of bits. The top two wires in the circuit belong to Alice, and the bottom wire belongs to Bob. Initially, the unknown qubit

$$|\psi\rangle = \alpha\,|0\rangle + \beta\,|1\rangle, \qquad\qquad (10.22)$$

is on the top wire, and Bob and Alice share the entangled state $|\beta_{00}\rangle$ on the middle wire and the bottom wire. It can be seen that Bob's qubit is unchanged until it reaches an X or Z gate that may or may not be activated. Alice sends the qubits on the top and middle wire through gates that the reader will recognize is an disentanglement operation shown in figure 10.28.

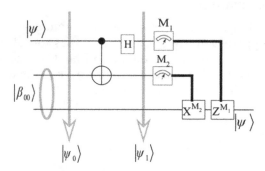

Figure 10.28. A teleportation circuit.

The initial state can be written as

$$|\psi_0\rangle = 1/\sqrt{2}\,(\alpha|000\rangle + \alpha|011\rangle + \beta|100\rangle + \beta\,|111\rangle), \qquad\qquad (10.23)$$

and with some manipulation this becomes

$$|\psi_0\rangle = 1/2\big[\,|\beta_{00}\rangle(\alpha|0\rangle + \beta\,|\,1\rangle) + |\beta_{10}\rangle(\alpha|0\rangle - \beta\,|1\rangle) \\ + |\beta_{01}\rangle(\alpha|\,1\rangle + \beta|0\rangle) + |\beta_{11}\rangle(\alpha|1\rangle - \beta|0\rangle)\big] \qquad (10.24)$$

The qubits in the top two wires have been rewritten in terms of recognizable entangled states. When these vectors are run through the disentanglement circuit, the result is

$$|\psi_0\rangle = 1/2\big[|\,00\rangle(\alpha|0\rangle + \beta|1\rangle) + |10\rangle(\alpha|0\rangle - \beta\,|1\rangle) \\ + |01\rangle(\alpha\,|1\rangle + \beta|0\rangle) + |11\rangle(\alpha|1\rangle - \beta|0\rangle)\big] \qquad (10.25)$$

Alice now runs the qubits on the top two wires through measuring devices. If the device senses the qubit $|0\rangle$ it sends out the bit 0, and if it senses $|1\rangle$ it sends out 1. As shown in the figure, the bits from the measuring devices are sent by ordinary wires to the boxes on Bob's quantum wire. The notation X^{M_2} means that, if the M_2 measuring device sends a 0, then the X gate is not activated and the Bob's qubit is not changed. If the M_2 measuring device sends a 1, then the X gate is activated and Bob's qubit is changed according to the rules in equation (10.13). Similarly, if the M_1 measuring device sends a 0, then the Z gate is not activated, while, if the M_1 measuring device sends a 1, then the Z gate is activated.

Looking at the qubit in the bottom wire from equation (10.25) and recalling the effect of the X and Z gates from equation (10.13), it can be seen that the qubit that accompanies a given pair of bits in the top two wires will be converted into the qubit in equation (10.22). Therefore, Alice can ensure that the final qubit that Bob has on his wire, $|\psi\rangle$, is the same as the one that was initially on her top wire. It should be noted that at no time does either she or Bob know what that qubit is.

10.20 Superdense coding

If Alice has one bit of information she wants to communicate to Bob, she can just send a $|0\rangle$ or $|1\rangle$. Suppose, however, Alice is in possession of two classical bits of information, b_1 and b_1, that she wants to send to Bob. Can she achieve this sending just one qubit? The answer is yes, if she uses a process called superdense coding. The superdense coding circuit is shown in figure 10.29.

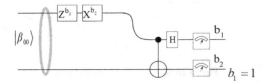

Figure 10.29. Superdense coding circuit.

Initially, Alice and Bob share two wires, and the entangled state $|\beta_{00}\rangle$ is on those two wires. Alice modifies the top wire according to the information that she wants to send. She then relinquishes her wire to Bob. Bob sends the two wires through a disentangling circuit, and does a measurement on them. As expected, if the measuring device senses the qubit $|a\rangle$ it converts it to the bit a.

Alice sends the qubit on the top wire through gates according to the following rules. The Z gate is activated if $b_1 = 1$ and it is not if $b_1 = 0$, and the X gate is activated if $b_2 = 1$ and it is not if $b_2 = 0$. Since the Z gate changes a $|0\rangle$ to a $|0\rangle$ and a $|1\rangle$ to a $-|1\rangle$, it changes $|\beta_{00}\rangle$ to $|\beta_{10}\rangle$. The disentangling circuit changes $|\beta_{10}\rangle$ to $|10\rangle$ and the measuring devices produce the bits 1 and 0. This and other operations are shown in table 10.2.

Table 10.2. The table of bits sent by Alice to Bob, and the gates that are activated.

b_1	b_2		
0	0	I	$|\beta_{00}\rangle$
1	0	Z	$|\beta_{10}\rangle$
0	1	X	$|\beta_{01}\rangle$
1	1	ZX	$|\beta_{11}\rangle$

A normal measurement of a qubit gives one of two values, i. e., 0 or 1. This seems to say that the qubit doesn't carry more information than a classical bit. Superdense coding is a way to send more information with a qubit, but it requires an entangled state. The entangled state is sometimes referred to as a resource that lets the circuit do something that would otherwise be impossible.

10.21 Deutsch's algorithm

Another example of a quantum computer being able to do something that a classical computer cannot is called quantum parallelism. It will be seen in the next chapter that this is one of the most important features of a quantum computer. Deutsch's algorithm is the simplest example of this.

A person has some black balls and some white balls. There are two cups on a table. He will put one ball under each cup according to one of these two rules. Rule 1: the two balls must be the same. Rule 2: the two balls must have different colors as shown in figure 10.30. Is it possible to tell which rule he is following by looking under just one cup?

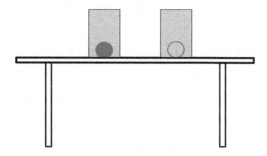

Figure 10.30. Illustration of Deutsch's algorithm.

In order to analyze this problem mathematically, assume there is a function $f(x)$ that has the values 0 or 1 for $x = 0$ or 1. There are four possible functions that can be constructed this way, as shown in table 10.3.

Table 10.3. Four possible functions.

	$x = 0$	$x = 1$	
f_1	0	0	rule 1
f_2	1	1	
f_3	0	1	rule 2
f_4	1	0	

Letting the colors white and black be represented by the numbers 0 and 1, the first two functions satisfy rule 1 and the last two satisfy rule 2. These functions are used in the circuit in figure 10.31.

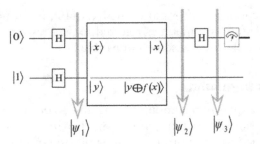

Figure 10.31. Circuit for Deutsch's algorithm.

The operator in the box can read the incoming bits and knows the function $f(x)$. When the operator sees a state $|xy\rangle$ it converts it to $|x(y \oplus f(x))\rangle$. The advantage to speaking theoretically or even hypothetically is that it is not necessary to specify how such a box might be constructed. That is a problem for experimentalists, but there is no reason to believe that it is impossible.

Given the rules for Hadamard gates, the state after the bits $|0\rangle$ and $|1\rangle$ pass those gates is

$$|\psi_1\rangle = \frac{|0\rangle + |1\rangle}{\sqrt{2}} \frac{|0\rangle - |1\rangle}{\sqrt{2}} = \frac{|00\rangle - |01\rangle + |10\rangle - |11\rangle}{2}. \tag{10.26}$$

If $f = f_1$, the function box has no effect whatever, so the state just beyond the box is

$$\begin{array}{c} f = f_1 \\ |\psi_2\rangle = |\psi_1\rangle \end{array}. \tag{10.27}$$

If $f = f_2$, the bits in the lower wire are switched from 0 to 1 and 1 to 0 to obtain

$$|\psi_2\rangle = \frac{|01\rangle - |00\rangle + |11\rangle - |10\rangle}{2} = -|\psi_1\rangle. \tag{10.28}$$

If $f = f_3$, the bits in the lower wire that follow a zero are unchanged and those that follow a one are flipped

$$|\psi_2\rangle = \frac{|00\rangle - |01\rangle - |10\rangle + |11\rangle}{2} = \frac{|0\rangle - |1\rangle}{\sqrt{2}} \frac{|0\rangle - |1\rangle}{\sqrt{2}}. \tag{10.29}$$

If $f = f_4$, the bits in the lower wire that follow a zero are flipped and those that follow a one are unchanged

$$|\psi_2\rangle = \frac{-|00\rangle + |01\rangle + |10\rangle - |11\rangle}{2} = -\frac{|0\rangle - |1\rangle}{\sqrt{2}} \frac{|0\rangle - |1\rangle}{\sqrt{2}}. \tag{10.30}$$

When the various forms of $|\psi_2\rangle$ are sent through the last Hadamard gate, the result is

$$|\psi_3\rangle = |a\rangle \frac{|0\rangle - |1\rangle}{\sqrt{2}}. \tag{10.31}$$

If $f = f_1$ or $f = f_2$, that is rule 1 is satisfied, $|a\rangle = \pm|0\rangle$. If $f = f_3$ or $f = f_4$, that is rule 2 is satisfied, $|a\rangle = \pm|1\rangle$. Quantum states are unchanged if a wave function is multiplied by a phase factor of magnitude one, so the final measuring device in the top wire does not see the minus signs. If rule 1 is satisfied, the measuring device sends out the bit 0, while if rule 2 is satisfied, the measuring device sends out the bit 1. This is the mathematical equivalent to the question about the balls and cups stated above. One measurement, the bit from the measuring device, is sufficient to distinguish between the cases that the balls under the cups are the same color or different colors.

10.22 Deutsch–Jozsa algorithm

The Deutsch algorithm is generalized by defining functions with 2^n arguments, each of which can have the value 0 or 1. The functions can only be of two kinds. One is a constant function for which all of the arguments are either 0 or 1, and the other is a balanced function for which exactly half of the arguments are 0 and the other half are 1. There are only 2 constant functions, but the number of balanced functions is $2^n!/(2^{(n-1)}!)^2$, which is very large. The problem is to decide if a function is constant or balanced. To do this classically, ½ of the functions arguments plus 1 would have to be determined. This would require $2^{(n-1)} + 1$ evaluations. It will be shown that a quantum computer can solve the problem with n evaluations, so this is a case where the quantum computer is exponentially faster than a classical one.

The circuit for the Deutsch–Jozsa (D-J) algorithm is shown in figure 10.32.

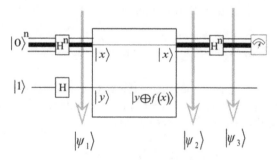

Figure 10.32. Deutsch–Joscza circuit.

There are $n + 1$ wires going into the box and the same number coming out. The top n wires initially carry qubit $|0\rangle$, and will be called the active wires. The bottom wire carries qubit $|1\rangle$, and will be called the passive wire. After passing through the Hadamard gates that precede the box, 2^n states are generated in the active wires

$$|\psi\rangle = \left(\frac{|0\rangle + |1\rangle}{\sqrt{2}}\right)^n. \tag{10.32}$$

If they are interpreted as binary numbers, they represent the integers from 0 to $2^n - 1$. For the case $n = 4$, the state looks like this

$$|\psi\rangle = \frac{1}{4}[|0000\rangle + |0001\rangle + |0010\rangle + |0011\rangle + |0100\rangle + |0101\rangle + |0110\rangle$$
$$+ |0111\rangle + |1000\rangle + |1001\rangle + |1010\rangle + |1011\rangle + |1100\rangle \qquad (10.33)$$
$$+ |1101\rangle + |1110\rangle + |1111\rangle]$$

The wave function can be written in a concise form

$$|\psi\rangle = 2^{-n/2} \sum_{i=0}^{2^n-1} |x_i\rangle. \qquad (10.34)$$

The wave function $|\psi_1\rangle$ includes the qubit in the bottom wire, and is written

$$|\psi_1\rangle = 2^{-n/2} \sum_{i=0}^{2^n-1} |x_i\rangle \frac{|0\rangle - |1\rangle}{\sqrt{2}}. \qquad (10.35)$$

The mathematical statement of the question to be investigated is to introduce the function $f(x_i)$ that takes on values of 0 or 1 for the 2^n arguments x_i. For rule 1, the function is the same for all x_i, either 0 or 1. For rule 2, the function is 0 for half of the x_i and 1 for the other half. The question is whether a function the has 2^n arguments can be shown to follow rule 1 or 2 making only n measurements.

The wave function is now sent through the box, and the state $|x_i y\rangle$ is converted to $|x_i(y + f(x_i))\rangle$. Noting the effect of the value of $f(x_i)$ on the qubit in the bottom wire, it is seen that if $f(x_i) = 0$, the term $|x_i\rangle \frac{|0\rangle - |1\rangle}{\sqrt{2}}$ is unchanged while if $f_i(x_i) = 1$ it is $-|x_i\rangle \frac{|0\rangle - |1\rangle}{\sqrt{2}}$. The wave function that comes out of the box is then

$$|\psi_2\rangle = 2^{-n/2} \sum_{i=0}^{2^n-1} (-1)^{f(x_i)} |x_i\rangle \frac{|0\rangle - |1\rangle}{\sqrt{2}}. \qquad (10.36)$$

Alice is going to send the output from the active wires through Hadamard gates after they come out of the box. To understand the outcome, consider what happens when a pure state qubit $|0\rangle$ or $|1\rangle$ is sent through H,

$$H|a\rangle = \frac{1}{\sqrt{2}} \sum_{b=0}^{1} (-1)^{ab} |b\rangle. \qquad (10.37)$$

The qubit $|x_i\rangle$ is, of course, a collection of n 0s and 1s $|a_1^i a_2^i a_3^i \cdots a_n^i\rangle$. Therefore, the Hadamard gates that are on the active wires after passing the box have the effect

$$H^n |a_1^i a_2^i a_3^i \cdots a_n^i\rangle = 2^{-n/2} \sum_{j=0}^{2^n-1} (-1)^{a_1^i b_1^j + a_2^i b_2^j + a_3^i b_3^j + \cdots a_n^i b_n^j} |b_1^j b_2^j b_3^j \cdots b_n^j\rangle$$
$$= 2^{-n/2} \sum_{j=0}^{2^n-1} (-1)^{x_i \cdot z_j} |z_j\rangle \qquad (10.38)$$

Combining this equation with equation (10.36) leads to

$$|\psi_3\rangle = 2^{-n}\sum_{i=0}^{2^n-1}(-1)^{f(x_i)}\sum_{j=0}^{2^n-1}(-1)^{x_i \cdot z_j}|z_j\rangle\frac{|0\rangle - |1\rangle}{\sqrt{2}}.$$
(10.39)

The equation above seems very complicated, but the required information can be obtained quite simply. The crux of the argument is that the state vectors are normalized throughout the process, so, if any term has a coefficient of magnitude one, all the other terms must have coefficients of zero. The coefficient of the state vector

$$|z_0\rangle = |000\cdots0\rangle = |0\rangle\,|0\rangle\,|0\rangle\ldots|0\rangle,$$
(10.40)

is

$$c_0 = 2^{-n}\sum_{i=0}^{2^n-1}(-1)^{f(x_i)}.$$
(10.41)

If $f(x_i)$ is 0 for all i, then c_0 is one, while, if $f(x_i)$ is 1 for all i, c_0 is minus one. In either case, $|c_0|^2 = 1$. Thus, a measurement of all n wires that finds 0 in every wire is a necessary and sufficient proof that rule 1 is being followed and $f(x_i)$ is constant.

If a measurement of the active wires gives as much as a single 1 in any wire proves that $|c_0|^2 \neq 1$ and rule 1 is not satisfied. The agreement was that the $f(x_i)$ would be chosen according to rule 1 or rule 2, so it follows the measurement of a 1 is a necessary and sufficient proof that the values of $f(x_i)$ are balanced.

This shows that at most n measurements with a quantum computer can give information that it would take $2^{(n-1)} + 1$ measurements on a classical computer, an example of an exponential speedup in computing. Notice that there is no entanglement operation in this experiment. It will be seen in the next chapter that the ability of a quantum computer to do certain operations exponentially faster than a classical computer is a more important resource than entanglement.

10.23 Four-level Deutsch–Jozsa experiment

The analysis of the D-J theory in the previous section is correct, but it is useful to apply it to an example. The wires that are initiated with the qubit $|0\rangle$ will be called the active wires, and the bottom wire that is initiated with $|1\rangle$ is called the passive wire. A D-J circuit with one active and one passive wire is the Deutsch circuit. Four active wires is enough to illustrate the D-J theory, but not so many as to be unmanageable. The narrative that accompanies a D-J experiment with four active wires is that there are sixteen balls under sixteen cups. Rule 1 is that the balls under the cups will be all black or all white. Rule 2 is that 8 of the balls will be white and 8 will be black. This case is illustrated in figure 10.33. The challenge is to make four measurements that will distinguish if the balls are distributed under rule 1 or 2.

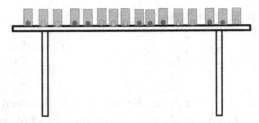

Figure 10.33. An illustration of the narrative for a four-level Deutsch–Josca experiment.

The circuit for a four-level D-J experiment is shown in figure 10.34.

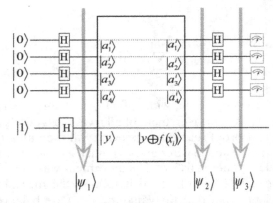

Figure 10.34. Circuit for the four-level Deutsch–Jozsa circuit.

The wave vector $|x_i\rangle$ can be written in a base 2 notation as $|a_1^i a_2^i a_3^i a_4^i\rangle$, where the a_j^i are 1s and 0s. The function $|\psi_1\rangle$ that describes the state after the qubits have passed through the Hadamard gates is

$|\psi_1\rangle$

$$= \frac{1}{16}\Big[|0000\rangle + |0001\rangle + |0010\rangle + |0011\rangle + |0100\rangle + |0101\rangle + |0110\rangle + |0111\rangle$$

$$+ |1000\rangle + |1001\rangle + |1010\rangle + |1011\rangle + |1100\rangle + |1101\rangle + |1110\rangle + |1111\rangle\Big] \times \frac{|0\rangle - |1\rangle}{\sqrt{2}} \quad .(10.42)$$

$$= \frac{1}{16}\sum_{i=0}^{15}|x_i\rangle \times \frac{|0\rangle - |1\rangle}{\sqrt{2}}$$

This function passes through the box. The mathematical statement that rule 1 is operating is that all of the $f(x_i)$ are one or zero. There are 12,870 ways to choose half of the $f(x_i)$ to be one and the other half to be zero, but any one of them corresponds to rule 2.

If all of the $f(x_i)$ are zero, the box has no effect and $|\psi_2\rangle = |\psi_1\rangle$. If they are all one, the qubits in the active wires are multiplied by -1 and the one in the passive wire is unchanged

$$|\psi_2\rangle = -\frac{1}{16}\sum_{i=0}^{15}|x_i\rangle \times \frac{|0\rangle - |1\rangle}{\sqrt{2}}. \tag{10.43}$$

Since the Hadamard gate is involutory, which means it is its own inverse, passing either of these wave functions through the last set of Hadamard matrices leads to

$$|\psi_3\rangle = \pm|0000\rangle \times \frac{|0\rangle - |1\rangle}{\sqrt{2}}. \tag{10.44}$$

The measuring devices are unaffected by the plus or minus signs. The conclusion is that rule 1 is operating if the output of all four devices is a 0.

For rule 2, $|\psi_3\rangle$ is a very complicated state and successive measurements will give different results. However, if a single 1 is returned by the measuring devices, that is proof that rule 1 is not operating and therefore rule 2 is. This constitutes the proof that the way the balls are distributed under the 16 cups shown in figure 10.33 can be found by making 4 measurements.

10.24 Discrete Fourier transform

There is nothing new about the discrete Fourier transform (DFT), but it turns out that it is important in quantum information systems. The ordinary Fourier transform is

$$y(k) = \frac{1}{\sqrt{2\pi}}\int_{-\infty}^{\infty} e^{ikt}x(t)dt, \quad x(t) = \frac{1}{\sqrt{2\pi}}\int_{-\infty}^{\infty} e^{-ikt}y(k)dk. \tag{10.45}$$

Inserting the function

$$x(t) = \sqrt{\frac{2\pi}{N}}\sum_{j=0}^{N-1}x_j\delta\left(t - \frac{2\pi j}{N}\right), \tag{10.46}$$

into the above equation leads to

$$y_k = \frac{1}{\sqrt{N}}\sum_{j=0}^{N-1}x_je^{i2\pi jk/N}. \tag{10.47}$$

This is the DFT. The inverse is found by

$$\frac{1}{\sqrt{N}}\sum_{k=0}^{N-1}y_ke^{-i2\pi jk/N} = x_j. \tag{10.48}$$

It is easy to prove that the DFT is periodic in k $y_{k+N} = y_k$, and also in j $x_{j+N} = x_j$.

The DFT is more useful than the ordinary Fourier transform for analyzing data, which normally appear as a list. For example, a set of data calculated at 200 points (there are 30 rough cycles) is shown in figure 10.35.

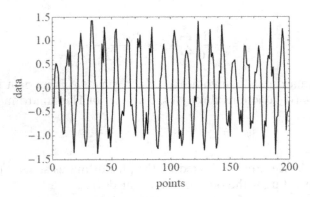

Figure 10.35. Noisy data evaluated for 200 points.

These data appear noisy or random, but there is an underlying pattern that is hard to spot with the naked eye. The DFT of these data appears in figure 10.36.

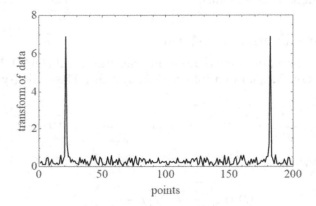

Figure 10.36. The DFT of the data above.

It shows a strong peak at 20 + 1, and a symmetric peak at 201–20, reflecting the frequency component of the major signal near 20/200. The underlying signal without the noise is shown in figure 10.37.

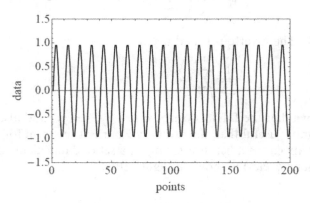

Figure 10.37. The underlying signal.

The formula for the function plotted in figure 10.37 is

$$f(n) = \sin\left(20\frac{2\pi n}{200}\right),$$ (10.49)

and its DFT has the two peaks shown in figure 10.36.

It will be seen in the next chapter when dealing with Shor's algorithm that it is particularly useful to know the DFT of the periodic sequence of 0s and 1s such as the one shown here.

$$0, 0, 1, 0, 0, 0, 0, 0, 0, 0, 0, 0, 1, 0, 0, 0, 0, 0, 0, 0,$$
$$0, 0, 1, 0, 0, 0, 0, 0, 0, 0, 0, 0, 1, 0, 0, 0, 0, 0, 0, 0,.$$ (10.50)
$$0, 0, 1, 0, 0, 0, 0, 0, 0, 0, 0, 0, 1, 0, 0, 0, 0, 0, 0$$

The first step in dealing with this class of functions is to change the origin of j and k so that equation (10.47) becomes

$$y_k = \frac{1}{\sqrt{N}} \sum_{j=1}^{N} x_j e^{i2\pi jk/N},$$ (10.51)

and k has the range 1, 2, ...N. Inserting the x_j like the one in equation (10.50) gives

$$y_k = \frac{1}{\sqrt{N}}\left(e^{i2\pi j_{min} k/N} + e^{i2\pi(r+j_{min})k/N} + e^{i2\pi(2r+j_{min})k/N} + ... + e^{i2\pi(nr+j_{min})k/N}\right),$$ (10.52)

where $nr + j_{min}$ is the last integer before N. This equation is equivalent to

$$y_k = \frac{1}{\sqrt{N}}e^{i2\pi j_{min} k/N}(1 + x + x^2 + ... + x^n).$$ (10.53)
$$x = e^{i2\pi rk/N}$$

The sum in equation (10.53) is

$$y_k = \frac{1}{\sqrt{N}}e^{i2\pi j_{min} k/N}\left(\frac{1 - x^{n+1}}{1 - x}\right).$$ (10.54)

Plots of the real and imaginary parts of y_k are very messy and of no interest, but the plot of

$$f(k) = \frac{|y_k|^2}{\sum_{l=1}^{N}|y_l|^2},$$ (10.55)

has an interpretation that will be given in the next chapter. The following plot shows $f(k)$ for integer values of k when the period is $r = 10$ and $N = 256$.

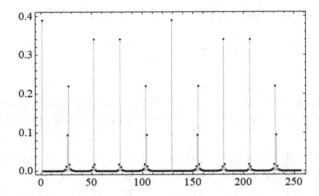

Figure 10.38. A plot of $f(k)$ from equation (10.55) for $r=10$ and $N=256$.

From equation (10.53), the peaks in $f(k)$ occur when $x \approx 1$ or

$$k \approx \frac{N}{r}m. \tag{10.56}$$

The notation means that k is the closest integer to the quantity on the right and m is an integer. The number of peaks is r because, if $m > r$, then $k > N$.

10.25 The quantum Fourier transform

The quantum version of the DFT is the discrete quantum Fourier transform (QFT). It starts from the idea that there is a basis set for the Hilbert space that is related to another basis set by a unitary transformation

$$U|j\rangle = \sum_{k=0}^{N-1} U_{jk}|k\rangle = \frac{1}{\sqrt{N}} \sum_{k=0}^{N-1} e^{i2\pi jk/N}|k\rangle. \tag{10.57}$$

The transform of a general vector is

$$U\sum_{j=0}^{N-1} x_j|j\rangle = \sum_{j=0}^{N-1}\sum_{k=0}^{N-1} x_j U_{jk}|k\rangle = \sum_{k=0}^{N-1} y_k|k\rangle, \tag{10.58}$$

and the y_k are the QFT of the x_j. Quantum QFTs can be found rapidly using quantum parallelism.

The formula for discrete QFT of a q-element basis function for n qubits is

$$\boxed{\text{DFT}_q}|a\rangle \rightarrow \frac{1}{\sqrt{q}} \sum_{c=0}^{q-1} e^{i2\pi\left(\frac{ac}{q}\right)}|c\rangle, \tag{10.59}$$

where $q = 2^n$ and the $|c\rangle$ are basis vectors. The q-element basis vector is a set of 0s and 1s, $|a\rangle = |a_{q-1}, a_{q-2}, \dots, a_2, a_1, a_0\rangle$, and has the ordinal number $a = \sum_{i=0}^{q-1} a_i 2^i$.

The ordinal number of the basis vector $|c\rangle$ is c. A set of n qubits $|b_i\rangle = b_i^0|0\rangle + b_i^1|1\rangle$ can be expressed as a linear combination of basis vectors $\sum_{i=0}^{2^n-1} c_i|i\rangle$.

If the QFT of the basis vectors is taken, then the QFT of any vector $|\psi\rangle = \sum_{a=0}^{q-1}\alpha_a|a\rangle$ is

$$\boxed{\text{QFT}_q}\,|\psi\rangle \rightarrow \frac{1}{\sqrt{q}}\sum_{a=0}^{q-1}\sum_{c=0}^{q-1}e^{i2\pi\left(\frac{ac}{q}\right)}\alpha_a\,|c\rangle = \sum_{c=0}^{q-1}\beta_c\,|c\rangle. \tag{10.60}$$

Ekert and Jozsa (E-J) showed that the QFT of a q-element basis vector can be evaluated using $\frac{q(q+1)}{2}$ gates operating on the q qubits. This is shown for the case of 4-element basis vector in figure 10.39.

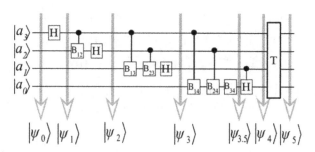

Figure 10.39. The E-J circuit for a 4-element basis vector.

The operators that are used in the QFT circuit are the Hadamard gate and a phase shift gate. From previous experience, it is known that when a $|0\rangle$ or $|1\rangle$ on a given wire passes through a Hadamard gate the output is on the same wire and is given by equation (10.37).

By looking at the phase shift gates in the drawing above, it is seen that they straddle two wires. There is a black dot on the reference wire, and an operation box on the lower wire. When a $|0\rangle$ or $|1\rangle$ on a given wire passes through the operation box, the output is on that same wire. The output, however, depends on the wire that the box is on and also the reference wire. It also depends on the qubit entering the box and the qubit on the reference wire.

$$B_{jk} = \begin{pmatrix} 1 & 0 & 0 & 0 \\ 0 & 1 & 0 & 0 \\ 0 & 0 & 1 & 0 \\ 0 & 0 & 0 & e^{i\pi/2^{k-j}} \end{pmatrix}. \tag{10.61}$$

The effect the phase shift gate has on the states is shown in table 10.4

Table 10.4. Operation of phase shift gate.

$\lvert \ldots a_j \ldots a_k \ldots \rangle$	$\lvert \ldots b_j \ldots b_k \ldots \rangle$
$\lvert \ldots 0 \ldots 0 \ldots \rangle$	$\lvert \ldots 0 \ldots 0 \ldots \rangle$
$\lvert \ldots 0 \ldots 1 \ldots \rangle$	$\lvert \ldots 0 \ldots 1 \ldots \rangle$
$\lvert \ldots 1 \ldots 0 \ldots \rangle$	$\lvert \ldots 1 \ldots 0 \ldots \rangle$
$\lvert \ldots 1 \ldots 1 \ldots \rangle$	$e^{i\pi/2^{k-j}}\lvert \ldots 1 \ldots 1 \ldots \rangle$

A 4-qubit case can be used as an example. The result of using formula (10.59) for the DFT is

$$\boxed{\text{DFT}_4}\,\lvert a \rangle \rightarrow \frac{1}{4}\Big[\lvert 0000 \rangle + e^{i\frac{2\pi a}{16}}\lvert 0001 \rangle + e^{i\frac{2\pi a}{16}2}\lvert 0010 \rangle + e^{i\frac{2\pi a}{16}3}\lvert 0011 \rangle$$

$$+ e^{i\frac{2\pi a}{16}4}\lvert 0100 \rangle + e^{i\frac{2\pi a}{16}5}\lvert 0101 \rangle + e^{i\frac{2\pi a}{16}6}\lvert 0110 \rangle + e^{i\frac{2\pi a}{16}7}\lvert 0111 \rangle \qquad (10.62)$$

$$+ e^{i\frac{2\pi a}{16}8}\lvert 1000 \rangle + e^{i\frac{2\pi a}{16}9}\lvert 1001 \rangle + e^{i\frac{2\pi a}{16}10}\lvert 1010 \rangle + e^{i\frac{2\pi a}{16}11}\lvert 1011 \rangle$$

$$+ e^{i\frac{2\pi a}{16}12}\lvert 1100 \rangle + e^{i\frac{2\pi a}{16}13}\lvert 1101 \rangle + e^{i\frac{2\pi a}{16}12}\lvert 1110 \rangle + e^{i\frac{2\pi a}{16}15}\lvert 1111 \rangle$$

where a is the ordinal number of $\lvert a \rangle = \lvert a_3, a_2, a_1, a_0 \rangle$.

A simple example is $\lvert a \rangle = \lvert 0001 \rangle$ with ordinal number $a = 1$. Formula (10.62) becomes

$$\boxed{\text{QFT}_4}\,\lvert 0001 \rangle \rightarrow \frac{1}{4}\Big[\lvert 0000 \rangle + e^{i\frac{2\pi}{16}}\lvert 0001 \rangle + e^{i\frac{2\pi}{16}2}\lvert 0010 \rangle + e^{i\frac{2\pi}{16}3}\lvert 0011 \rangle$$

$$+ e^{i\frac{2\pi}{16}4}\lvert 0100 \rangle + e^{i\frac{2\pi}{16}5}\lvert 0101 \rangle + e^{i\frac{2\pi}{16}6}\lvert 0110 \rangle + e^{i\frac{2\pi}{16}7}\lvert 0111 \rangle \qquad (10.63)$$

$$+ e^{i\frac{2\pi}{16}8}\lvert 1000 \rangle + e^{i\frac{2\pi}{16}9}\lvert 1001 \rangle + e^{i\frac{2\pi}{16}10}\lvert 1010 \rangle + e^{i\frac{2\pi}{16}11}\lvert 1011 \rangle$$

$$+ e^{i\frac{2\pi}{16}12}\lvert 1100 \rangle + e^{i\frac{2\pi}{16}13}\lvert 1101 \rangle + e^{i\frac{2\pi}{16}14}\lvert 1110 \rangle + e^{i\frac{2\pi}{16}15}\lvert 1111 \rangle.$$

The E-J circuit for a 4-element basis vector is shown in figure 10.39. The initial vector is obviously $\lvert \psi_0 \rangle = \lvert 0001 \rangle$. After this vector passes the first Hadamard gate, it becomes

$$\lvert \psi_1 \rangle = \frac{1}{\sqrt{2}}[\lvert 0001 \rangle + \lvert 1001 \rangle]. \qquad (10.64)$$

Because the second wire has a 0 on it, the phase shift operator acts like a unit operator and

$$\lvert \psi_2 \rangle = \frac{1}{2}[\lvert 0001 \rangle + \lvert 0101 \rangle + \lvert 1001 \rangle + \lvert 1101 \rangle]. \qquad (10.65)$$

The qubit pairs that could lead to a phase shift are on the first and third wires and the second and third wires. Neither of these have a 11 pair, so the phase shift operators have no effect. After $\lvert \psi_2 \rangle$ passes the next Hadamard gate the result is

$$|\psi_3\rangle = \frac{1}{2^{3/2}}[|0001\rangle + |0011\rangle + |0101\rangle + |0111\rangle + |1001\rangle + |1011\rangle$$
$$+ |1101\rangle + |1111\rangle]. \tag{10.66}$$

This vector is made up of 8 basis vectors, and will pass through three phase shift operators. The last four basis vectors in $|\psi_3\rangle$ have a 1 in the first and fourth wire, so the phase shift operator applies a shift of $\exp[i\pi/8]$ to them. The third, fourth, seventh, and eighth basis vectors have a 1 in the second and fourth wires, so they acquire shifts of $\exp[i\pi/4]$. Finally, the second, fourth, sixth, and eighth basis vectors have a 1 in the third and fourth wires, so they acquire $\exp[i\pi/2]$. Since the last vector has a 1 on every wire, it acquires a phase shift of

$$\exp\left[i\pi*\left(\frac{1}{8} + \frac{1}{4} + \frac{1}{2}\right)\right] = \exp\left[i\pi\frac{7}{8}\right]. \tag{10.67}$$

Similar calculations must be done for the other basis vectors, and the result is

$$|\psi_{3.5}\rangle = \frac{1}{2^{3/2}}[|0001\rangle + e^{i\pi/2}|0011\rangle + e^{i\pi/4}|0101\rangle + e^{i\pi 3/4}|0111\rangle$$
$$+ e^{i\pi/8}|1001\rangle + e^{i\pi 5/8}|1011\rangle + e^{i\pi 3/8}|1101\rangle + e^{i\pi 7/8}|1111\rangle] \tag{10.68}$$

Passing this vector through the final Hadamard gate leads to

$$|\psi_4\rangle = \frac{1}{4}[|0000\rangle - |0001\rangle + e^{i2\pi 4/16}|0010\rangle - e^{i2\pi 4/16}|0011\rangle$$
$$+ e^{i2\pi 2/16}|0100\rangle - e^{i2\pi 2/16}|0101\rangle + e^{i2\pi 6/16}|0110\rangle - e^{i2\pi 6/16}|0111\rangle$$
$$+ e^{i2\pi/16}|1000\rangle - e^{i2\pi/16}|1001\rangle + e^{i2\pi 5/16}|1010\rangle - e^{i2\pi 5/16}|1011\rangle$$
$$+ e^{i2\pi 3/16}|1100\rangle - e^{i2\pi 3/16}|1101\rangle + e^{i2\pi 7/16}|1110\rangle - e^{i2\pi 7/16}|1111\rangle] \tag{10.69}$$

To put this into a form that can be compared with equation (10.62) it is necessary to replace the minus signs with $e^{i2\pi(8/16)}$, which leads to

$$|\psi_4\rangle = \frac{1}{4}[|0000\rangle + e^{i2\pi(8/16)}|0001\rangle + e^{i2\pi(4/16)}|0010\rangle + e^{i2\pi(12/16)}|0011\rangle$$
$$+ e^{i2\pi(2/16)}|0100\rangle + e^{i2\pi(10/16)}|0101\rangle + e^{i2\pi(6/16)}|0110\rangle + e^{i2\pi(14/16)}|0111\rangle \tag{10.70}$$
$$+ e^{i2\pi(1/16)}|1000\rangle + e^{i2\pi(9/16)}|1001\rangle + e^{i2\pi(5/16)}|1010\rangle + e^{i2\pi(13/16)}|1011\rangle$$
$$+ e^{i2\pi(3/16)}|1100\rangle + e^{i2\pi(11/16)}|1101\rangle + e^{i2\pi(7/16)}|1110\rangle + e^{i2\pi(15/16)}|1111\rangle].$$

This is the desired result except that, as mentioned previously, the basis vectors come out in the form $|c_0c_1c_2c_3\rangle$ instead of $|c_3c_2c_1c_0\rangle$. For cosmetic reasons, if nothing else, the qubits can be put in the right order by a transposition operator $\mathbf{T}$, which has the effect $\mathbf{T}|c_0c_1c_2c_3\rangle = |c_3c_2c_1c_0\rangle$. Applying $\mathbf{T}$ to $|\psi_4\rangle$ gives

$$\mathbf{T}|\psi_4\rangle = |\psi_5\rangle = \boxed{\text{QFT}_4}\,|0001\rangle. \tag{10.71}$$

From

$$\boxed{\text{QFT}_q} \, |\psi\rangle \rightarrow \frac{1}{\sqrt{q}} \sum_{a=0}^{q-1} \sum_{c=0}^{q-1} e^{i2\pi\left(\frac{ac}{q}\right)} \alpha_a \, |c\rangle = \sum_{c=0}^{q-1} \beta_c \, |c\rangle \qquad (10.72)$$

it is seen that this looks like the classical DFT

$$\beta_c = \frac{1}{\sqrt{q}} \sum_{a=0}^{q-1} e^{i2\pi\left(\frac{ac}{q}\right)} \alpha_a. \qquad (10.73)$$

The gyrations that were necessary to find the QFT of one basis vector seem immense, but this is an emulation of the actions of a (hypothetical) quantum circuit. If the device is built, it could find the QFT exponentially faster than a classical computer.

Problems

P10.1 Suppose you have a computer that has a function Query(n) such that Query(n) = 1 if n is a prime number and Query(n) = 0 if n is not a prime number. Expand the pseudocode above to compute SumOfTwoPrimeNumbers and check the general statement in it.

P10.2 Expand the table in figure 10.11 to write the equivalent of the NOR, EX-OR, and EX-NOR gates in terms of fanouts and NAND gates.

P10.3 In order to win the Florida lottery it is necessary to guess 6 out of 53 numbers correctly. What is the information entropy for this guess?

P10.4 A crummy computer has only one megabyte of memory available for calculations and storage. It is necessary to calculate the function $f(x)$ for many values of x. It takes 100 flops to evaluate the function to acceptable accuracy for one x, and it is then stored in single precision. It takes fifty bytes of memory to carry out one flop. How many times can the function be evaluated before the computer starts heating up with Landauer entropy. If heating is ignored, how many times can the function be evaluated?

P10.5 The problem is to find a corporate password. The corporation has a rule that the password must contain n characters out of m legal characters. A step is trying the password to see if it works. What is the complexity of this problem?

P10.6 Suppose Alice has a qubit $|\psi\rangle$. Is there any way that she can tell Bob in a classical message what it is so that he can construct the identical qubit?

P10.7 For all of the two qubit gates statements are made like $|a\rangle$ is in the top wire and $|b\rangle$ is on the bottom one to create the state $|ab\rangle$. Are these statements consistent with the no cloning theorem?

P10.8 Which of these states was used by Einstein in the EPR paper?

P10.9 Prove equations (10.19) and (10.20).

P10.10 Do the algebra of this section for the special cases that Alice's initial state is $|\psi\rangle = |0\rangle$ or $|\psi\rangle = |1\rangle$?

P10.11 Why can this circuit not be used to teleport a qubit?

P10.12

The above notation appears in figure 10.32. Draw a larger picture to show exactly what it means.

P10.13 Work through a two level Deutsch–Jozsa experiment.

P10.14 Work through the above analysis for the basis function $|a\rangle = |0101\rangle$ with ordinal number $a = 5$.

Reference

[1] Nielsen M A and Chuang I L 2000 *Quantum Computation and Quantum Information* (Cambridge: Cambridge University Press)

IOP Publishing

Modern Quantum Mechanics and Quantum Information

J S Faulkner

Chapter 11

Quantum cryptography

11.1 The Caesar cipher

One of the simplest examples of a substitution cipher is the Caesar cipher, which is said to have been used by Julius Caesar to communicate with his army. Caesar decided that shifting each letter in the message would be his standard algorithm, and so he informed all of his generals of his decision and was then able to send them secured messages. Using the shift of three to the right, the message,

'RETURN TO ROME'

would be encrypted as,

'UHWXUA WR URPH'

In this example, 'R' is shifted to 'U', 'E' is shifted to 'H', and so on. Now, even if the enemy did intercept the message, it would be useless, since only Caesar's generals knew how to decrypt it.

Like many examples of ancient technology, this effort at encryption is very flimsy. The message can be decrypted with a laptop computer in seconds. Shifts from 1 to 25 can be tested until the message appears. The field of cryptography has evolved greatly over the years. The initial applications of the stronger methods were in the areas of business, diplomacy and warfare.

Today, cryptography plays an essential role in e-commerce. No one would buy an item unless they were certain the company had techniques to keep their credit card number secret. Most people do their banking and other business on the web. This would be impossible without the encryption of the personal data that pours through these systems. Physicists need to know about the elements of encryption because they are at the center of an industry focused on the creation of high strength codes and also the possible of breaking those codes.

doi:10.1088/978-0-7503-2167-9ch11

11.2 Symmetric key cryptography

Symmetric key cryptography makes use of keys that are shared between the sender and receiver of an encrypted message. The most famous historical example of symmetric key cryptography is the Enigma machine, which has electro-mechanical rotors and is used for the encryption and decryption of secret messages. The first Enigma was invented by a German engineer at the end of World War I. This model and its variants were used commercially from the early 1920s, and adopted by military and government services of several countries, most notably by Nazi Germany before and during World War II. A range of Enigma models was produced, but the German military model, the Wehrmacht Enigma, is the version most commonly discussed.

The machine has become notorious because Polish and then Allied cryptographers were able to decrypt a vast number of messages that had been encrypted using the Wehrmacht Enigma. One of the key figures in this effort was the British mathematician Alan Turing of Turing machine fame. The intelligence gleaned from this source, codenamed ULTRA by the British, was a substantial aid to the Allied war effort. The exact influence of ULTRA is debated, but an oft-repeated assessment is that decryption of German ciphers hastened the end of the European war by two years.

A more important version of symmetric key cryptography is called DES, an acronym for the Data Encryption Standard. DES is the name of the Federal Information Processing Standard (FIPS) 46–3, which describes the data encryption algorithm (DEA). The DEA is also defined in the ANSI standard X3.92. DEA is an improvement of the algorithm Lucifer developed by IBM in the early 1970s. DES was abandoned in about 1997. Over time, many implementations are expected to upgrade to the AES block cipher because it offers a 128, 192 and 256 bit keys. As of 2006, AES is one of the most popular algorithms used in symmetric key cryptography.

The National Security Agency (NSA) has affirmed that the design and strength of all key lengths of the AES algorithm (i.e., 128, 192 and 256) are sufficient to protect US Government classified information up to the SECRET level. TOP SECRET information will require use of either the 192 or 256 key lengths. It is also required by the standard specifying security mechanisms for wireless networks (Wi-Fi). High level block ciphers like AES require computers because the message bits are substituted and permuted many times in the encipherment process, but asymmetric key ciphers described below require even more computer resources.

11.3 Public-key cryptography (asymmetric cryptography)

In symmetric cryptography, it is necessary to have $N(N - 1)/2$ keys in order to give each pair of N persons their own keys. In asymmetric cryptography, there is no secret key. However, it is necessary that the decryption key D_k cannot be found from a knowledge of the public encryption key E_k.

Name	Public key (published)	Secret key (not published)
Alice	E_A	D_A
Bob	E_B	D_B

Sending a message: Bob has a message x that he wants to send to Alice. He looks up her public key and sends the encrypted message $y = E_A[x]$. Alice decrypts using her private key $D_A E_A[x] = x$. Clearly, $D_A = E_A^{-1}$. In order for this asymmetric code to be secure, the process of finding the inverse of the public key must be impossible as a practical matter.

Authentification: It is necessary to know you are talking to the right person. Alice can send Bob a message x that they agreed on by sending him $E_B D_A[x]$. Bob can strip off E_B with his secret key D_B, $D_B E_B D_A[x] = D_A[x]$. Then he can use Alice's public key to find x., $E_A D_A[x] = x$. Only Alice and Bob could carry out this set of operations. The operations must be such that, even if you know E_A, $D_A[x]$, and x, you still cannot find D_A.

The strongest codes in use today are asymmetric. Their strength relies on the existence of one way trapdoor functions. A one way trapdoor function is such that calculating $y = f(x)$ is easy, but finding $x = f^{-1}(y)$ is exponentially difficult. In order to understand how such a function can be generated, it is necessary to become familiar with some advanced topics in mathematics.

11.4 Modular arithmetic

A famous set of trapdoor functions is based on the area of number theory that deals with prime numbers and modular arithmetic. The ring $\mathbb{Z}_N$ is the set of integers $\{0,1,2,...N-1\}$, which is closed under addition and multiplication modulo N. The totient function $\phi(n)$, also called Euler's totient function, is defined as the number of positive integers $\leqslant n$ that are relatively prime to (i.e., do not contain any factor in common with) n, where 1 is counted as being relatively prime to all numbers.

Since a number less than or equal to and relatively prime to a given number is called a totative (or, more commonly, a unit), the totient function $\phi(n)$ can be simply defined as the number of totatives of n. For example, there are eight totatives of 24 (1, 5, 7, 11, 13, 17, 19, and 23), so $\phi(24) = 8$. The function $\phi(n)$ is always even for $n \geqslant 3$. By convention, $\phi(0) = 1$. For a prime p, $\phi(p) = p - 1$, since all numbers less than p are relatively prime to p.

The fundamental theorem of arithmetic states that any integer n can be written as a product of primes, $n = p_1{}^{m_1} p_2{}^{m_2} ... p_N{}^{m_N}$. The totient function for this n is

$$\phi(n) = p_1{}^{m_1-1}(p_1 - 1)p_2{}^{m_2-1}(p_2 - 1)...p_N{}^{m_N-1}(p_N - 1)$$

$$= \frac{n}{p_1 p_2 \cdots p_N}(p_1 - 1)(p_2 - 1)...(p_N - 1). \tag{11.1}$$

In the example above, $24 = 3 \times 2^3$ so $\phi(24) = \dfrac{24}{2 \cdot 3} 2 \cdot 1 = 8$, as had been calculated before. From this formula, the totient of the product of two primes p and q is

$$\phi(pq) = (p - 1)(q - 1), \tag{11.2}$$

so, for example, $\phi(15) = 2 \cdot 4 = 8$.

The set of units of $\mathbb{Z}_N$ with multiplication modulo N is a group $\mathbb{Z}_N^*$. Euler's theorem (also known as the Fermat–Euler theorem or Euler's totient theorem) states that if n is a positive integer and a is coprime to n, then

$$a^{\phi(n)} = 1(\mathrm{mod}\ n). \tag{11.3}$$

The previous statements are demonstrated for $n = 15$ and $\phi = 8$ in the following tables. $\mathbb{Z}_{15}^*$ is the multiplicative group of units, modulo 15. From the discussions in chapter 4, it can be seen that table 11.1 satisfies the requirements for a group multiplication table.

Table 11.1. Table of units for $\mathbb{Z}_{15}^*$.

	1	2	4	7	8	11	13	14
1	1	2	4	7	8	11	13	14
2	2	4	8	14	1	7	11	13
4	4	8	1	13	2	14	7	11
7	7	14	13	4	11	2	1	8
8	8	1	2	11	4	13	14	7
11	11	7	14	2	13	1	8	4
13	13	11	7	1	14	8	4	2
14	14	13	11	8	7	4	2	1

Table 11.2 illustrates Euler's theorem for $\mathbb{Z}_{15}^*$. The elements of $\mathbb{Z}_{15}$ for which Euler's theorem does not hold are not elements of $\mathbb{Z}_{15}^*$ because their greatest common denominator with 15 is not 1, $\gcd(x, 15) \neq 1$.

The purpose of this foray into modular arithmetic is to factor the large number n which is the product of two primes p and q. Euler's totient theorem says that if x is any number coprime with n, $x^{\phi(n)} = 1 \bmod n$, but the theorem does not say that $\phi(n)$ is the only number with this property. The simple example shown in the above table illustrates that there are numbers r such that $x^r = 1 \bmod n$. Such a number is called the order of x in the group $\mathbb{Z}_n^*$. With a little algebra, it can be shown that

$$(x^{r/2} + 1)(x^{r/2} - 1) = 0(\mathrm{mod}\ n), \tag{11.4}$$

when r is the order of x. From the first and last equalities in this paragraph, it appears that the product $(x^{r/2} + 1)(x^{r/2} - 1)$ must contain the product pq. Table 11.3 shows the possibilities. It follows that the prime factors of n can be obtained by

Table 11.2. Illustration of equation (11.3) for $\mathbb{Z}_{15}^*$.

					$n = 15\ \phi=8$								
X	(x^y)mod15								r	x^(r/2)−1		x^(r/2)+1	
1	1	1	1	1	1	1	1	1	1	0		2	
2	2	4	8	1	2	4	8	1	4	3		5	
3	3	9	12	6	3	9	12	6					
4	4	1	4	1	4	1	4	1	2	3		5	
5	5	10	5	10	5	10	5	10					
6	6	6	6	6	6	6	6	6					
7	7	4	13	1	7	4	13	1	4	3		5	
8	8	4	2	1	8	4	2	1	4	3		5	
9	9	6	9	6	9	6	9	6					
10	10	10	10	10	10	10	10	10					
11	11	1	11	1	11	1	11	1	2	10	2×5	12	4×3
12	12	9	3	6	12	9	3	6					
13	13	4	7	1	13	4	7	1	4	3		5	
14	14	1	14	1	14	1	14	1	2	13		15	3×5

Table 11.3. Table showing the possible implications of equation (11.4).

$(x^{r/2} - 1)$	$(x^{r/2} + 1)$
$k_1 p$	$k_2 q$
$k_2 q$	$k_1 p$
$k_1 pq = 0 \bmod n$	k_2
k_2	$k_1 pq = 0 \bmod n$

finding the order of an element in the group $\mathbb{Z}_n^*$. The previous statements are demonstrated for $n = 15$ and $\phi = 8$ in table 11.2.

The application of modular arithmetic to factor a prime number becomes more interesting as n becomes bigger. For example, the powers of 8^m in $\mathbb{Z}_{15}^*$ are shown in figure 11.1. From the data in figure 11.1, it can be seen that $r = 4$. Since $x^{r/2}(\bmod n) = 8^2(\bmod 15) = 4$, the factors of 15 are found from $(x^{r/2} - 1)(\bmod n) = 3$ and $(x^{r/2} + 1)(\bmod n) = 5$.

In figure 11.2 the powers of 38^m in $\mathbb{Z}^*_{77}$ are plotted. From the data in figure 11.2, it can be seen that $r = 30$. Since $38^{15}(\bmod 77) = 34$, the factors of 77 are found from $(x^{r/2} - 1)(\bmod n) = 33 = 3 \times 11$ and $(x^{r/2} + 1)\ (\bmod n) = 35 = 5 \times 7$. From these results it is easy to get the correct factors of 77.

Finally, the powers of 72^m in $\mathbb{Z}^*_{143}$ are shown in figure 11.3. From the data in this figure, it can be seen that $r = 60$. Since $72^{30}(\bmod 143) = 12$, the factors of 143 are found to be $(x^{r/2} - 1)(\bmod n) = 11$ and $(x^{r/2} + 1)\ (\bmod n) = 13$.

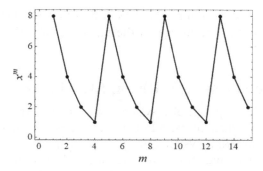

Figure 11.1. The powers of $8^m(\text{mod }15)$ as a function of m for $1 \leqslant m \leqslant 15$.

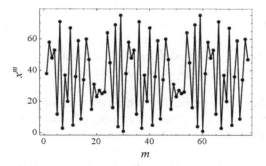

Figure 11.2. The powers of $38^m(\text{mod }77)$ as a function of m for $1 \leqslant m \leqslant 77$.

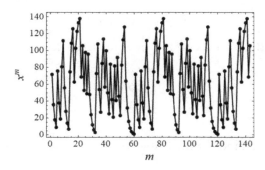

Figure 11.3. The powers of $72^m(\text{mod }143)$ as a function of m for $1 \leqslant m \leqslant 143$.

From these examples, it might be thought that this is a good way to reduce a number n into its prime factors for any size of n. It will be seen in the next section that, for the purpose of modern cryptography, the numbers that must be treated have 260 digits in the standard base 10 notation or 862 1s and 0s in the binary notation. For numbers of that size, the method described in the preceding paragraph is not practical. It will be seen in the final section of this chapter that the techniques of modular arithmetic will reemerge as the basis for factoring a number using a quantum computer.

11.5 RSA public key system. Rivest, Shamir, Adleman

The RSA asymmetric cryptography system was put forward by Ron Rivest, Adi Shamir, and Leonard Adleman, who publicly described the algorithm in 1977. It is widely used, and competing systems are very similar to it. It is conventional to describe the system by imagining two characters, Bob and Alice, who want to communicate securely. Bob chooses two large primes p and q, each having approximately 130 decimal digits or 430 binary digits. He calculates $n = p \times q$. He also calculates $\phi(n) = (p - 1) \cdot (q - 1)$. Bob chooses an integer s from the group of units of the ring $\mathbb{Z}_{\phi(n)}$, $s \in \mathbb{Z}^*_{\phi(n)}$. It is not necessary to study $\mathbb{Z}^*_{\phi(n)}$ in detail, but only that s the and $\phi(n)$ are that are chosen relatively prime, $\gcd(s, \phi(n)) = 1$. The requirement on s guarantees that it has an inverse, mod $\phi(n)$. Bob calculates the inverse t that satisfies $s \cdot t = 1(\text{mod } \phi(n))$. When dealing with very large numbers, it is not difficult to guess s and check it because the probability is high that a randomly chosen number will suffice.

Bob can now publish his public key, which consists of the two integers n, s. That means that he keeps p, q, and t secret. Of course, a hacker could break this code if n could be factored into it's two prime numbers. The RSA Laboratories has put forward a set of numbers made up of products as a challenge to the public. So far, no one has been able to factor a number with 260 decimal digits.

When Alice wants to send Bob a message, she first converts her message into a string of decimal digits or binary bits, x. She then looks up Bob's public key and creates the number

$$y = E_B(x) = x^s \text{ mod } n, \tag{11.5}$$

which she sends to Bob. To decrypt Alice's message, Bob operates on the string she sent him

$$D_B(y) = y^t \text{ mod } n = x. \tag{11.6}$$

Proof: Since $s \cdot t = 1(\text{mod } \phi(n))$, it follows that $s \cdot t = 1 + k\phi(n)$, where k is any integer $k \in \mathbb{Z}$. Then

$$y^t \text{ mod } (n) = x^{st} \text{ mod } (n) = x(x^{\phi(n)})^k. \tag{11.7}$$

If x is coprime with n, Euler's totient theorem says $x^{\phi(n)} = 1 \text{ mod } (n)$ and, of course, any integer power of it will be 1. Inserting this into equation (11.7) proves the decryption. It is overwhelmingly likely that x will be coprime with n, because that only means that x is not a multiple of p or q.

Below is an example in which the message is one decimal digit long and the number n is two digits. It takes a flight of imagination to translate to an example in which the message and n have more than 260 digits.

Example:

Alice	Bob
Alice wants to send message $x = 2$	Public parameters $n = 15$ and $s = 3$
She sends y, where	Private $t = 3$, $\phi = 8$
$y = E_B(x) = x^s \bmod n = 2^3 \bmod 15 = 8$	Bob decrypts with $D_B(y) = y^t \bmod n = 8^3 \bmod 15$
	$= 512 \bmod 15 = 2 = x$

11.6 Diffie–Hellman key exchange

RSA is a relatively slow algorithm, and because of this, it is less commonly used to directly encrypt user data. More often, RSA passes encrypted shared keys for symmetric key cryptography which in turn can perform bulk encryption–decryption operations at much higher speed. Key exchange is still a major aspect of cryptography. The Diffie–Hellman key agreement protocol (also called exponential key agreement) was developed by Diffie and Hellman in 1976 and published in the ground-breaking paper 'New Directions in Cryptography'. The protocol allows two users to exchange a secret key over an insecure medium without any prior secrets. As with RSA, the security of the protocol relies on the use of very large numbers.

The protocol has two system parameters p and g. They are both public and may be used by all the users in a system. Parameter p is a prime number and parameter g (usually called a generator) is an integer less than p with the following property: for every number n between 1 and $p-1$ inclusive, there is a power k of g such that $n = g^k \bmod p$.

Suppose Alice and Bob want to agree on a shared secret key using this protocol. They proceed as follows: First, Alice generates a random private integer a and Bob generates a random private integer b. Then they derive their public values using parameters p and g and their private values. Alice's public value is $g^a \bmod p$ and Bob's public value is $g^b \bmod p$. They then exchange their public values. Finally, Alice uses her secret integer b to compute $g^{ab} \bmod p = K$, and Bob uses his secret integer to compute $g^{ba} \bmod p = K$. Alice and Bob now have a shared secret key K.

The protocol depends on the discrete logarithm problem for its security. Breaking the Diffie–Hellman protocol is equivalent to computing discrete logarithms under certain assumptions. The same mathematical principal that makes the factorization of a large integer into its component primes impossible as a practical matter makes the solution of the discrete logarithm problem impossible for large integers g and p.

11.7 Discrete logarithm problem

For a group element g of the group $\mathbb{Z}_p^*$ and a number n, let g^n denote the element obtained by multiplying g by itself n times modulo p

$$g^2 = g*g \bmod p, \; g^3 = g*g^2 \bmod p, \; g^4 = g*g^3 \bmod p..., \qquad (11.8)$$

and so on. The discrete logarithm problem is as follows: given an element g in $\mathbb{Z}_p^*$ and another element h contained in the same group, find the integer x such that $g^x = h$.

Table 11.4 shows the multiplication table for the group $\mathbb{Z}_{23}^*$ which is of order 22. It shows the integers 1–22 taken to powers 1–22 mod 23. Using this table, a discrete logarithm problem based on this group can easily be solved. For example, given $g = 5$ and $h = 15$, it is seen from the table that $x = 17$. The table will be used to give an example of the Diffie–Hellman protocol.

The following is an example of the Diffie–Hellman key exchange protocol.

Table 11.4. This is the multiplication table for the group $\mathbb{Z}_{23}^*$, made by exponentiating the row number to the column number. Every element is a unit. The totient is 22.

y=1	2	3	4	5	6	7	8	9	10	11	12	13	14	15	16	17	18	19	20	21	22
1	1	1	1	1	1	1	1	1	1	1	1	1	1	1	1	1	1	1	1	1	1
2	4	8	16	9	18	13	3	6	12	1	2	4	8	16	9	18	13	3	6	12	1
3	9	4	12	13	16	2	6	18	8	1	3	9	4	12	13	16	2	6	18	8	1
4	16	18	3	12	2	8	9	13	6	1	4	16	18	3	12	2	8	9	13	6	1
5	2	10	4	20	8	17	16	11	9	22	18	21	13	19	3	15	6	7	12	14	1
6	13	9	8	2	12	3	18	16	4	1	6	13	9	8	2	12	3	18	16	4	1
7	3	21	9	17	4	5	12	15	13	22	16	20	2	14	6	19	18	11	8	10	1
8	18	6	2	16	13	12	4	9	3	1	8	18	6	2	16	13	12	4	9	3	1
9	12	16	6	8	3	4	13	2	18	1	9	12	16	6	8	3	4	13	2	18	1
10	8	11	18	19	6	14	2	20	16	22	13	15	12	5	4	17	9	21	3	7	1
11	6	20	13	5	9	7	8	19	2	22	12	17	3	10	18	14	16	15	4	21	1
12	6	3	13	18	9	16	8	4	2	1	12	6	3	13	18	9	16	8	4	2	1
13	8	12	18	4	6	9	2	3	16	1	13	8	12	18	4	6	9	2	3	16	1
14	12	7	6	15	3	19	13	21	18	22	9	11	16	17	8	20	4	10	2	5	1
15	18	17	2	7	13	11	4	14	3	22	8	5	6	21	16	10	12	19	9	20	1
16	3	2	9	6	4	18	12	8	13	1	16	3	2	9	6	4	18	12	8	13	1
17	13	14	8	21	12	20	18	7	4	22	6	10	9	15	2	11	3	5	16	19	1
18	2	13	4	3	8	6	16	12	9	1	18	2	13	4	3	8	6	16	12	9	1
19	16	5	3	11	2	15	9	10	6	22	4	7	18	20	12	21	8	14	13	17	1
20	9	19	12	10	16	21	6	5	8	22	3	14	4	11	13	7	2	17	18	15	1
21	4	15	16	14	18	10	3	17	12	22	2	19	8	7	9	5	13	20	6	11	1
22	1	22	1	22	1	22	1	22	1	22	1	22	1	22	1	22	1	22	1	22	1

Example:

Alice	Bob
Public parameters $p = 23$ and $g = 14$	Public parameters $p = 23$ and $g = 14$
Private $a = 8$	Private $b = 12$
Alice sends $g^a \bmod p = 14^8 \bmod 23 = 13$	Bob sends $g^b \bmod p = 14^{12} \bmod 23 = 9$
$K = 9^8 \bmod 23 = 13$	$K = 13^{12} \bmod 23 = 13$

While this example shows how the protocol works, it must be kept in mind that the numbers used in practice have more than 260 decimal digits.

11.8 ElGamal

The ElGamal encryption system is an asymmetric key encryption algorithm for public-key cryptography which is based on the Diffie–Hellman key exchange. It was described by Taher Elgamal in 1985. ElGamal encryption is used in the free software foundation's GNU Privacy Guard software, recent versions of pretty good privacy (PGP), and other cryptosystems. The Digital Signature Algorithm (DSA) is a variant of the ElGamal signature scheme. The Digital Signature Algorithm (DSA) is a Federal Information Processing Standard for digital signatures. DSA is patented but NIST has made this patent available worldwide royalty-free.

ElGamal encryption can be defined over any cyclic group G, most commonly the multiplicative group of integers modulo p, $\mathbb{Z}_p^*$. Its security depends upon the difficulty of solving the discrete logarithm problem for large integers p. The ElGamal system consists of both encryption and key signature variants. It is similar in nature to the Diffie–Hellman key agreement protocol.

The system parameters for the ElGamal cryptosystem are a prime p and an integer g. The powers of g modulo p generate elements of $\mathbb{Z}_p^*$ (it is not necessary for g to be a generator of the group $\mathbb{Z}_p^*$; but it is ideal). Alice has a private key a and a public key y and g, where

$$y = g^a \bmod p. \tag{11.9}$$

Suppose Bob wishes to send a message m (length less than p) to Alice. Bob first generates a random number b less than p. He then computes

$$\begin{aligned} y_1 &= g^b \bmod p \\ y_2 &= my^b \bmod p \end{aligned} \tag{11.10}$$

Bob sends the pair (y_1, y_2) to Alice. Alice computes $y_1^{-a}y_2 \bmod p$, which is equal to m

$$y_1^{-a}y_2 \bmod p = g^{-ab}mg^{ab} \bmod p = m. \tag{11.11}$$

It is helpful to apply ElGamal to an actual message using the multiplication table for $\mathbb{Z}_{23}^*$ in table 11.4.

Example:

Alice	Bob
Public parameters $p = 23$ and $g = 14$	Private $b = 12$
Private $a = 8$	Bob has a message $m=17$ to send.
Alice makes public $y = g^a \bmod p = 14^8 \bmod 23 = 13$	Bob sends
Alice calculates $9^{-1} \bmod 23$ from $9x = 1 \bmod 23$.	$y_1 = g^b \bmod p = 14^{12} \bmod 23 = 9$
By brute force, $x = 9^{-1} \bmod 23 = 18$. She then calculates	$y_2 = y^b m \bmod p = 13^{12} \cdot 17 \bmod 23 = 14$.
$18^8 \cdot 14 \bmod 23 = 16 \cdot 14 \bmod 23 = 17 = m$	

Now imagine doing a calculation like this for a prime number p with more than 260 decimal digits.

11.9 Elliptic curves

Elliptic curve cryptography (ECC) was proposed by Victor Miller and Neal Koblitz in the mid 1980s. The NSA decided to move to elliptic curve based public-key cryptography in the late 1990s. The full package of specified cryptography algorithms was called 'Suite B'. The argument for ECC is based on a comparison of the size of keys. More recently, the NSA seems to be moving away from ECC because the quantum methods that promise to successfully break the RSA can also be used on the ECC. Their are other reasons to prefer the RSA over the ECC. One is that more systems use the RSA, even today.

Whole books are written on the mathematics of elliptic curves, but to discuss that theory in this book would be moving too far afield. There are even more modern efforts to develop cryptographic protocols, such as the use of non-commutative braid groups. To understand this requires even more advanced mathematics.

11.10 The Vernam cipher

Many cryptography schemes are based on the *one time pad* or *Vernam cipher*. The idea is simple. Number the letters in the alphabet from 1 to 26. Alice and Bob both have a list of random numbers. Alice adds the numbers from her pad to the numbers representing letters in her message, modulo 26. In the example below, Alice's message is QUANTUM. The middle row, is the set of random numbers. The message received by Bob is mcttbiw. Since he has the identical set of random numbers, he simply subtracts them from the numbers corresponding to that message, and thus obtains Alice's message.

Q	U	A	N	T	U	M
17	21	1	14	20	21	13
22	8	19	6	17	20	5
13	3	20	20	11	15	18
m	c	t	t	b	i	w

Claude Shannon proved that this code is unbreakable as long as Alice and Bob use a new sheet of numbers for each message. The problem is distributing the sheets of numbers, or keys as they are called.

Physically handing a one time pad to an operative is not a practical scheme in most cases. Several classical schemes for exchanging keys securely over public communication lines have been described. One is the Diffie–Hellman key exchange protocol, and another is ElGamal. Methods that use the unique features of quantum mechanics are described below.

11.11 Quantum key distribution

A functioning quantum computer (QC) will make the classical public key distribution methods that are presently in use obsolete. What QCs take away with one hand, they return with another. The best known quantum key distribution scheme is BB84 [1]. It uses the quantum concept of conjugate variables.

Alice generates a random string of 0s and 1s length n called a and another of the same length called b. Using these strings, she sends a string of n qubits $| \psi_{ab} \rangle$ to Bob as illustrated in the following table.

a	1	0	0	1	0	1	1	1	0
state	$\lvert \psi_{11} \rangle$	$\lvert \psi_{00} \rangle$	$\lvert \psi_{01} \rangle$	$\lvert \psi_{10} \rangle$	$\lvert \psi_{01} \rangle$	$\lvert \psi_{10} \rangle$	$\lvert \psi_{10} \rangle$	$\lvert \psi_{10} \rangle$	$\lvert \psi_{01} \rangle$
b	1	0	1	0	1	0	0	0	1

The qubits corresponding to a given $| \psi_{ab} \rangle$ are

$$| \psi_{00} \rangle = |0\rangle \; | \psi_{10} \rangle = |1\rangle$$
$$| \psi_{01} \rangle = \frac{|0\rangle + |1\rangle}{\sqrt{2}} \; | \psi_{11} \rangle = \frac{|0\rangle - |1\rangle}{\sqrt{2}} . \tag{11.12}$$

The b are called basis bits, and the states with basis bit 1 are created with Hadamard gates. Bob will create his own string of basis bits b' and will put a qubit through a Hadamard gate before measuring it if the corresponding bit in his list is a 1. At the end of the process, Alice and Bob publicly compare their strings of basis bits b and b'. Any qubit for which the bit in b and b' differ is discarded. Statistically, this would be about half of the qubits that Alice sent. The bits a in states for which Bob and Alice chose the same basis bits should be the same. This gives both of them a string of 1s and 0s that can be used as a key in any symmetric cryptographic system.

To test for tampering, they choose a random subset of the a that have the same basis bits and announce them publicly. From there they can compute the error rate (i.e., the fraction of points where their values disagree). If the error rate is unreasonably high, above around say 10%, they throw away all the data (and perhaps try again later). The reason for this is that an evesdropper, Eve, that is intercepting and rebroadcasting Alice's qubits has no way of knowing which of them were sent through Hadamard gates. If she guesses wrong, she will cause an error. If the error rate is acceptably small, Alice and Bob assume that their communication was not breached.

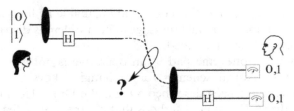

Figure 11.4. An illustration of the BB84 quantum key distribution algorithm.

They then perform error correction and also 'privacy amplification' to distill out a shorter string of bits which will act as the secret key. These steps essentially ensure that their keys agree, are random, and are unknown to Eve (Figure 11.4).

One practical implementation consists in the transmission of $0°$, $90°$, $45°$ and $135°$ linear polarizations by Alice over optical fiber. This is possible by polarization scrambling or polarization modulation. The reason that BB84 is difficult is that the the laser source must be intense enough so that the signal will propagate over a reasonable distance, but weak enough so that only one photon will be counted in a given pulse. If two photons are sent as a packet, Eve can intercept one of them and gain information without interfering with Bob and Alice. Thus, she can eavesdrop without being detected.

Other quantum key distribution schemes have also been proposed. For example, Artur Ekert (of Oxford) proposed one based on quantum mechanically correlated (i.e., entangled) photons and using Bell inequalities as a check of security [2].

There are ways to create entangled states experimentally

$$\Xi_x = 1/\sqrt{2}(|x; \uparrow\rangle \, |x; \downarrow\rangle - |x; \downarrow\rangle \, |x; \uparrow\rangle)$$
$$\Xi_y = 1/\sqrt{2}(|y; \uparrow\rangle \, |y; \downarrow\rangle - |y; \downarrow\rangle \, |y; \uparrow\rangle) \qquad (11.13)$$

Such entangled states can be used in cryptography for key distribution. They have an additional advantage that no one can eavesdrop on their communication without destroying it. If Eve intercepts the electron destined for Bob and makes a measurement, the spin wave function becomes $\Xi = |\uparrow\rangle \, |\downarrow\rangle$ or $\Xi = |\downarrow\rangle \, |\uparrow\rangle$. There is no way that Eve can reconstitute the spin function into Ξ_α and send it on to Bob. It follows that there will be no correlation between the spin directions measured by Bob and Alice. Ekert's proposal is similar to BB84 in that Bob and Alice have equipment that can analyze the states in an x or y plane. Alice uses polarization in the two planes randomly as does Bob. When they get through, they tell each other which planes they measured in over an unsecure line.

The most effective use for quantum key distribution at the time of this writing is a Chinese project that makes use of ground stations and a satellite. The stations are connected to China's Quantum Science Satellite, nicknamed Mozi, which was launched in August 2016. A key was relayed via Mozi between the ground station in Jinan and a fixed station in Shanghai. The Industrial and Commercial Bank of China (ICBC) uses the quantum key distribution system to insure the privacy or their commercial communications.

11.12 Shor factoring algorithm

In the year 2020, it was announced that the United States Department of Energy would spend \$625 M dollars to create five quantum information science research centers at the Argonne National Laboratory, Brookhaven National Laboratory, Fermi National Accelerator Laboratory, Lawrence Berkeley National Laboratory, and the Oak Ridge National Laboratory. This is a mere fraction of the funding for research on quantum information and quantum computing being spent by the United States government and also private companies. Similar efforts are being made by all of the other technologically advanced countries in the world. A prime motivation for this massive research effort stems from a mathematical paper entitled '*Polynomial-Time Algorithms for Prime Factorization and Discrete Logarithms on a Quantum Computer*' published in 1994 by Peter W Shor [3]. The RSA asymmetric cryptography system introduced in 1977 revolutionized the field of cryptography that is pivotal to modern e-commerce, and Shor's factoring algorithm promises to undo that. All asymmetric cryptography systems rely on the fact that it is much harder to reduce a large number into its prime factors than to multiply two primes together. Even with the fastest modern computers, it would take years to factor a very large number using the most advanced techniques. Shor has provided an algorithm that uses the features of a QC to factor the same number in a day. In this section, that algorithm will be described and analyzed.

It was shown in section 11.4 that the factors of an integer n can be found by obtaining the power r of an element x of the group $\mathbb{Z}_n^*$ such that $x^r = 1 \bmod n$. Finding the power r for which this is true is known as solving the discrete logarithm problem, and in general that problem is just as difficult as factoring. However, using quantum parallelism, Shor showed how it can be done using the parallelism feature of a QC.

The Shor calculation pictures a quantum device with two registers called register 1 and register 2. The combination of these two registers is $|\text{reg}\rangle = |\text{reg1}, \text{reg2}\rangle$. Register 1 is loaded with the integers $i = 0,1,2,...,a-1$, where $a = 2^m$ and m is the number of qubits (wires) that go into the register. The use of Hadamard gates for doing this was explained in the discussion of the Deutsch–Jozsa algorithm. For example, for $m = 4$,

$$|\text{reg1}\rangle = \left(\frac{|0\rangle + |1\rangle}{\sqrt{2}}\right)^4 = \frac{1}{4}(c_0 |0000\rangle + c_1 |0001\rangle + c_2 |0010\rangle$$
$$+ c_3 |0011\rangle + c_4 |0100\rangle + c_5 |0101\rangle + c_6 |0110\rangle + c_7 |0111\rangle \quad (11.14)$$
$$+ c_8 |1000\rangle + c_9 |1001\rangle + c_{10} |1010\rangle + c_{11} |1011\rangle + c_{12} |1100\rangle$$
$$+ c_{13} |1101\rangle + c_{14} |1110\rangle + c_{15} |1111\rangle)$$

The c_i are all 1 to start off with. Register 2 contains the powers $x^i \bmod n$, where x is chosen by the operator and n is the number to be factored. These powers can be generated in a QC using a method devised by Shor. This leads to the result $|\text{reg2}\rangle = \sum_{i=0}^{a-1} c_i |x^i \bmod n\rangle$ and the c_i are all the same. The full register is then

$|\text{reg}\rangle = \frac{1}{\sqrt{a}} \sum_{i=0}^{a-1} |i, x^i \bmod n\rangle$. For the case that $n = 21$, $m = 7$, $a = 128$, and $x = 10$, the contents of register 2 are shown in figure 11.5.

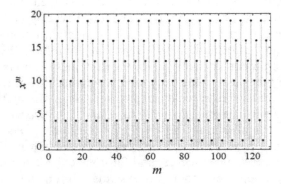

Figure 11.5. The powers of x mod$n$ as a function of for the case that $n = 21$, $m = 7$, $a = 128$, and $x = 10$. The parameter m ranges from 1 to 128.

When register 2 is measured, a number k is found. It is, of course, in the range 0 $-(n-1)$. It must be emphasized that this is an internal measurement and the operator of the QC does not know that number. By the process of measuring k the quantum state of both registers is changed. The set of integers j such that $x^j \bmod n = k$ is projected out in register 1. That is, register 1 is now $|\text{reg1}\rangle = \sum_{j=0}^{a-1} c_j |j\rangle =$ the c_j are zero except for the integers j described above. For the values of n, m, a, and x listed above and also $k = 4$, register 1 now has the contents shown in figure 11.6.

0, 0, 0, 1, 0, 0, 0, 0, 0, 1, 0, 0, 0, 0, 0, 1, 0, 0, 0, 0, 0, 1, 0, \
0, 0, 0, 0, 1, 0, 0, 0, 0, 0, 1, 0, 0, 0, 0, 0, 1, 0, 0, 0, 0, 0, 1, \
0, 0, 0, 0, 0, 1, 0, 0, 0, 0, 0, 1, 0, 0, 0, 0, 0, 1, 0, 0, 0, 0, 0, \
1, 0, 0, 0, 0, 0, 1, 0, 0, 0, 0, 0, 1, 0, 0, 0, 0, 0, 1, 0, 0, 0, 0, \
0, 1, 0, 0, 0, 0, 0, 1, 0, 0, 0, 0, 0, 1, 0, 0, 0, 0, 0, 1, 0, 0, 0, \
0, 0, 1, 0, 0, 0, 0, 0, 1, 0, 0, 0, 0

Figure 11.6. The contents of register 1 for the case that $n = 21$, $m = 7$, $a = 128$ and a measurement of register 2 has given the value $k = 4$.

At this stage, the QC contains enough information to calculate r, since it is just the period of the array of ones and zeroes seen above. For the parameters of the example, $r = 6$. However, there is no way to secure this information. Only one measurement can be made on a QC because that measurement projects the computer into a specific state. Further measurements on the same computer simply gives the same result. If the computer is restarted, it can give another result. A measurement on register 1 at this stage will most likely give zero and less likely give one. Repeatedly restarting the computer hundreds of times will still not give r.

For this reason, Shor proposes that the discrete Fourier transform (DFT) of register 1 shall be taken, and the result placed in register 1. The DFT of the set of integers c_j in register 1 gives the results $|reg1\rangle = \sum_{i=0}^{a-1} d_i |i\rangle$ with $d_i = \frac{1}{\sqrt{a}} \sum_{j=0}^{a-1} e^{i2\pi i c_j / a} |j\rangle$. Register 1 is now measured, and the probability of finding it in state i is $p_i = |d_i|^2$. It was shown in the preceding chapter that the DFT of a periodic array of 1s and 0s is such that the probabilities p_i take the form of a series of very sharp peaks. For the parameters that were introduced above, the probabilities of measuring state i are:

The data shown in figure 11.7 are consistent with the earlier analysis in that the number of peaks is $r = 6$ and the average spacing between the peaks is $a/r = 21.333$. The actual spacings between the peaks shown is $21, 22, 21, 21, 22$, because they must be integers.

Of course, the ordinary manipulations that lead to a DFT are not possible in a QC so Shor proposed a QC circuit to achieve the same goal. This circuit for producing quantum Fourier transforms (QFTs) was described in the last section of the previous chapter.

After the contents of register 1 have been put into the form of peaks that have a well-defined relationship to the period r like the one in figure 11.7, a single measurement on the QC can give a useful result. To understand this, it is useful to consider a classical model. Suppose you are throwing darts at a target as seen in figure 11.8. The rules of the game are that if the dart lands in any one of the white areas it is a win. Only if it lands in a black area is it a loss. Analogously, a measurement of register 1 after the QFT is far more likely to give the position of one of the peaks than it is to give any other result. This is Shor's solution to the conundrum of how to get a useful result from a QC when quantum mechanics requires that the outcome of measurements must be random.

It is easy to create a computer code that emulates a quantum computer. The inputs are n, m, and x, and the output is the position of a peak. Below is such a code written in Mathematica:

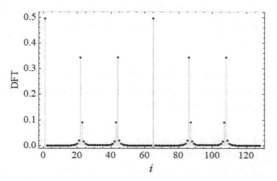

Figure 11.7. The contents of register 1 after the DFT of the original contents of the register has been taken. The parameter i ranges from 1 to 128.

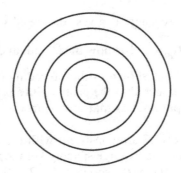

Figure 11.8. A hypothetical sketch of the target.

This is a simple simulation of Shor' s factoring algorithm using the Mathematica DFT,

```
ClearAll
time1=TimeUsed[];
Print['The following numbers are the input to the simulation.
&-&-&-&&-&-&-&&-&-&-&&-&-&-&&-&-&-&&-&-&-&&-&-&-&']
n=21;
Print['The number to be factored, n, is ', n]
m=Ceiling[Log[2,4 n]];
Print['The number of qubits in the top register and bottom register, m, is ', m ]
Print['This can be considered to be the numbers of wires that are in each register.']
a=2^m;
Print['The number of binary basis vectors, a=2^m, is ', a ]
x=2 Round[n/4];
Print['The number whose powers generate the contents of register 2, x, is ', x]
Print['&-&-&-&&-&-&-&&-&-&-&&-&-&-&&-&-&-&&-&-&-&&-&-&-&']
Print['The following numbers are generated in the QC and are not knowable
&-&-&-&&-&-&-&&-&-&-&&-&-&-&&-&-&-&&-&-&-&&-&-&-&']
Print['The ',a,' numbers in register 2 are generated']
y=Table[ Mod[x^i,n],{i,1,a}];
k=RandomChoice[y];
Print['The QC measures register 2 and finds a number k= ', k]
Print['The measurement of k projects out a set of 1s and 0s in register 1.']
Do[If[y[[i]]==k,y[[i]]=1,y[[i]]=0],{i,a}];
ones=List[];
Do[If[y[[i]]==1,AppendTo[ones,i]],{i,a}];
r=ones[[2]]-ones[[1]]; period=N[a/r];
Print['The results of the DFT of the qubits in register 1 are put back into register 1']
f=Fourier[y];
SetOptions[ListLinePlot,Frame->True,Axes->True,
LabelStyle->{FontFamily->'Times New Roman',FontSize->14},
```

```
PlotStyle->Black,ImageSize->Medium];
QFTVec=Abs[f]^2;
xnorm=Norm[QFTVec];
prob=QFTVec/xnorm;
Print['A measure of the wires in register 1 will give an ',m,' digit binary number.
The base 10 number this corresponds to is between 1 and ',a]
y=Ordering[prob,-r];
peak=RandomChoice[y];
Print['The peak measured by the QC is at i= ', peak]
```

The output of this code is the following:

```
ClearAll
  The following numbers are the input to the simulation.
&-&-&-&&-&-&-&&-&-&-&&-&-&-&&-&-&-&&-&-&-&&-&-&-&
  The number to be factored, n, is  21
  The number of qubits in the top register and bottom
register, m, is  7
  This can be considered to be the numbers of wires that are
in each register.
  The number of binary basis vectors, a=2^m, is  128
  The number whose powers generate the contents of register
2, x, is  10
  &-&-&-&&-&-&-&&-&-&-&&-&-&-&&-&-&-&&-&-&-&&-&-&-&
  The following numbers are generated in the QC and are not
knowable
  &-&-&-&&-&-&-&&-&-&-&&-&-&-&&-&-&-&&-&-&-&&-&-&-&
  The  128  numbers in register 2 are generated
  The QC measures register 2 and finds a number k=   13
  The measurement of k projects out a set of 1s and 0s in
register 1.
  The results of the DFT of the qubits in register 1 are put
back into register 1
  A measure of the wires in register 1 will give an  7  digit
binary number.
  The base 10 number this corresponds to is between 1 and  128
  The peak measured by the QC is at i=   44
```

It is unlikely that a single run of a QC will give enough information to specify r. Therefore, it must be run several times with the same input. The quantum factoring algorithm takes asymptotically O((log n)2(log log n)(log log log n)) steps on a QC. The post-processing required to find r from the finite number of peak positions obtained by repeating the QC calculations takes polynomial time.

A number of schemes have been proposed for post-processing. An example worked out by the author of this book uses the fact that since the peaks in register 1 are separated by an approximately constant spacing, as seen in figure 11.7, it should be possible to predict the positions of all peaks by a least-squares fit of the measured ones to a periodic function. First, place the known peak positions on the x-axis. Take a trial function

$$f(x) = \sin(Pkx), \tag{11.15}$$

where

$$P = \frac{\pi}{\tau_0}, \tag{11.16}$$

and τ_0 is a first guess at the spacing between peaks. A least-squares fit gives the optimum value for k, called K, and hence the best value for the period of the peak positions is

$$\tau = \frac{\pi}{PK}. \tag{11.17}$$

There are r peaks in the range from 1 to a, so

$$r = \text{Round}\left(\frac{a}{\tau}\right) \tag{11.18}$$

where the Round function gives the nearest integer value.

The following gives a simple Mathematica code that will do this calculation. It is applied to the factoring of 143.

```
ClearAll;
n=143;
Print['The number to be factored, n, is ', n,
' This is the only input to the QC.']
m=Ceiling[Log[2,4 n]];
Print['The number of qubits in the top register and bottom register, m, is ', m ]
a=2^m;
Print['The number of binary basis vectors, a, is ', a ]
x=2 Round[n/4];
Print['The number whose powers generate the contents of register 2, x, is ', x]
data = {{189,0},{223,0},{240,0},{291,0},{428,0}};
P=N[π/17];
model = Sin[ P k t ];
Print['The function that the data is fitted to is model = Sin[ P k t ]']
fit=FindFit[data,model,{k},t];
K=N[k/.fit];
Print['The value of k found from the least squares fit is ', K]
Evaluate[model/.fit];
modelf=Function[{t},Evaluate[model/.fit]];
```

```
SetOptions[Plot,Frame->True,Axes->True,LabelStyle->{FontFamily->'Times
New Roman',FontSize->14},
    PlotStyle->Black,ImageSize->Large];
    Plot[modelf[t],{t,1.0,1024.0},Epilog->{PointSize[Large],Point[data]}]
    period=N[π/(P K)];
    rp=N[a/period];
    r=Round[rp];
    Print['The calculated average separation between peaks is ',period, ' and the
    resulting value of r is ', rp]
    Print['Rounding this value gives the r = ', r]
    Print['The factors of n are obtained from y=FactorInteger[Mod[x^{r/2}±1,n]]. For this
    case they are ']
    y1=FactorInteger[Mod[x^{r/2}–1,n]]
    y2=FactorInteger[Mod[x^{r/2}+1,n]]
    Print['****************************************']
```

The output of this code is:

```
The number to be factored, n, is  143  This is the only
input to the QC.
   The number of qubits in the top register and bottom
register, m, is  10
   The number of binary basis vectors, a, is  1024
   The number whose powers generate the contents of register
2, x, is  72
   The function that the data is fitted to is model = Sin[ P k t ]
   The value of k found from the least squares fit is  0.992 287
   The calculated average separation between peaks is
17.1321  and the resulting value of r is  59.7707
   Rounding this value gives the r =  60
   The factors of n are obtained from y=FactorInteger[Mod[x^{r/
2±1},n]]. For this case they are
   11
   13
****************************************
```

This post-processing method works well for values of n that are small enough so that a classical emulation of quantum processes is practical. Just as there is a lot of work to be done to make a quantum computer a reality, there is also room for improvements in post-processing. The calculations described here are shown in figure 11.9.

Because of the importance of Shor's factoring algorithm, there are many sources that endeavor to explain it. Perhaps the clearest and easiest to follow is the original paper by Shor [3], which is readily available from arXiv. As pointed out by Shor, similar methods

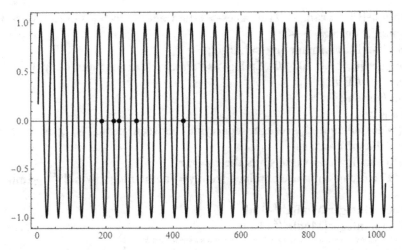

Figure 11.9. A plot of fitted function sin PKx for $1 \leqslant x \leqslant 1024$. The dots show the positions of the five peak positions used in the fitting procedure. The intersections of the sine function with the x-axis show the approximate positions of all 60 of the peaks in register 1.

can be used to solve the discrete logarithm problem. This paper should be comprehensible to anyone familiar with the last two chapters in this book.

Problems

P11.1 Encrypt HAPPYFALLSEMESTER with a Caesar's cipher. See how long it takes your friends to decrypt it.

P11.2 What was done at Bletchley Park during World War II?

P11.3 In what sense is the cryptography in section 11.2 symmetric and in section 11.3 asymmetric?

P11.4 What is the totient $\phi(21)$ of Z_{21}^*?

P11 5 Create table 1b, which is the table of units for Z_{21}^*. It is suggested that you use Mathematica, Matlab, Python, or Excel.

P11.6 Create table 2b, which is the collection of powers x^y for $1 \leqslant x \leqslant 20$ and $1 \leqslant y \leqslant \phi(21)$. It is suggested that you use Mathematica, Matlab, Python, or Excel.

P11.7 Use table 2b to find a set of rs for factoring 21.

P11.8 Bob and Alice have agreed to use RSA. Bob publishes $n = 21$ and $s = 5$. Alice wants to send the message 3. She sends Bob $y = 3^5$ mod 21. Bob knows that $\phi(21) = 12$ so he can calculate t. Calculate y^t mod 21 to see if Bob got the right message.

P11.9 Alice and Bob want exchange keys so that they can do symmetric cryptography. They exchange public parameters $p = 17$ and $g = 9$. Alice's private key is 7 and Bob's is 9. What is the key K that they exchange?

P11.10 Alice and Bob agree to use the ElGamal encryption system. They exchange public parameters $p = 17$ and $g = 9$. Alice's private key is 7 and Bob's is 9. Bob wants to send the message 9. Show how it is encrypted and decrypted.

P11.11 The codes in section 11.12 are Mathematica notebooks. They can be easily modified for Matlab or Python. Write the programs and run them for $n = 35$.

References

[1] Bennett C H and Brassard G 1984 *Proc. IEEE International Conference on Computers, Systems, and Signal Processing* (Los Alamitos, CA: IEEE Press)

[2] Ekert A 1991 *Phys. Rev. Lett.* **67** 661

[3] Shor P W 1996 arXiv:quant-ph/9508027v2

Chapter 12

Many particle systems

12.1 The Schrödinger equation

The easiest system to consider when describing the fundamental equations of quantum mechanics is a single particle moving in a potential field. In applications of quantum mechanics to the study of solids and molecules, many electrons and nuclei are involved. Although the principles are the same, a number of technical problems arise in dealing with many particle systems. The non-relativistic version of the theory is emphasized in the following discussion even though the Dirac theory is more often used in modern calculations. It is easy enough to convert to the relativistic version, and it saves a lot of time to present the development in the simpler formalism.

Let us first consider the Hamiltonian that will be used

$$E = T + V + J + U, \tag{12.1}$$

where T is the kinetic energy

$$T = \sum_{i=1}^{N} \frac{p_i^2}{2m_i} = -\sum_{i=1}^{N} \frac{\hbar^2}{2m_i} \nabla_i^2. \tag{12.2}$$

The external Coulomb potential function is

$$V = -\sum_{i=1}^{N} \sum_{\alpha=1}^{N_\alpha} \frac{Z_\alpha e^2}{|\mathbf{r}_i - \mathbf{R}_\alpha|}, \tag{12.3}$$

where N is the number of electrons, N_α is the number of nuclei, and Z_α is the atomic number. The electron–electron interaction is

$$J = \frac{1}{2} \sum_{i=1}^{N} \sum_{\substack{j=1 \\ j \neq i}}^{N} \frac{e^2}{|\mathbf{r}_i - \mathbf{r}_j|}. \tag{12.4}$$

doi:10.1088/978-0-7503-2167-9ch12

The Coulomb interaction from the nuclei is

$$U = \sum_{\alpha=1}^{N_\alpha} \sum_{\substack{\beta=1 \\ \beta \neq \alpha}}^{N_\alpha} \frac{Z_\alpha Z_\beta e^2}{|\mathbf{R}_\alpha - \mathbf{R}_\beta|}. \tag{12.5}$$

The wave function for this general system is

$$\Psi = \Psi(\mathbf{r}_1, \sigma_1, \mathbf{r}_2, \sigma_2, \mathbf{r}_3, \sigma_3, ..., \mathbf{r}_N, \sigma_N, t), \tag{12.6}$$

where the $\mathbf{r}_i$ are the positions and the σ_i the spins of the particles. The Schrödinger equation is

$$H\Psi = i\hbar \frac{\partial \Psi}{\partial t}. \tag{12.7}$$

The primary focus will be on energy eigenfunctions, which lead to

$$H\Psi = E\Psi, \tag{12.8}$$

and

$$\Psi = \Psi_E(\mathbf{r}_1, \sigma_1, \mathbf{r}_2, \sigma_2, \mathbf{r}_3, \sigma_3, ..., \mathbf{r}_N, \sigma_N)e^{-i\frac{Et}{\hbar}}. \tag{12.9}$$

The total energy is

$$E = \int ... \int \Psi_E H \Psi_E dv_1 dv_2 dv_3 ... dv_N \tag{12.10}$$

where the differential dv_i indicates and integration over space and sum over spin variables.

As a general rule, the motion of the nuclei is ignored and the $\mathbf{R}_\alpha$ are simply parameters. If the crystal structure or the molecular structure is being calculated, the calculation is repeated for several choices of the nuclear positions.

It is convenient to use dimensionless units in calculations. Distance is measured in Bohr radii a_0, where

$$a_0 = \frac{\hbar^2}{me^2} = 0.529177 \text{Angstroms}$$
$$= 0.529177 \times 10^{-8} \text{cm} = 0.529177 \times 10^{-10} \text{m} \tag{12.11}$$

Energy measured in Rydbergs,

$$R = \frac{me^4}{2\hbar^2} = \frac{e^2}{2a_0} = 13.60535 \text{eV}. \tag{12.12}$$

It follows that $\frac{\hbar^2}{2m} \rightarrow 1$, $e^2 \rightarrow 2$, and

$$H = \sum_{i=1}^{N} \left[-\nabla_i^2 - \sum_{\alpha=1}^{N_\alpha} \frac{2Z_\alpha}{|\mathbf{r}_i - \mathbf{R}_\alpha|} + \sum_{j=1}^{N} \frac{1}{|\mathbf{r}_i - \mathbf{r}_j|} \right]. \tag{12.13}$$

12.2 Hartree theory

It may seem strange but, in the earliest effort to solve the Schrödinger equation for an atom with $Z > 1$ by D R Hartree, the assumption was made that, to a first approximation, the electron–electron interaction can be replaced by an average. Then the Hamiltonian can be written as a sum of N single-particle Hamiltonians

$$H = \sum_{i=1}^{N} H_i, \tag{12.14}$$

with

$$H_i = -\frac{\hbar^2}{2m}\nabla_i^2 + V_i. \tag{12.15}$$

The one-particle potentials V_i including the averaging process will be defined in detail as the theory develops.

Using the separation of variables, the wave function may be written

$$\Psi_E = \psi_{\varepsilon_1}(\mathbf{r}_1, \sigma_1)\psi_{\varepsilon_2}(\mathbf{r}_2, \sigma_2)\psi_{\varepsilon_3}(\mathbf{r}_3, \sigma_3)...\psi_{\varepsilon_N}(\mathbf{r}_N, \sigma_N), \tag{12.16}$$

and the form of the energy becomes

$$E = \varepsilon_1 + \varepsilon_2 + \varepsilon_3 + ... + \varepsilon_N. \tag{12.17}$$

The one-electron wave functions and energies are solutions of the one-electron equations

$$H_i\psi_{\varepsilon_i}(\mathbf{r}_i, \sigma_i) = \varepsilon_i\psi_{\varepsilon_i}(\mathbf{r}_i, \sigma_i). \tag{12.18}$$

It follows from this equation that the wave functions corresponding to different energy eigenvalues are orthonormal

$$\int \psi_{\varepsilon_i}^*(\mathbf{r}, \sigma)\psi_{\varepsilon_j}(\mathbf{r}, \sigma)dv = \delta_{ij}. \tag{12.19}$$

Using the Hartree approximation for the wave function, equation (12.16), and the Hamiltonian from equation (12.13), the total energy from equation (12.10) leads to

$$E(\Psi_E) = \sum_{a=1}^{N} h_a + \sum_{a\neq b=1}^{N} j_{ab}, \tag{12.20}$$

with

$$h_a = \int \psi_a^*(\mathbf{r}, \sigma)[-\nabla^2 + V(\mathbf{r})]\psi_a(\mathbf{r}, \sigma)dv, \tag{12.21}$$

and

$$V(\mathbf{r}) = -\sum_{\alpha=1}^{N_\alpha} \frac{2Z_\alpha}{|\mathbf{r} - \mathbf{R}_\alpha|}. \tag{12.22}$$

The quantities j_{ab} are

$$j_{ab} = \int \psi_a^*(\mathbf{r}, \sigma)\psi_b^*(\mathbf{r}', \sigma')\frac{1}{|\mathbf{r} - \mathbf{r}'|}\psi_a(\mathbf{r}, \sigma)\psi_b(\mathbf{r}', \sigma')dvdv'. \qquad (12.23)$$

A factor of 2 is absorbed because dimensionless units $e^2/2 \rightarrow 1$ are being used.

The calculus of variations will be used to find the functions that minimize the energy functional in equation (12.20). Taking the functional derivative of $E(\Psi_E)$ with respect to $\psi_a^*(\mathbf{r}, \sigma)$ and using Lagrange multipliers [1] to include the restrictions $\int \psi_a^*(\mathbf{r}, \sigma)\psi_a(\mathbf{r}, \sigma)dv - 1 = 0$ leads to

$$\frac{\delta\left\{ E - \varepsilon_a\left[\int \psi_a^*(\mathbf{r}', \sigma')\psi_a(\mathbf{r}', \sigma')dv' - 1 \right] \right\}}{\delta\psi_a^*(\mathbf{r}, \sigma)} = 0, \qquad (12.24)$$

where the Lagrange multipliers are called ε_a. Using the functional form E from equation (12.20) in this equation leads to the Hartree equations

$$[-\nabla^2 + V(\mathbf{r})]\psi_a(\mathbf{r}, \sigma) +$$
$$\left\{ \sum_{b\neq a=1}^{N} \int \psi_b^*(\mathbf{r}', \sigma')\frac{2}{|\mathbf{r} - \mathbf{r}'|}\psi_b(\mathbf{r}', \sigma')dv' \right\}\psi_a(\mathbf{r}, \sigma) = \varepsilon_a\psi_a(\mathbf{r}, \sigma). \qquad (12.25)$$

Note that the 2 has appeared because the double sum has been changed to a single sum.

From the preceding equation, the one-electron effective potential is

$$V_a(\mathbf{r}) = -\sum_{a=1}^{N_a}\frac{2Z_a}{|\mathbf{r} - \mathbf{R}_a|} + \sum_{b\neq a=1}^{N}\int \psi_b^*(\mathbf{r}', \sigma')\frac{2}{|\mathbf{r} - \mathbf{r}'|}\psi_b(\mathbf{r}', \sigma')dv'. \qquad (12.26)$$

Clearly this potential depends on the function that is being solved for and, indeed, the Hartree equation is different for each function $\psi_a(\mathbf{r}, \sigma)$. The physical meaning of the potential is that the electron a sees the Coulomb potential from the nuclear charges at the positions $\mathbf{R}_a$ screened by the charges of all of the electrons other than a.

Multiplying equation (12.25) by $\psi_a^*(\mathbf{r}, \sigma)$ and integrating over $\mathbf{r}, \sigma$ leads to

$$\varepsilon_a = h_a + 2\sum_b j_{ab}. \qquad (12.27)$$

It follows from this and equation (12.20) that

$$E = \sum_{a=1}^{N}\varepsilon_a - \sum_{a\neq b=1}^{N} j_{ab}. \qquad (12.28)$$

The sum over b is called the double counting term, which must appear to cancel the fact that that j_{ab} appears twice in the sum over ε_a. The existence of the double counting makes it difficult to consider the ε_a to be one-electron energies.

Since the Hamiltonian doesn't contain spin, the spin-orbitals $\psi_a(\mathbf{r}, \sigma)$ can be written, $\psi_a(\mathbf{r}, \sigma) = \phi_a(\mathbf{r})\chi_a(\sigma)$, where $\chi_a(\sigma) = \alpha(\sigma)$ for spin up, or $\chi_a(\sigma) = \beta(\sigma)$ for spin down. The 'integration over spin' is a formal device to describe the inner product between spin vectors, $(\alpha, \alpha) = \int |\alpha|^2 \, d\sigma = 1$, $(\beta, \beta) = \int |\beta|^2 \, d\sigma = 1$, and $(\alpha, \beta) = \int \alpha\beta d\sigma = 0$. The part of the electronic density that comes from the bth orbital is $\rho_b(\mathbf{r}) = \int \psi_b^*(\mathbf{r}, \sigma)\psi_b(\mathbf{r}, \sigma)d\sigma = \phi_b^*(\mathbf{r})\phi_b(\mathbf{r})$, so the Hartree equation in equation (12.25) can be rewritten,

$$[-\nabla^2 + V(\mathbf{r})]\phi_a(\mathbf{r}) + \left\{ \sum_{b(\neq a)=1}^{N} \int \frac{2\rho_b(\mathbf{r}')}{|\mathbf{r} - \mathbf{r}'|} d\mathbf{r}' \right\} \phi_a(\mathbf{r}) = \varepsilon_a \phi_a(\mathbf{r}) \qquad (12.29)$$

This is the eigenvalue equation for a spinless electron moving in the electric field of the nuclei and all the other electrons, and is the actual equation used by Hartree.

There is no mathematical reason that, in Hartree's theory, the electrons cannot all be put in the orbital corresponding to the lowest one-electron energy ε_a. The exclusion principal that allows no more than 2 electrons in each orbital is added in an ad hoc manner in the Hartree theory.

12.3 Hartree–Fock theory

One of the most important developments in quantum theory was the realization by Prof. Wolfgang Pauli that the N-particle wave function must be antisymmetric under the interchange of two position and spin arguments

$$\begin{aligned} \Psi_E(\mathbf{r}_1, \sigma_1, \mathbf{r}_2, \sigma_2, ..., \mathbf{r}_i, \sigma_i, ..., \mathbf{r}_j, \sigma_j, ..., \mathbf{r}_N, \sigma_N) \\ = -\Psi_E(\mathbf{r}_1, \sigma_1, \mathbf{r}_2, \sigma_2, ..., \mathbf{r}_j, \sigma_j, ..., \mathbf{r}_i, \sigma_i, ..., \mathbf{r}_N, \sigma_N) \end{aligned} \qquad (12.30)$$

Among other things, this principle leads to the exclusion principle and the uncertainty relations. The minimum extension to the Hartree theory that includes the Pauli principle is to create a determinantal wave function from the one-electron wave functions

$$\begin{aligned} &\Psi_E(\mathbf{r}_1, \sigma_1, \mathbf{r}_2, \sigma_2, \mathbf{r}_3, \sigma_3, ..., \mathbf{r}_N, \sigma_N) \\ &= \frac{1}{\sqrt{N!}} \begin{vmatrix} \psi_{e_1}(\mathbf{r}_1, \sigma_1) & \psi_{e_1}(\mathbf{r}_2, \sigma_2) & \psi_{e_1}(\mathbf{r}_3, \sigma_3) & \cdots & \psi_{e_1}(\mathbf{r}_N, \sigma_N) \\ \psi_{e_2}(\mathbf{r}_1, \sigma_1) & \psi_{e_2}(\mathbf{r}_2, \sigma_2) & \psi_{e_2}(\mathbf{r}_3, \sigma_3) & \cdots & \psi_{e_2}(\mathbf{r}_N, \sigma_N) \\ \psi_{e_3}(\mathbf{r}_1, \sigma_1) & \psi_{e_3}(\mathbf{r}_2, \sigma_2) & \psi_{e_3}(\mathbf{r}_3, \sigma_3) & \cdots & \psi_{e_3}(\mathbf{r}_N, \sigma_N) \\ \cdots & \cdots & \cdots & \cdots & \cdots \\ \psi_{e_N}(\mathbf{r}_1, \sigma_1) & \psi_{e_N}(\mathbf{r}_2, \sigma_2) & \psi_{e_N}(\mathbf{r}_3, \sigma_3) & \cdots & \psi_{e_N}(\mathbf{r}_N, \sigma_N) \end{vmatrix}. \end{aligned} \qquad (12.31)$$

It is well known that the interchange of two columns of this determinant, the equivalent to equation (12.30), changes the sign. It is also known that, if two columns of the determinant are equal, the determinant is zero, which leads to the exclusion principle.

Another way to write the wave function is

$$\Psi_E(\mathbf{r}_1, \sigma_1, \mathbf{r}_2, \sigma_2, \mathbf{r}_3, \sigma_3, ..., \mathbf{r}_N, \sigma_N)$$

$$= \frac{1}{\sqrt{N!}} A \psi_{e_1}(\mathbf{r}_1, \sigma_1)\psi_{e_2}(\mathbf{r}_2, \sigma_2)\psi_{e_3}(\mathbf{r}_3, \sigma_3)...\psi_{e_N}(\mathbf{r}_N, \sigma_N), \tag{12.32}$$

where A is a linear operator that makes a wave function of N identical electrons antisymmetric under the exchange of the coordinates of any pair of them [2]. It can be shown that this antisymmetrizing operator is such that

$$A^2 = N!A. \tag{12.33}$$

Using this wave function to get the expectation functional

$$E(\Psi_E) = \iint ... \int ... \int \Psi_E^* H \Psi_E dv_1 dv_2 ... dv_N, \tag{12.34}$$

leads to

$$E(\Psi_E) = \sum_{a=1}^{N} h_a + \sum_{a \neq b=1}^{N} (j_{ab} - k_{ab}), \tag{12.35}$$

where h_a and j_{ab} are given in equations (12.21) and (12.23). The new quantity that arises because of the antisymmetry of the wave function is called the exchange integral

$$k_{ab} = \int \psi_a^*(\mathbf{r}, \sigma)\psi_b^*(\mathbf{r}', \sigma') \frac{1}{|\mathbf{r} - \mathbf{r}'|} \psi_b(\mathbf{r}, \sigma)\psi_a(\mathbf{r}', \sigma') dv dv'. \tag{12.36}$$

Recall that dimensionless units in which $e^2/2 \rightarrow 1$ are being used.

The calculus of variations is used to find the functions that minimize the energy functional in equation (12.36). Taking the functional derivative of $E(\Psi_E)$ and using Lagrange multipliers to include the restrictions $\int \psi_a^*(\mathbf{r}, \sigma)\psi_a(\mathbf{r}, \sigma)dv - 1 = 0$

$$\frac{\delta \left\{ E - \varepsilon_a \left[\int \psi_a^*(\mathbf{r}', \sigma')\psi_a(\mathbf{r}', \sigma')dv' - 1 \right] \right\}}{\delta \psi_a^*(\mathbf{r}, \sigma)} = 0, \tag{12.37}$$

leads to the Hartree–Fock (H-F) equations

$$[-\nabla^2 + V(\mathbf{r})]\psi_a(\mathbf{r}, \sigma) + \left\{ \sum_{b \neq a=1}^{N} \int \psi_b^*(\mathbf{r}', \sigma') \frac{2}{|\mathbf{r} - \mathbf{r}'|} \psi_b(\mathbf{r}', \sigma')dv' \right\} \psi_a(\mathbf{r}, \sigma)$$

$$- \left\{ \sum_{b \neq a=1}^{N} \int \psi_b^*(\mathbf{r}', \sigma') \frac{2}{|\mathbf{r} - \mathbf{r}'|} \psi_a(\mathbf{r}', \sigma')dv' \right\} \psi_b(\mathbf{r}, \sigma) = \varepsilon_a \psi_a(\mathbf{r}, \sigma). \tag{12.38}$$

The second integral on the left is called the exchange term. Since the solution $\psi_a(\mathbf{r}, \sigma)$ appears inside of the integral, the H-F equation is an integro-differential equation.

Using the same argument that led to equation (12.28), it can be seen that

$$E = \sum_{a=1}^{N} \varepsilon_a - \sum_{a \neq b = 1}^{N} (j_{ab} - k_{ab}). \tag{12.39}$$

The double sum is called the double counting term. If it were not for this term, the energy would be the sum of the one-electron energies ε_a. Ignoring the double counting term leads to Koopman's approximation. Using this approximation, exciting an electron from state a to state b requires an energy ε_b minus ε_a. The correct H-F method would be to calculate the total energy of the system with the electron in state a and subtract that from the energy with the electron in state b.

Since the Hamiltonian doesn't contain spin, the spin-orbitals can be written $\psi_a(\mathbf{r}, \sigma) = \phi_a(\mathbf{r})\chi_a(\sigma)$. The part of the electronic density that comes from the bth orbital is $\rho_b(\mathbf{r}) = \int \psi_b^*(\mathbf{r}, \sigma)\psi_b(\mathbf{r}, \sigma)d\sigma = \phi_b^*(\mathbf{r})\phi_b(\mathbf{r})$, so the above can be rewritten,

$$[-\nabla^2 + V(\mathbf{r})]\phi_a(\mathbf{r}) + \left\{ \sum_{b=1}^{N} \int \frac{2\rho_b(\mathbf{r}')}{|\mathbf{r} - \mathbf{r}'|} d\mathbf{r}' \right\} \phi_a(\mathbf{r})$$

$$- \left\{ \sum_{b=1}^{N} \int \phi_b^*(\mathbf{r}') \frac{2}{|\mathbf{r} - \mathbf{r}'|} \phi_a(\mathbf{r}') d\mathbf{r}' \delta_{\sigma_b \sigma_a} \right\} \phi_b(\mathbf{r}) = \varepsilon_a \phi_a(\mathbf{r}) \tag{12.40}$$

This is the eigenvalue equation for an electron moving in the electric field of the nuclei and all the other electrons, but with an additional exchange term that only appears when the spins of the a and b electrons point in the same direction. It is not necessary to exclude $b = a$ from the sums because the two integrals cancel for that case.

The spin plays an important role in the H-F equations, so the occupancy of the orbitals $\phi_a(\mathbf{r})$ is not arbitrary, as it is in the Hartree formulation. The orbitals are filled in order of increasing one-electron energies ε_a with two electrons per orbital, one with spin up and the other with spin down. This is called the exclusion principle, and it provides a first-principles explanation of atomic structure. The exchange energy that appears in H-F theory has many other ramifications in the calculation of properties of molecules and solids using quantum mechanics.

A look at the Hartree and H-F equations, equations (12.29) and (12.40), makes it clear that the coefficients in the equations depend on the solutions to the equations. The only way to solve them is to make an initial guess for the solutions, calculate the coefficients, and then solve for more accurate solutions. The more accurate solutions are then used to calculate new coefficients. The iteration process is repeated until there is no difference between the current and the former solutions. This technique for solving the Hartree and H-F equations is called the self-consistent field (SCF) method. The H-F equations give total energies and charge densities for atoms that are in excellent agreement with experiment.

The NIST Multiconfiguration Hartree–Fock and Multiconfiguration Dirac–Hartree–Fock Database was developed by Charlotte Froese Fischer, a student of Hartree. These computer programs are freely available to the public, and make it possible for anyone with the appropriate skills to calculate the energies and wave functions of any atom. Many other computer packages have been developed and made available to the community that can be used for Hartree and H-F calculations.

12.4 Configuration interaction (CI) calculations

The H-F equation produces an infinite number of solutions $\psi_i(\mathbf{r}, \sigma)$. These solutions can be used to create an infinite number of determinantal wave functions, which are called configurations. The configuration that corresponds to the ground state in the H-F approximation is the one that is used to define the charge density in the H-F equation. The other configurations can be pictured as describing states in which electrons are promoted out of this ground state into excited one-electron states. It should be obvious that the configurations are an orthonormal set

$$\int \Psi_{E'}(\mathbf{r}_1, \sigma_1, \mathbf{r}_2, \sigma_2, ..., \mathbf{r}_N, \sigma_N)\Psi_{E''}(\mathbf{r}_1, \sigma_1, \mathbf{r}_2, \sigma_2, ..., \mathbf{r}_N, \sigma_N)dv_1dv_2...dv_N = \delta_{E'E''}. \quad (12.41)$$

A more general antisymmetric function can be written as a linear combination of N_C configurations

$$\Psi(\mathbf{r}_1, \sigma_1, \mathbf{r}_2, \sigma_2, \mathbf{r}_3, \sigma_3, ..., \mathbf{r}_N, \sigma_N)$$
$$= \sum_{E'=1}^{N_C} C_{E'}\Psi_{E'}(\mathbf{r}_1, \sigma_1, \mathbf{r}_2, \sigma_2, \mathbf{r}_3, \sigma_3, ..., \mathbf{r}_N, \sigma_N). \quad (12.42)$$

The integrals

$$\int \Psi_{E'}(\mathbf{r}_1, \sigma_1, \mathbf{r}_2, \sigma_2, ..., \mathbf{r}_N, \sigma_N)H\Psi_{E''}(\mathbf{r}_1, \sigma_1, \mathbf{r}_2, \sigma_2, ..., \mathbf{r}_N, \sigma_N)dv_1dv_2...dv_N$$
$$= H_{E'E''} \quad (12.43)$$

with H the exact Hamiltonian in equation (12.13), are used in the Rayleigh–Ritz variational equation

$$\begin{pmatrix} H_{11} - E_g & H_{12} & ... & H_{1N_c} \\ H_{21} & H_{22} - E_g & ... & H_{2N_c} \\ ... & ... & ... & ... \\ H_{N_c1} & H_{N_c2} & ... & H_{N_cN_c} - E_g \end{pmatrix}\begin{pmatrix} C_1 \\ C_2 \\ ... \\ C_{N_c} \end{pmatrix} = 0 \quad (12.44)$$

to find an improved ground state energy E_g. With a reasonable choice of configurations, E_g will be a better approximation to the exact ground state energy than the one obtained from the simple H-F equation.

In principle, the CI approach can be used to find exact solutions for any many-electron problem. They quickly become unmanageable, however, when N_C becomes large. Suppose, for example, M orbitals that correspond to the lowest one-electron eigenvalues ε_a are chosen. The number of configurations that can be constructed is

$$N_C = \frac{(2M)!}{(2M - N)!N!}. \tag{12.45}$$

In a system with 20 electrons, including only the first three states above the H-F ground configuration leads to 230,230 configurations. Methods for using the CI method more effectively are lumped into the categories called post H-F. Rules are developed for including a subset of the possible configurations. Specific versions are the coupled cluster method and Møller–Plesset perturbation theory.

12.5 The electron gas in the Hartree–Fock approximation

The expressions for the total energy in the H-F approximation can be summed up

$$E(\Psi_E) = T + V + J + K + N \tag{12.46}$$

with

$$T = \sum_{a=1}^{N} \int \psi_a^*(\mathbf{r}, \sigma) \left[-\frac{\hbar^2}{2m} \nabla^2 \right] \psi_a(\mathbf{r}, \sigma) dv \tag{12.47}$$

and

$$V = -\sum_{a=1}^{N} \sum_{\alpha=1}^{N_\alpha} \int \psi_a^*(\mathbf{r}, \sigma) \frac{e^2 Z_\alpha}{|\mathbf{r} - \mathbf{R}_\alpha|} \psi_a(\mathbf{r}, \sigma) dv. \tag{12.48}$$

The electron–electron Coulomb interaction is

$$J = \frac{1}{2} \sum_{a=1}^{N} \sum_{b=1}^{N} \int \psi_a^*(\mathbf{r}, \sigma) \psi_b^*(\mathbf{r}', \sigma') \frac{e^2}{|\mathbf{r} - \mathbf{r}'|} \psi_a(\mathbf{r}, \sigma) \psi_b(\mathbf{r}', \sigma') dv dv', \tag{12.49}$$

and the exchange integral is

$$K = -\frac{1}{2} \sum_{a=1}^{N} \sum_{b=1}^{N} \int \psi_a^*(\mathbf{r}, \sigma) \psi_b^*(\mathbf{r}', \sigma') \frac{e^2}{|\mathbf{r} - \mathbf{r}'|} \psi_b(\mathbf{r}, \sigma) \psi_a(\mathbf{r}', \sigma') dv dv'. \tag{12.50}$$

The Coulomb interaction between the nuclei is

$$N = \frac{1}{2} \sum_{\alpha=1}^{N_\alpha} \sum_{\beta=1}^{N_\beta} \frac{Z_\alpha Z_\beta e^2}{|\mathbf{R}_\alpha - \mathbf{R}_\beta|}. \tag{12.51}$$

The charge density of the electrons is

$$\rho_e(\mathbf{r}) = e \sum_{a=1}^{N} \psi_a^*(\mathbf{r}) \psi_a(\mathbf{r}). \tag{12.52}$$

Consider a system in which the positive charge density

$$\rho_p(\mathbf{r}) = e\sum_{\alpha=1}^{N_\alpha} Z_\alpha \delta(\mathbf{r} - \mathbf{R}_\alpha) \tag{12.53}$$

can be smeared out into a uniform positive background. The system has been named 'jellium' by the theorists who work in this field. In jellium, the sum of the Coulomb potentials is

$$V + J + N = \int_{-\infty}^{\infty}\int_{-\infty}^{\infty} \frac{\left[\rho_e(\mathbf{r}) - \rho_p(\mathbf{r})\right]\left[\rho_e(\mathbf{r}') - \rho_p(\mathbf{r}')\right]}{|\mathbf{r} - \mathbf{r}'|} d\mathbf{r}d\mathbf{r}'. \tag{12.54}$$

Since the background is a constant, the electron gas will be uniform and, by charge neutrality $\rho_e(\mathbf{r}) = \rho_p(\mathbf{r})$. It follows that the total Coulomb energy for this jellium model is zero

$$V + J + N = 0. \tag{12.55}$$

As a consequence, the only contributions remaining to the energy are from the kinetic and exchange terms

$$E = T + K. \tag{12.56}$$

The H-F orbitals for the electrons in jellium must be plane waves

$$\psi_\mathbf{k}(\mathbf{r}) = \frac{1}{\sqrt{\Omega}}e^{i\mathbf{k}\cdot\mathbf{r}}\chi(\sigma). \tag{12.57}$$

For simplicity, it is assumed that they satisfy periodic boundary conditions and are normalized within a box with dimensions L_x, L_y, and L_z and volume $\Omega = L_xL_yL_z$. The one-electron energy is

$$\varepsilon_0(\mathbf{k}) = \frac{k^2}{2m}, \tag{12.58}$$

in units where $\hbar = 1$.

The kinetic energy is

$$T = 2\sum_\mathbf{k}\varepsilon_0(\mathbf{k})n(\mathbf{k}). \tag{12.59}$$

The step function is

$$\begin{aligned}n(\mathbf{k}) &= 1 \text{ if } \mathbf{k} \text{ is occupied}\\ n(\mathbf{k}) &= 0 \text{ if } \mathbf{k} \text{ is not occupied}\end{aligned}. \tag{12.60}$$

The 2 is due to the spin.

To obtain the exchange energy, equation (12.50) is evaluated with

$$\psi_a(\mathbf{r}, \sigma) = \psi_{\mathbf{k}}(\mathbf{r}, \sigma) = \frac{1}{\sqrt{\Omega}} e^{i\mathbf{k}\cdot\mathbf{r}} \chi_a(\sigma)$$

$$\psi_b(\mathbf{r}, \sigma) = \psi_{\mathbf{k}'}(\mathbf{r}, \sigma) = \frac{1}{\sqrt{\Omega}} e^{i\mathbf{k}'\cdot\mathbf{r}} \chi_b(\sigma)$$

(12.61)

which gives

$$K = -\frac{1}{2\Omega^2} \sum_{\mathbf{k}} n(\mathbf{k}) \sum_{\mathbf{k}'} n(\mathbf{k}') \int e^{-i(\mathbf{k}-\mathbf{k}')\cdot(\mathbf{r}-\mathbf{r}')} \frac{e^2}{|\mathbf{r}-\mathbf{r}'|} d^3\mathbf{r}\, d^3\mathbf{r}'.$$

(12.62)

The sums over $\mathbf{k}$ are not multiplied by 2 because the contributions include only terms in which the a and b spins are parallel. Since the integrand in the above equation depends only on the difference between $\mathbf{r}$ and $\mathbf{r}'$, the first integral over $\mathbf{r}'$ merely introduces a factor of Ω leading to

$$K = -\frac{1}{2\Omega} \sum_{\mathbf{k}} n(\mathbf{k}) \sum_{\mathbf{k}'} n(\mathbf{k}') \int e^{-i(\mathbf{k}-\mathbf{k}')\cdot\mathbf{r}} \frac{e^2}{r} d^3\mathbf{r}.$$

(12.63)

Referring back to the discussion of the Born approximation, it can be seen that the integral above is just a constant times the matrix element of the potential

$$(2\pi)^3 \langle \mathbf{k}' | V | \mathbf{k} \rangle = \frac{4\pi e^2}{|\mathbf{k}-\mathbf{k}'|^2},$$

(12.64)

so

$$K = -\frac{1}{2\Omega} \sum_{\mathbf{k}} n(\mathbf{k}) \sum_{\mathbf{k}'} n(\mathbf{k}') \frac{4\pi e^2}{|\mathbf{k}-\mathbf{k}'|^2}.$$

(12.65)

It will be seen to be convenient to write the exchange contribution to the energy as

$$K = \frac{1}{2} \sum_{\mathbf{k}} n(\mathbf{k}) \Sigma_x(\mathbf{k}),$$

(12.66)

where the self-energy is

$$\Sigma_x(\mathbf{k}) = -\frac{e^2}{\Omega} \sum_{\mathbf{k}'} \frac{4\pi}{|\mathbf{k}-\mathbf{k}'|^2} n(\mathbf{k}')$$

$$\rightarrow -\frac{1}{(2\pi)^3} \int_0^\pi \int_0^\infty \frac{4\pi e^2}{k^2 + (k')^2 - 2kk'\cos\theta} 2\pi n(k') \sin\theta\, d\theta (k')^2 dk'$$

(12.67)

The volume Ω was replaced by $(2\pi)^3$ because the sum was converted to an integral. The step function $n(k')$ has the effect of limiting this integral over k' to an integral from zero to a Fermi momentum k_F that is related to the Fermi energy by

$$k_F = \sqrt{2m\varepsilon_F}.$$

(12.68)

The self-energy becomes

$$
\begin{aligned}
\Sigma_x(\mathbf{k}) &= -\frac{e^2}{2\pi k} \int_0^{k_F} \ln \frac{|k+k'|}{|k-k'|} k'\, dk' \\
&= -\frac{e^2 k_F}{2\pi y} \int_0^1 \ln \frac{|y+y'|}{|y-y'|} y'\, dy' = \frac{e^2 k_F}{2\pi} S(y)
\end{aligned}
\tag{12.69}
$$

where $y = k/k_F$ and

$$
S(y) = -\left[1 + \frac{1-y^2}{2y} \ln\left(\frac{|1+y|}{|1-y|}\right) \right].
\tag{12.70}
$$

It is conventional for this special case to define a one-particle energy such that

$$
E = \sum_{\mathbf{k}} \varepsilon(\mathbf{k}) n(\mathbf{k}),
\tag{12.71}
$$

and an effective mass m_* so that

$$
\varepsilon(\mathbf{k}) = \varepsilon_0(\mathbf{k}) + \Sigma_x(\mathbf{k}) = \frac{k^2}{2m_*} = \left(\frac{m}{m_*}\right) \varepsilon_0(\mathbf{k}).
\tag{12.72}
$$

The most useful expression for the effective mass m^* is

$$
\frac{m}{m_*} = \frac{\partial \varepsilon(\mathbf{k})}{\partial \varepsilon_0(\mathbf{k})} = 1 + \frac{\partial \Sigma_x(\mathbf{k})}{\partial \varepsilon_0(\mathbf{k})}.
\tag{12.73}
$$

Using the preceding definitions and the laws of differential calculus, this expression becomes

$$
\begin{aligned}
\frac{m}{m_*} &= 1 + \frac{m}{\hbar^2 k} \frac{\partial \Sigma_x(k)}{\partial k} \\
&= 1 + \frac{m}{\hbar^2 k_F^2 y} \frac{\partial \Sigma_x(k)}{\partial y} = 1 + \frac{me^2}{\hbar^2 k_F 2\pi y} \frac{\partial S(y)}{\partial y}.
\end{aligned}
\tag{12.74}
$$

It is easy enough to calculate the function $S(y)$ using Mathematica or the equivalent. A plot of that function is shown in figure 12.1.

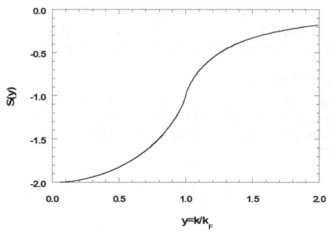

Figure 12.1. A plot of the dimensionless quantity $S(y)$ versus y.

The important feature to note about the function $S(y)$ is that for particles at the Fermi energy, $y = 1$, the slope of the function is infinity. This predicts an effective mass of $m^* = 0$ for electrons with $\varepsilon = \varepsilon_F$. The implications of this are large because, in condensed matter physics, theoretical predictions for many experimentally measured quantities are based on the concept of quasiparticles. These are electrons with energies approximately equal to the Fermi energy and masses equal to m^*. The quasiparticle theory explains the electron contribution to the specific heat, the paramagnetic susceptibility, and the mobility of electrons [3]. The H-F approximation gives the unphysical values for all these quantities. The Hartree approximation, which predicts $\varepsilon(\mathbf{k}) = \varepsilon_0(\mathbf{k})$, is actually better for extended system like the electron gas.

12.6 Critique of the H-F approximation

As pointed out in section 12.3, H-F calculations on atoms give excellent results for the optical spectrum and the energies of atoms. It was shown above that the application of H-F theory to the electron gas gives a very bad answer. An extrapolation of this observation is that, when the orbitals of the electrons are restricted to a small region of space, H-F is a good approximation. As the range of the orbitals increases, H-F becomes less good. This can be checked by doing H-F calculations on larger systems, such as large molecules and nanoparticles.

In principle, more accurate solutions of the many-electron Schrödinger equation for an electron gas can be obtained using the techniques of many-body physics [4]. Time-dependent perturbation theory is used to develop an infinite-order perturbation theory. That, in turn, is analyzed using diagrams like the ones originally developed by Feynman in quantum electrodynamics. The diagrams that correspond to the H-F approximation are known, and it was realized immediately that summing them did not lead to a good theory. As more diagrams are included, some are seen to cancel the H-F $\Sigma_\mathbf{x}(\mathbf{k})$ and still others must be summed to replace it with something more sensible. Many-body theory provides an understanding of electronic states, but it does not lead to a formalism that can be programmed into a computer to calculate the properties of molecules and solids.

There is an approximation that is more useful for extended systems. Walter Kohn was awarded the Nobel prize for the development of the density functional theory (DFT) and the local density approximation (LDA), which will now be discussed.

12.7 Density matrices

The Hamiltonian for an N-body system is written in equation (12.13) and the properly antisymmetric wave function in equation (12.30). It is frequently necessary to find the expectation value of observables which are written in the position representation as

$$F_1 = \sum_{i=1}^{N} f_1(\mathbf{r}_i, \sigma_i),$$
(12.75)

or

$$F_2 = \sum_{i=1}^{N}\sum_{j=1}^{N} f_2(\mathbf{r}_i, \sigma_i, \mathbf{r}_j, \sigma_j). \tag{12.76}$$

There are very few observables that depend on the positions of more than two particles.

Because of the antisymmetry of $|\Psi_E\rangle$, the expectation of the operator $\langle\Psi_E|F_1|\Psi_E\rangle$ is

$$\sum_{i=1}^{N}\int \Psi_E(\mathbf{r}_1, \sigma_1, \mathbf{r}_2, \sigma_2, ..., \mathbf{r}_N, \sigma_N)f_1(\mathbf{r}_i, \sigma_i)\Psi_E^*(\mathbf{r}_1, \sigma_1, \mathbf{r}_2, \sigma_2, ..., \mathbf{r}_N, \sigma_N)dv_1dv_2...dv_N$$
$$= N\int \Psi_E(\mathbf{r}, \sigma, \mathbf{r}_2, \sigma_2, ..., \mathbf{r}_N, \sigma_N)f_1(\mathbf{r}, \sigma)\Psi_E^*(\mathbf{r}, \sigma, \mathbf{r}_2, \sigma_2, ..., \mathbf{r}_N, \sigma_N)dvdv_2...dv_N. \tag{12.77}$$

Observables that do not depend on spin can be averaged by defining a charge density matrix

$$\rho(\mathbf{r}, \mathbf{r}') = N\int \Psi_E(\mathbf{r}, \sigma, \mathbf{r}_2, \sigma_2, ..., \mathbf{r}_N, \sigma_N)\Psi_E^*(\mathbf{r}', \sigma, \mathbf{r}_2, \sigma_2, ..., \mathbf{r}_N, \sigma_N)d\sigma dv_2...dv_N \tag{12.78}$$

and writing

$$\langle\Psi_E|F_1|\Psi_E\rangle = <F>_E = \int_{\infty} \rho(\mathbf{r})f_1(\mathbf{r})d\mathbf{r}, \tag{12.79}$$

where $\rho(\mathbf{r})$ is a diagonal element of $\rho(\mathbf{r}, \mathbf{r}')$. It is pictured as the density of the electrons. Defining the function

$$\rho_2(\mathbf{r}, \mathbf{r}',) = \frac{N(N-1)}{2}\int \Psi_E(\mathbf{r}, \sigma, \mathbf{r}', \sigma', ..., \mathbf{r}_N, \sigma_N)$$
$$\Psi_E^*(\mathbf{r}, \sigma, \mathbf{r}', \sigma', ..., \mathbf{r}_N, \sigma_N)d\sigma d\sigma'dv_3...dv_N \tag{12.80}$$

the expectation value of F_2 is

$$\langle\Psi_E|F_2|\Psi_E\rangle = \iint f_2(\mathbf{r}, \mathbf{r}')\rho_2(\mathbf{r}, \mathbf{r}')d\mathbf{r}d\mathbf{r}'. \tag{12.81}$$

In section 12.1 the Hamiltonian for a N-particle system was written without any approximation as $E = T + V + J + U$, where U is the nuclear contribution and will frequently be ignored in the derivations. Inserting the kinetic energy terms into equation (12.77) leads to

$$T = -\frac{\hbar^2}{2m}\iint \nabla'^2\rho(\mathbf{r}, \mathbf{r}')\delta(\mathbf{r} - \mathbf{r}')d\mathbf{r}d\mathbf{r}'. \tag{12.82}$$

This expression means that the differential operator is applied to the density function and then $\mathbf{r}$ is set equal to $\mathbf{r}'$. The potential from the nuclei from equation (12.3) and equation (12.79) is

$$V = -\int \sum_{\alpha=1}^{N_\alpha} \frac{Z_\alpha e^2}{|\mathbf{r} - \mathbf{R}_\alpha|}\rho(\mathbf{r})d\mathbf{r}'. \tag{12.83}$$

From equations (12.4) and (12.81), the electron–electron interaction is

$$J = e^2 \iint \frac{\rho_2(\mathbf{r}, \mathbf{r}')}{|\mathbf{r} - \mathbf{r}'|} d\mathbf{r} d\mathbf{r}'. \tag{12.84}$$

The Coulomb interaction among the nuclei, U, is ignored at this point.

Condensed matter theorists and quantum chemists developed the notation in the preceding paragraph many years ago. Their goal was to write everything in terms of the charge density $\rho(\mathbf{r})$. As a step toward this goal, the two-particle density is written

$$\rho_2(\mathbf{r}, \mathbf{r}') = \frac{\rho(\mathbf{r})\rho(\mathbf{r}')}{2}[1 - h(\mathbf{r}, \mathbf{r}')], \tag{12.85}$$

where $h(\mathbf{r}, \mathbf{r}')$ is called the two-particle correlation function. Inserting this into equation (12.84) leads to $J = J_{ee} + K$, with

$$J_{ee} = \frac{e^2}{2} \iint \frac{\rho(\mathbf{r})\rho(\mathbf{r}')}{|\mathbf{r} - \mathbf{r}'|} d\mathbf{r} d\mathbf{r}', \tag{12.86}$$

and

$$K = -\frac{e^2}{2} \iint \frac{\rho(\mathbf{r})h(\mathbf{r}, \mathbf{r}')\rho(\mathbf{r}')}{|\mathbf{r} - \mathbf{r}'|} d\mathbf{r} d\mathbf{r}'. \tag{12.87}$$

The contribution J_{ee} is the same as the Coulomb interaction in the Hartree approximation. The second term includes the interaction caused by the antisymmetry of the wave function. It pushes electrons apart and leads to an exchange-correlation hole. The obvious fact that

$$\int \rho_2(\mathbf{r}, \mathbf{r}') d\mathbf{r}' = \frac{N - 1}{2}\rho(\mathbf{r}), \tag{12.88}$$

leads to the interesting result that

$$\int \rho(\mathbf{r}')[1 - h(\mathbf{r}, \mathbf{r}')] d\mathbf{r}' = N - 1 = N - \int \rho(\mathbf{r}')h(\mathbf{r}, \mathbf{r}') d\mathbf{r}', \tag{12.89}$$

and hence

$$\int \rho(\mathbf{r}')h(\mathbf{r}, \mathbf{r}') d\mathbf{r}' = 1. \tag{12.90}$$

The meaning of this is that the exchange-correlation hole contains one electron.

The preceding equations are exact, but they are not very useful because they all assume that the wave function is known.

12.8 Single configuration approximation

When the eigenfunction can be approximated by one determinant,

$$
\begin{aligned}
&\Psi_E(\mathbf{r}_1,\, \sigma_1,\, \mathbf{r}_2,\, \sigma_2,\, \mathbf{r}_3,\, \sigma_3,\, ...,\, \mathbf{r}_N,\, \sigma_N) \\[4pt]
&= \frac{1}{\sqrt{N!}}
\begin{vmatrix}
\psi_{\varepsilon_1}(\mathbf{r}_1,\sigma_1) & \psi_{\varepsilon_1}(\mathbf{r}_2,\sigma_2) & \psi_{\varepsilon_1}(\mathbf{r}_3,\sigma_3) & \cdots & \psi_{\varepsilon_1}(\mathbf{r}_N,\sigma_N) \\
\psi_{\varepsilon_2}(\mathbf{r}_1,\sigma_1) & \psi_{\varepsilon_2}(\mathbf{r}_2,\sigma_2) & \psi_{\varepsilon_2}(\mathbf{r}_3,\sigma_3) & \cdots & \psi_{\varepsilon_2}(\mathbf{r}_N,\sigma_N) \\
\psi_{\varepsilon_3}(\mathbf{r}_1,\sigma_1) & \psi_{\varepsilon_3}(\mathbf{r}_2,\sigma_2) & \psi_{\varepsilon_3}(\mathbf{r}_3,\sigma_3) & \cdots & \psi_{\varepsilon_3}(\mathbf{r}_N,\sigma_N) \\
\cdots & \cdots & \cdots & \cdots & \cdots \\
\psi_{\varepsilon_N}(\mathbf{r}_1,\sigma_1) & \psi_{\varepsilon_N}(\mathbf{r}_2,\sigma_2) & \psi_{\varepsilon_N}(\mathbf{r}_3,\sigma_3) & \cdots & \psi_{\varepsilon_N}(\mathbf{r}_N,\sigma_N)
\end{vmatrix},
\end{aligned}
\tag{12.91}
$$

it is known that

$$
\begin{aligned}
J &= \frac{e^2}{2}\sum_{a=1}^{N}\sum_{b=1}^{N}\int \frac{\psi_a^*(\mathbf{r},\sigma)\psi_a(\mathbf{r},\sigma)\psi_b^*(\mathbf{r}',\sigma')\psi_b(\mathbf{r}',\sigma')}{|\mathbf{r}-\mathbf{r}'|}\,dvdv' \\[4pt]
&= \frac{e^2}{2}\sum_{a=1}^{N}\sum_{b=1}^{N}\int \frac{\phi_a^*(\mathbf{r})\phi_a(\mathbf{r})\phi_b^*(\mathbf{r}')\phi_b(\mathbf{r}')}{|\mathbf{r}-\mathbf{r}'|}\,d\mathbf{r}d\mathbf{r}' \\[4pt]
&= \frac{e^2}{2}\int \frac{\rho(\mathbf{r})\rho(\mathbf{r}')}{|\mathbf{r}-\mathbf{r}'|}\,d\mathbf{r}d\mathbf{r}'
\end{aligned}
\tag{12.92}
$$

With an obvious definition for the particle density. The exchange term is

$$
\begin{aligned}
K &= -\frac{e^2}{2}\sum_{a=1}^{N}\sum_{b=1}^{N}\int \psi_a^*(\mathbf{r},\sigma)\psi_b^*(\mathbf{r}',\sigma')\frac{1}{|\mathbf{r}-\mathbf{r}'|}\psi_b(\mathbf{r},\sigma)\psi_a(\mathbf{r}',\sigma')\,dvdv' \\[4pt]
&= -\frac{e^2}{2}\sum_{a=1}^{N}\sum_{b=1}^{N}\int \frac{\phi_a^*(\mathbf{r})\phi_a(\mathbf{r}')\phi_b^*(\mathbf{r}')\phi_b(\mathbf{r})}{|\mathbf{r}-\mathbf{r}'|}\,d\mathbf{r}d\mathbf{r}'\,\delta_{\chi_a\chi_b} \\[4pt]
&= -\frac{e^2}{4}\int \frac{\rho(\mathbf{r},\mathbf{r}')\rho(\mathbf{r}',\mathbf{r})}{|\mathbf{r}-\mathbf{r}'|}\,d\mathbf{r}d\mathbf{r}'.
\end{aligned}
\tag{12.93}
$$

In these expressions for J and K, the formula for the density matrix based on orbitals has been introduced,

$$
\rho(\mathbf{r},\mathbf{r}') = \sum_{a=1}^{N}\phi_a(\mathbf{r})\phi_a^*(\mathbf{r}') = 2\sum_{a=1}^{N/2}\phi_a(\mathbf{r})\phi_a^*(\mathbf{r}'),
\tag{12.94}
$$

where $\phi_a(\mathbf{r})$ is a normalized orbital, and the factor of 2 takes into account the spin.

From the more general expression for K in equation (12.93), it can be seen that, for the one-configuration case, the two-particle correlation function can be calculated if the density matrix $\rho(\mathbf{r},\mathbf{r}')$ is known. This will be done for a specific model below. The first thing that can be seen is that K will cancel one half of J at $\mathbf{r} = \mathbf{r}'$ because the cancellation is only effective when the spins of the electrons are aligned.

12.9 The Thomas–Fermi and Thomas–Fermi–Dirac theories

The definition of the density matrix $\rho(\mathbf{r}, \mathbf{r}')$ above assumes that the orbitals $\phi_a(\mathbf{r})$ are known. The idea of the Thomas–Fermi (T-F) and Thomas–Fermi–Dirac (TFD) theories is to write the total energy entirely as functionals of $\rho(\mathbf{r}) = \rho(\mathbf{r}, \mathbf{r})$ with no prior information. The contributions V and J_{ee} are already in the correct form in equations (12.83) and (12.86).

The jellium model that has already been developed can be used to make the kinetic energy T a functional of $\rho(\mathbf{r})$. The free-electron orbital is

$$\phi_{\mathbf{k}}(\mathbf{r}) = \langle \mathbf{r}|\mathbf{k}\rangle = \frac{1}{\sqrt{\Omega}}e^{i\mathbf{k}\cdot\mathbf{r}}, \tag{12.95}$$

when they are normalized in the periodically reproduced volume $\Omega = L^3$. Then

$$\rho(\mathbf{r}, \mathbf{r}') = \frac{2}{\Omega}\sum_{\mathbf{k}} e^{i\mathbf{k}\cdot(\mathbf{r}-\mathbf{r}')}n(\mathbf{k}), \tag{12.96}$$

where $n(\mathbf{k})$=1if $\phi_{\mathbf{k}}(\mathbf{r})$ is occupied and 0 otherwise. With this definition,

$$\nabla^2\rho(\mathbf{r}, \mathbf{r}') = -\frac{2}{\Omega}\sum_{\mathbf{k}} k^2 e^{i\mathbf{k}\cdot(\mathbf{r}-\mathbf{r}')}n(\mathbf{k}). \tag{12.97}$$

The sum can be converted to an integral as $L \to \infty$

$$\frac{1}{\Omega}\sum_{\mathbf{k}} \to \frac{1}{(2\pi)^3}\int d\mathbf{k}. \tag{12.98}$$

It follows that

$$\int \nabla^2\rho(\mathbf{r}, \mathbf{r}')\delta(\mathbf{r} - \mathbf{r}')d\mathbf{r}d\mathbf{r}' = -\frac{2\Omega}{(2\pi)^3}\int k^2 n(\mathbf{k})(4\pi k^2)dk$$
$$= -\frac{\Omega}{\pi^2}\int_0^{k_F} k^4 dk = -\frac{\Omega}{5\pi^2}k_F^5 \tag{12.99}$$

The number of electrons that can be put in states with $|\mathbf{k}| \leqslant k_F$ is

$$\rho = 2 \times \frac{4}{3}\pi k_F^3 \times \frac{\Omega}{(2\pi)^3} = \frac{k_F^3}{3\pi^2}, \tag{12.100}$$

and the Fermi radius is

$$k_F = (3\pi^2\rho)^{1/3}. \tag{12.101}$$

It follows that the kinetic energy of a free-electron gas in the volume Ω is

$$-\frac{\hbar^2}{2m}\int \nabla^2\rho(\mathbf{r}, \mathbf{r}')\delta(\mathbf{r} - \mathbf{r}')d\mathbf{r}d\mathbf{r}' = \frac{\hbar^2}{2m}\frac{\Omega}{5\pi^2}(3\pi^2\rho)^{5/3}$$
$$= \frac{\hbar^2}{2m}\frac{3}{5}(3\pi^2)^{2/3}\rho^{5/3}\Omega \tag{12.102}$$

At this stage the assumption is introduced that the electron density depends on $\mathbf{r}$, but the variation is on a macroscopic scale. Then the volume element can be replaced by an infinitesimal $\Omega \to d\mathbf{r}$ and the kinetic energy is a functional

$$T[\rho] = \frac{\hbar^2}{2m}\frac{3}{5}(3\pi^2)^{2/3}\int_\Omega \rho^{5/3}(\mathbf{r})d\mathbf{r}, \tag{12.103}$$

as was to be shown.

The next step is to make K a functional of $\rho(\mathbf{r})$. The density matrix for the electron gas is written as in equation (12.96), but the integral form is

$$\rho(\mathbf{r}, \mathbf{r}') = \frac{2}{(2\pi)^2}\int_0^{k_F}\int_0^\pi e^{iks\cos\theta}\sin\theta d\theta k^2 dk \tag{12.104}$$

where $s = |\mathbf{r} - \mathbf{r}'|$. Since

$$\int_0^\pi e^{iks\cos\theta}\sin\theta d\theta = \int_{-1}^1 e^{iksw}dw = \frac{2}{ks}\sin ks, \tag{12.105}$$

it follows that

$$\rho(\mathbf{r}, \mathbf{r}') = \rho(s) = \frac{2}{(2\pi)^2}\int_0^{k_F}\frac{2}{ks}\sin ks k^2 dk = \frac{k_F^3}{\pi^2}\frac{\sin t - t\cos t}{t^3}, \tag{12.106}$$

where $t = k_F s$. Using equation (12.101), this equation can be rewritten

$$\rho(\mathbf{r}, s) = 3\rho(\mathbf{r})\frac{\sin t - t\cos t}{t^3}, \tag{12.107}$$

where $t = (3\pi^2\rho)^{1/3}s$. Expanding the trigonometric functions in a Taylor series for small t, it is seen that the function defined by this equation approaches

$$\rho(\mathbf{r}, \mathbf{r}') = \rho(\mathbf{r}), \tag{12.108}$$

as it should. Because of the t^3 term in the denominator in equation (12.107), it follows that $\rho(\mathbf{r}, \mathbf{r}') \to 0$ as $t \propto |\mathbf{r} - \mathbf{r}'| \to \infty$. This is the expected behavior because, from a comparison of equations (12.87) and (12.93),

$$\rho(\mathbf{r}, \mathbf{r}')\rho(\mathbf{r}', \mathbf{r}) = 2\rho(\mathbf{r})h(\mathbf{r}, \mathbf{r}')\rho(\mathbf{r}'). \tag{12.109}$$

The asymptotic behavior from the free-electron model shows that $h(\mathbf{r} - \mathbf{r}') \to \frac{1}{2}$, as $|\mathbf{r} - \mathbf{r}'| \to 0$, because of spin. There is no correlation between electrons, and hence no exchange contribution, when $|\mathbf{r} - \mathbf{r}'| \to \infty$.

Referring back to the definition in equation (12.93),

$$K = -\frac{e^2}{4} \int \frac{\rho(\mathbf{r}, \mathbf{r}')\rho(\mathbf{r}', \mathbf{r})}{|\mathbf{r} - \mathbf{r}'|} d\mathbf{r} d\mathbf{r}' = -\frac{e^2}{4} \int \frac{|\rho(\mathbf{r}, s)|^2}{s} d\mathbf{r} ds$$
$$= -\frac{9e^2}{4} \int \int \int_0^\infty \frac{f(t)^2}{t} \frac{4\pi}{k_F^2} t^2 dt \rho(\mathbf{r})^2 d\mathbf{r} \qquad (12.110)$$

where

$$f(t) = \frac{\sin t - t \cos t}{t^3}. \qquad (12.111)$$

The definite integral is

$$\int_0^\infty f(t)^2 t \, dt = \frac{1}{4}, \qquad (12.112)$$

so

$$K = -\frac{9e^2}{4} \int \frac{\pi}{k_F^2} \rho(\mathbf{r})^2 d\mathbf{r} = -\frac{9\pi e^2}{4} \int \frac{1}{(3\pi^2 \rho(\mathbf{r}))^{2/3}} \rho(\mathbf{r})^2 d\mathbf{r}. \qquad (12.113)$$

It follows that K is a functional of the density,

$$K[\rho] = -\frac{3e^2}{4} \left(\frac{3}{\pi}\right)^{1/3} \int \rho(\mathbf{r})^{4/3} d\mathbf{r}, \qquad (12.114)$$

in this approximation. This formula has many applications, and is attributed to P A M Dirac.

Summing up the derivations in the preceding section, to the approximations stated, the energy can be written as a functional of the charge density of the electrons

$$E[\rho] = \frac{\hbar^2}{2m} \frac{3}{5} (3\pi^2)^{2/3} \int_\Omega \rho^{5/3}(\mathbf{r}) d\mathbf{r} + \int v(\mathbf{r})\rho(\mathbf{r}) d\mathbf{r}'$$
$$+ \frac{e^2}{2} \int \frac{\rho(\mathbf{r})\rho(\mathbf{r}')}{|\mathbf{r} - \mathbf{r}'|} d\mathbf{r} d\mathbf{r}' - \frac{3e^2}{4} \left(\frac{3}{\pi}\right)^{1/3} \int \rho(\mathbf{r})^{4/3} d\mathbf{r} + \sum_{\alpha=1}^{N_p} \sum_{\substack{\beta=1 \\ \beta \neq \alpha}}^{N_p} \frac{Z_\alpha Z_\beta e^2}{|\mathbf{R}_\alpha - \mathbf{R}_\beta|}, \qquad (12.115)$$

which can be written

$$E[\rho] = C_T \int_\Omega \rho^{5/3}(\mathbf{r}) d\mathbf{r} + \int v(\mathbf{r})\rho(\mathbf{r}) d\mathbf{r}'$$
$$+ \frac{e^2}{2} \int \frac{\rho(\mathbf{r})\rho(\mathbf{r}')}{|\mathbf{r} - \mathbf{r}'|} d\mathbf{r} d\mathbf{r}' - C_K \int \rho(\mathbf{r})^{4/3} d\mathbf{r} + U. \qquad (12.116)$$

The normalization condition on the density is

$$\int \rho(\mathbf{r}) d\mathbf{r} = N, \qquad (12.117)$$

where N is the number of electrons. The function $\rho(\mathbf{r})$ that minimizes the energy is the solution one for which the functional derivative

$$\frac{\delta\left\{E[\rho] - \mu\left(\int \rho(\mathbf{r}')d\mathbf{r}' - N\right)\right\}}{\delta\rho(\mathbf{r})} = 0, \tag{12.118}$$

which leads to the TFD equation

$$\frac{5}{3}C_T\rho(\mathbf{r})^{2/3} + v(\mathbf{r}) + e^2 \int \frac{\rho(\mathbf{r}')}{|\mathbf{r} - \mathbf{r}'|}d\mathbf{r}' - \frac{4}{3}C_K\rho(\mathbf{r})^{1/3} = \mu, \tag{12.119}$$

with

$$v(\mathbf{r}) = -\sum_{\alpha=1}^{N_\alpha} \frac{Z_\alpha e^2}{|\mathbf{r} - \mathbf{R}_\alpha|}. \tag{12.120}$$

The effects of exchange and correlation on the electrons is in the term

$$V_{xc}(\mathbf{r}) = -\frac{4}{3}C_K\rho(\mathbf{r})^{1/3} = -e^2\left(\frac{3}{\pi}\right)^{1/3}\rho(\mathbf{r})^{1/3}. \tag{12.121}$$

The T-F equation is obtained by setting that term equal to zero

$$\frac{5}{3}C_T\rho(\mathbf{r})^{2/3} + v(\mathbf{r}) + e^2 \int \frac{\rho(\mathbf{r}')}{|\mathbf{r} - \mathbf{r}'|}d\mathbf{r}' = \mu. \tag{12.122}$$

This integral equation has a unique solution, and μ is determined in the process. As with the Hartree and H-F equations, the equation is solved self-consistently.

The T-F equation generated a great deal of interest, and was used as a simple method to calculate the electronic structure of atoms. It was later applied to calculations of the equation of state of solids. However, it was proved by Teller in 1962 and, more clearly, by Lieb and Simon in 1977 that the energy of a collection of atoms is reduced by moving the atoms farther apart. This is called the negative pressure theorem. It follows that the energy of the atoms is minimized by separating them, which means that a molecule or solid has no bound states in the T-F approximation. Going from the T-F to the TFD equation, $C_K \neq 0$, makes matters worse, not better.

In spite of its failure, the TFD theory contains the germ of an idea that has proved useful. The H-F equation can be used to calculate the electronic states of an atom, but it is much more difficult to apply to molecules and it cannot be used for calculations on solids. For that reason, John Slater developed a theory that leads to a set of one-electron equations similar to the Hartree theory, but supplemented by the addition of the exchange-correlation (XC) potential in equation (12.121). Using arguments put forward by Eugene Wigner, John Slater suggested that the XC potential should be

$$V_{\text{Slater}}(\mathbf{r}) = -\frac{3e^2}{2}\left(\frac{3\rho(\mathbf{r})}{\pi}\right)^{1/3}. \tag{12.123}$$

He later improved this theory by making the multiplicative coefficient adjustable. Physicists of the day referred to this as the $X\alpha$ or the 'rho to the one third' approximation.

The successes of Thomas, Fermi and Slater led Walter Kohn to develop the most widely used theory for treating many-electron systems, the density functional theory. That will be discussed in the next section.

12.10 The density functional theory (DFT)

The DFT is based on two simple but profound theorems proved by Hohenberg and Kohn [5]. The first Hohenberg–Kohn (H-K) theorem states:

There is a unique relationship between the external potential $v(\mathbf{r})$ and the charge density $\rho(\mathbf{r})$.

Proof:

Assume that

$$H\Psi = E_0\Psi, \tag{12.124}$$

with E_0 the ground state eigenvalue, and

$$H = T_{op} + V_{ee} + U + V_{ext}(\mathbf{r}_1, \mathbf{r}_2, ...\mathbf{r}_N). \tag{12.125}$$

As usual

$$T_{op} = -\sum_{i=1}^{N}\frac{\hbar^2}{2m}\nabla_i^2$$

$$V_{ee} = \frac{1}{2}\sum_{i=1}^{N}\sum_{\substack{j=1\\j\neq i}}^{N}\frac{e^2}{|\mathbf{r}_i - \mathbf{r}_j|}. \tag{12.126}$$

$$U = \sum_{\alpha=1}^{N_p}\sum_{\substack{\beta=1\\\beta\neq\alpha}}^{N_p}\frac{Z_\alpha Z_\beta e^2}{|\mathbf{R}_\alpha - \mathbf{R}_\beta|}$$

The external potential can be caused by the positive nuclei and/or external fields

$$V_{ext}(\mathbf{r}_1, \mathbf{r}_2, ...\mathbf{r}_N) = -\sum_{i=1}^{N}\sum_{\alpha=1}^{N_\alpha}\frac{Z_\alpha e^2}{|\mathbf{r}_i - \mathbf{R}_\alpha|} + F(\mathbf{r}_1, \mathbf{r}_2, ...\mathbf{r}_N). \tag{12.127}$$

The charge density is given in terms of the eigenfunction

$$\rho(\mathbf{r}) = N\int\Psi(\mathbf{r}, \sigma, \mathbf{r}_2, \sigma_2, ..., \mathbf{r}_N, \sigma_N)\Psi^*(\mathbf{r}, \sigma, \mathbf{r}_2, \sigma_2, ..., \mathbf{r}_N, \sigma_N)d\sigma dv_2...dv_N \tag{12.128}$$

Consider another Hamiltonian that differs only in that it has a different external potential

$$H' = T_{op} + V_{ee} + U + V'_{ext}(\mathbf{r}_1, \mathbf{r}_2, ...\mathbf{r}_N) \tag{12.129}$$

It has a ground state eigenvalue and eigenfunction

$$H'\Psi' = E'_0\Psi' \tag{12.130}$$

and charge density

$$\rho'(\mathbf{r}) = N\int \Psi'(\mathbf{r}, \sigma, \mathbf{r}_2, \sigma_2, ..., \mathbf{r}_N, \sigma_N)\Psi'^*(\mathbf{r}, \sigma, \mathbf{r}_2, \sigma_2, ..., \mathbf{r}_N, \sigma_N)d\sigma dv_2...dv_N \tag{12.131}$$

Question:
   Can $\rho(\mathbf{r}) = \rho(\mathbf{r}')$ even though $V_{ext}(\mathbf{r}_1, \mathbf{r}_2, ...\mathbf{r}_N) \neq V'_{ext}(\mathbf{r}_1, \mathbf{r}_2, ...\mathbf{r}_N)$?
   Find the expectation value of H in the state described by the wave function Ψ'

$$\langle\Psi'|\,H\,|\,\Psi'\rangle = \int \Psi'(\mathbf{r}_1, \sigma_1, ..., \mathbf{r}_N, \sigma_N)H\Psi'^*(\mathbf{r}_1, \sigma_1, ..., \mathbf{r}_N, \sigma_N)dv_1 dv_2...dv_N. \tag{12.132}$$

This equation can be manipulated as follows

$$\langle\Psi'|\,H\,|\,\Psi'\rangle = \langle\Psi'|\,H'\,|\,\Psi'\rangle + \langle\Psi'|\,(H - H')\,|\,\Psi'\rangle. \tag{12.133}$$

By the rules of the variational principle

$$\langle\Psi'|\,H\,|\,\Psi'\rangle > E_0, \tag{12.134}$$

so

$$E_0 < E'_0 + \int \rho'(\mathbf{r})[V_{ext}(\mathbf{r}) - V'_{ext}(\mathbf{r})]d\mathbf{r}. \tag{12.135}$$

Similarly,

$$\langle\Psi|\,H'|\Psi\rangle = \langle\Psi|\,H\,|\Psi\rangle + \langle\Psi|\,(H' - H)\,|\Psi\rangle, \tag{12.136}$$

so

$$E'_0 < E_0 - \int \rho(\mathbf{r})[V_{ext}(\mathbf{r}) - V'_{ext}(\mathbf{r})]d\mathbf{r}. \tag{12.137}$$

Adding the two equations together,

$$E_0 + E'_0 < E_0 + E'_0 + \int [\rho'(\mathbf{r}) - \rho(\mathbf{r})][V_{ext}(\mathbf{r}) - V'_{ext}(\mathbf{r})]d\mathbf{r}. \tag{12.138}$$

If $\rho'(\mathbf{r}) = \rho(\mathbf{r})$, then the math leads to the result that $E_0 + E'_0 < E_0 + E'_0$, which is impossible. This completes the proof of the first H-K theorem, which can be restated:
   There is a unique relationship between the external potential $V_{ext}(\mathbf{r})$ and the charge density $\rho(\mathbf{r})$. Two different potentials $V_{ext}(\mathbf{r}) \neq V'_{ext}(\mathbf{r})$ cannot lead to the same charge density.
   Since this theorem proves that $\rho(\mathbf{r})$ determines $V_{ext}(\mathbf{r})$, and it is obvious that it determines the number of electrons in the system $N = \int \rho(\mathbf{r})d\mathbf{r}$, it determines everything about the system including the energy. It follows from the theorem that E is a functional of $\rho(\mathbf{r})$,

$$E = E[\rho]. \tag{12.139}$$

The second H-K theorem states:

The energy is variational in the density. That is, the functional $E[\rho]$ is minimized by the correct $\rho(\mathbf{r})$, and the minimum value is the ground state energy E_0.

Proof:

Consider an arbitrary $\rho'(\mathbf{r})$. By H-K I it determines a $V'_{ext}(\mathbf{r})$ and hence a Hamiltonian H' that is used in a Schrödinger equation. The Ψ' obtained from that equation can be used in the calculation of the expectation value of H, which is the functional $E[\rho']$. By the Rayleigh–Ritz theorem, the expectation value of H in the state Ψ' satisfies the inequality

$$\langle \Psi' | H | \Psi' \rangle = E[\rho'] \geqslant E_0, \tag{12.140}$$

thus proving the second theorem H-K II.

The mathematics of H-K I and II is so simple that they were not believed at first, but they have withstood every challenge. By construction, the energy has been shown to be a functional of $\rho(\mathbf{r}, \mathbf{r}')$ and $\rho_2(\mathbf{r}, \mathbf{r}')$, or, alternatively, $\rho(\mathbf{r})$ and the two-particle correlation function $h(\mathbf{r}, \mathbf{r}')$. Massive and unsuccessful efforts were made for many years to show that the energy can be made a functional of something simpler. H-K I and II solved the problem completely. It follows that T and K must be functionals of $\rho(\mathbf{r})$, although not the simple functionals used in the T-F and TFD theories.

These theorems were made into a practical approach for calculating the electronic states in condensed matter by Kohn and Sham [6]. They propose that the exact Hamiltonian

$$H = \sum_{i=1}^{N}\left[-\frac{\hbar^2}{2m}\nabla_i^2 + \frac{1}{2}\sum_{\substack{j=1\\j\neq i}}^{N}\frac{e^2}{|\mathbf{r}_i - \mathbf{r}_j|} \right] + V_{ext}(\mathbf{r}_1, \mathbf{r}_2, ...\mathbf{r}_N), \tag{12.141}$$

can be turned into an effective Hamiltonian

$$H = \sum_{i=1}^{N}\left[-\frac{\hbar^2}{2m}\nabla_i^2 + \widehat{v}(\mathbf{r}_i) \right], \tag{12.142}$$

where $\widehat{v}(\mathbf{r})$ is a best effective potential to be determined. From equation (12.31), it is known that the state vector can be written

$$\Psi_E(\mathbf{r}_1, \sigma_1, \mathbf{r}_2, \sigma_2, \mathbf{r}_3, \sigma_3, ..., \mathbf{r}_N, \sigma_N)$$

$$= \frac{1}{\sqrt{N!}}\begin{vmatrix} \psi_{\varepsilon_1}(\mathbf{r}_1, \sigma_1) & \psi_{\varepsilon_1}(\mathbf{r}_2, \sigma_2) & \psi_{\varepsilon_1}(\mathbf{r}_3, \sigma_3) & \cdots & \psi_{\varepsilon_1}(\mathbf{r}_N, \sigma_N) \\ \psi_{\varepsilon_2}(\mathbf{r}_1, \sigma_1) & \psi_{\varepsilon_2}(\mathbf{r}_2, \sigma_2) & \psi_{\varepsilon_2}(\mathbf{r}_3, \sigma_3) & \cdots & \psi_{\varepsilon_2}(\mathbf{r}_N, \sigma_N) \\ \psi_{\varepsilon_3}(\mathbf{r}_1, \sigma_1) & \psi_{\varepsilon_3}(\mathbf{r}_2, \sigma_2) & \psi_{\varepsilon_3}(\mathbf{r}_3, \sigma_3) & \cdots & \psi_{\varepsilon_3}(\mathbf{r}_N, \sigma_N) \\ \cdots & \cdots & \cdots & \cdots & \cdots \\ \psi_{\varepsilon_N}(\mathbf{r}_1, \sigma_1) & \psi_{\varepsilon_N}(\mathbf{r}_2, \sigma_2) & \psi_{\varepsilon_N}(\mathbf{r}_3, \sigma_3) & \cdots & \psi_{\varepsilon_N}(\mathbf{r}_N, \sigma_N) \end{vmatrix} \tag{12.143}$$

where

$$\psi_a(\mathbf{r}, \sigma) = \phi_a(\mathbf{r})\chi_a(\sigma),$$ (12.144)

and

$$\left[-\frac{\hbar^2}{2m}\nabla^2 + \widehat{v}(\mathbf{r})\right]\phi_a(\mathbf{r}) = \varepsilon_a\phi_a(\mathbf{r}).$$ (12.145)

The charge density is

$$\rho(\mathbf{r}) = 2\sum_{a=1}^{N/2}\phi_a(\mathbf{r})\phi_a^*(\mathbf{r}).$$ (12.146)

According to Kohn and Sham, the best choice of $\widehat{v}(\mathbf{r})$ is the one that makes this charge density equal to the exact one.

The kinetic energy functional is

$$\widehat{T}[\rho] = -\frac{\hbar^2}{m}\sum_{a=1}^{N/2}[\nabla^2\phi_a(\mathbf{r})]\phi_a^*(\mathbf{r}),$$ (12.147)

and the total energy functional is

$$E[\rho] = \widehat{T}[\rho] + \int \widehat{v}(\mathbf{r})\rho(\mathbf{r})d\mathbf{r}.$$ (12.148)

Write

$$\int \widehat{v}(\mathbf{r})\rho(\mathbf{r})d\mathbf{r} = \int V_{ext}(\mathbf{r})\rho(\mathbf{r})d\mathbf{r} + \frac{1}{2}\int V_{cou}(\mathbf{r})\rho(\mathbf{r})d\mathbf{r} + E_{xc}[\rho]$$ (12.149)

where

$$V_{cou}(\mathbf{r}) = e^2\int \frac{\rho(\mathbf{r}')}{|\mathbf{r} - \mathbf{r}'|}d\mathbf{r}'.$$ (12.150)

The quantity $E_{xc}[\rho]$ is called the exchange-correlation functional. Writing the total energy in the Kohn–Sham way $E = \widehat{T} + V + J + E_{xc}$ and the usual way $E = T + V + J + K$ leads to a formal definition

$$E_{xc}[\rho] = T[\rho] - \widehat{T}[\rho] + K[\rho].$$ (12.151)

Taking the functional derivative gives

$$\widehat{v}(\mathbf{r}) = V_{ext}(\mathbf{r}) + V_{cou}(\mathbf{r}) + \mu_{xc}(\mathbf{r}),$$ (12.152)

where

$$\mu_{xc}(\mathbf{r}) = \frac{\delta E_{xc}[\rho]}{\delta\rho(\mathbf{r})}.$$ (12.153)

This is an exact expression for the effective potential $\widehat{v}(\mathbf{r})$, except for the fact that $E_{xc}[\rho]$ is not known unless the problem has been solved exactly to start off with. Nonetheless, it turns out that this is an excellent starting place for approximate theories of $\widehat{v}(\mathbf{r})$.

12.11 The local density approximation (LDA)

The idea of the LDA is to calculate, as exactly as possible, the exchange-correlation energy per particle for a uniform electron gas of density ρ, $\varepsilon_{xc}(\rho)$. Then the total exchange-correlation (XC) energy is the sum of the contributions from each volume element

$$E_{xc}[\rho] = \int \rho(\mathbf{r})\varepsilon_{xc}[\rho(\mathbf{r})]d\mathbf{r}, \tag{12.154}$$

and the contribution to the potential is

$$\mu_{xc}(\mathbf{r}) = \varepsilon_{xc}[\rho(\mathbf{r})] + \rho(\mathbf{r})\frac{\delta\varepsilon_{xc}[\rho]}{\delta\rho(\mathbf{r})}. \tag{12.155}$$

Many to calculate $\varepsilon_{xc}[\rho]$ have been developed using modern many-body techniques such as the quantum Monte-Carlo method. An early formula that is still widely used is the Hedin–Lundquist potential [7]

$$\mu_{xc}(\mathbf{r}) = e^2\left[\frac{3\rho(\mathbf{r})}{\pi}\right]^{1/3}\left[1 + 0.7734X \ln\left(1 + \frac{1}{X}\right)\right], \tag{12.156}$$

with

$$X(\mathbf{r}) = \frac{[4\pi\rho(\mathbf{r})]^{1/3}}{21}. \tag{12.157}$$

A simpler formula is the Kohn–Sham exchange that is simply the exchange contribution for the free-electron gas

$$\mu_{xc}^{KS}(\mathbf{r}) = e^2\left[\frac{3\rho(\mathbf{r})}{\pi}\right]^{1/3}. \tag{12.158}$$

After the work of Kohn and Sham, Slater suggested that the coefficient before $\rho^{1/3}$ could be made adjustable by choosing a parameter α between 1 and $\frac{2}{3}$

$$\mu_{xc}^{X\alpha}(\mathbf{r}) = -\alpha\frac{3e^2}{2}\left(\frac{3\rho(\mathbf{r})}{\pi}\right)^{1/3}. \tag{12.159}$$

This 'Xα method was used in many calculations. More recent suggestions for $\mu_{xc}(\mathbf{r})$ include the gradient of ρ, and are called generalized gradient approximations (GGA).

Today there are large collections of XC potentials that are used for many specialized purposes. One of these is called LibXC [8].

12.12 Beyond the density functional theory

Modern DFT calculations of the total energy of solids and large molecules are reliable and very useful for explaining the properties of such systems. They have the great computational advantage that they only require the solution of one-electron differential equations that are actually simpler than ordinary H-F equations. DFT also gives a good description of Fermi surfaces and effective masses in metals.

The limitation on DFT is that there are phenomena that only exist when electrons get together and behave in a cooperative way. A simple example of this is plasmons in metals. These are collective excitations in an electron gas, and will be discussed in detail below. They are well-known experimentally and have important technological applications. Other intrinsically many-body phenomena are super-conductivity, metal–insulator transitions, heavy Fermions, itinerant electron magnetism, etc.

12.13 Infinite-order perturbation theory and Feynman diagrams

The lowest order perturbation terms are limited in that they cannot describe a transition between two states that are qualitatively different, such as a set of plane waves and a bound state wave function. Complex problems require that an infinite number of terms in a perturbation expansion must be summed. Richard Feynman faced this problem in the development of his theory of quantum electrodynamics (QED). He realized that a brute-force approach is doomed to failure. He demonstrated that the terms in a perturbation expansion could be visualized by diagrams [9], and certain types of diagrams turn out to be more important than others. By summing just those terms to infinity, he was able to calculate the fine structure constant in various physical systems to great accuracy.

Shortly after Feynman's success, his diagrammatic technique was adopted by quantum field theorists for their studies of high-energy physics and fundamental particle theory. It was also adopted by condensed matter theorists to treat the many-body phenomena described above.

Large books [4] are written on applications of Feynman diagrams to the study of many-body problems, and it would be impractical to attempt to reproduce all of the details of that approach. The easiest way to introduce the basic ideas is to describe a problem that has been solved in terms of diagrams.

In chapter 6 it was shown that Green's function for a system containing one scatterer can be written in the momentum representation

$$
\langle \mathbf{k} | G(E) | \mathbf{k} \rangle = G(E, \mathbf{k}) = G_0(E, \mathbf{k}) + G_0(E, \mathbf{k}) V_{\mathbf{kk}} G_0(E, \mathbf{k})
$$
$$
+ \int G_0(E, \mathbf{k}) V_{\mathbf{kk'}} G_0(E, \mathbf{k'}) V_{\mathbf{k'k}} G_0(E, \mathbf{k}) d\mathbf{k'}
$$
$$
+ \int\int G_0(E, \mathbf{k}) V_{\mathbf{kk'}} G_0(E, \mathbf{k'}) V_{\mathbf{k'k''}} G_0(E, \mathbf{k''}) V_{\mathbf{k''k}} G_0(E, \mathbf{k}) d\mathbf{k'} d\mathbf{k''} + \ldots
$$
.(12.160)

This diagrammatic picture of this equation is

where the heavy line with an arrow represents the total Green's function $G(\mathbf{k})$ and the light lines with an arrow represents $G_0(\mathbf{k})$. The dotted lines represent the interactions with the scattering potential V. It was also pointed out that the preceding equation can be obtained by iterating the equation

$$G(E, \mathbf{k}) = G_0(E, \mathbf{k}) + G_0(E, \mathbf{k})V_{\mathbf{kk}}G(E, \mathbf{k}), \tag{12.161}$$

which leads to the set of diagrams

where the lines have the same meaning as before. This kind of equation is called a Dyson equation after the theorist Freeman Dyson who showed the connection between the Feynman's work on QED with that of Sin-Itiro Tomonaga and Julian Schwinger. Tomonaga, Schwinger, and Feynman were awarded the Nobel prize jointly.

The picture of a Green function as a directed arrow makes more sense when it is considered as a function of time rather than energy

$$G_0(t - t') = e^{-iH_0(t-t')}\theta(t - t'), \tag{12.162}$$

where the step function $\theta(x)$ is zero for negative values of the argument and one for positive values. In the $\mathbf{k}$ representation,

$$\langle \mathbf{k}| \, G_0(t) \, |\mathbf{k}\rangle = e^{-i\frac{k^2}{2m}t}\theta(t). \tag{12.163}$$

The Green's functions using the energy argument are obtained by a Fourier transform

$$G_0(E) = \int_{-\infty}^{\infty} e^{iEt}G_0(t)dt. \tag{12.164}$$

In the units that will be used in this chapter,

$$\langle \mathbf{k}| \, G_0(E) \, |\mathbf{k}\rangle = \frac{1}{E - \dfrac{k^2}{2m} + i\delta}. \tag{12.165}$$

The positive $i\delta$ in the denominator guarantees that Green's function describes the propagation of a fermion in the positive time direction.

12-27

12.14 Dielectric function of a degenerate electron gas

Diagrams become much more interesting when used to calculate the properties of a gas made up of many electrons in a uniform positive background or jellium. This model has been studied from different points of view in the preceding sections. The electrons fill all of the states with energies less than the Fermi energy E_F, and this state is called a Fermi sea. If a particular electron is excited into a state with energy greater than E_F, it behaves like a quasiparticle with an effective mass m^* defined by

$$\varepsilon_{\mathbf{k}} = \frac{k^2}{2m^*} = \frac{k^2}{2m} - E_F. \tag{12.166}$$

When an electron is excited out of the Fermi sea, it leave behind a hole. The hole acts like a quasiparticle, but it has a charge of $-e$. The energy of the hole is measured relative to E_F, so it is also negative. The Green function for the hole is similar to the one for an electron but $G_0(t)$ propagates backward in time. In the energy representation, $i\delta$ is negative. The diagrams that describe electrons and holes are similar to the diagrams for electrons and positrons studied by Feynman, including the fact that electrons and positrons can annihilate. If an electron has the correct kind of interaction it can lose energy and fill up the hole. In that sense, the electrons and holes also annihilate.

The dielectric function has a very long history in electromagnetics and optics, and appears in the Maxwell equations. The electric field $\mathbf{E}$ and the displacement field $\mathbf{D}$ are both zero in jellium in its ground state. If a charged impurity is inserted into the system, the displacement field is related to it by

$$\nabla \cdot \mathbf{D}(\mathbf{r},\, t) = 4\pi\hat{\rho}_i(\mathbf{r},\, t), \tag{12.167}$$

where $\hat{\rho}_i(\mathbf{r},\, t)$ is the charge density of the impurity. The most general assumption is that it can depend on both the position and time. It is obvious that the impurity will induce a screening charge in the jellium $\hat{\rho}_s(\mathbf{r},\, t)$, and Maxwells formalism relates the total charge to the electric field by

$$\nabla \cdot \mathbf{E}(\mathbf{r},\, t) = 4\pi\big(\hat{\rho}_i(\mathbf{r},\, t) + \hat{\rho}_s(\mathbf{r},\, t)\big). \tag{12.168}$$

It will be assumed that the impurity charge density is positive $\hat{\rho}_i(\mathbf{r},\, t) = e\rho_i(\mathbf{r},\, t)$ and the screening charge density is negative $\hat{\rho}_s(\mathbf{r},\, t) = -e\rho_s(\mathbf{r},\, t)$. Clearly the displacement and electric fields are related, and historically this relation is expressed using a dielectric function

$$\mathbf{D}(\mathbf{r},\, t) = \varepsilon(\mathbf{r},\, t)\mathbf{E}(\mathbf{r},\, t). \tag{12.169}$$

Inserting this definition into equation (12.167) leads to

$$[\nabla\varepsilon(\mathbf{r},\, t)] \cdot \mathbf{E}(\mathbf{r},\, t) + \varepsilon(\mathbf{r},\, t)4\pi\big[\hat{\rho}_i(\mathbf{r},\, t) + \hat{\rho}_s(\mathbf{r},\, t)\big] = 4\pi\hat{\rho}_i(\mathbf{r},\, t). \tag{12.170}$$

Assuming that the dielectric function varies slowly in space and time means that the first term in this equation can be ignored, which is called the linear screening approximation, and it leads to

$$\varepsilon(\mathbf{r}, t) = \frac{\hat{\rho}_i(\mathbf{r}, t)}{\hat{\rho}_i(\mathbf{r}, t) + \hat{\rho}_s(\mathbf{r}, t)}. \tag{12.171}$$

It was pointed out that the Thomas–Fermi (T-F) formulas are unsuccessful in predicting the binding of molecules and condensed matter, but the derivations of the theory are not without merit. Assuming that there exists a potential function such that $\mathbf{E} = -\nabla V$ leads to

$$-\nabla^2 V(\mathbf{r}, t) = 4\pi\big(\hat{\rho}_i(\mathbf{r}, t) + \hat{\rho}_s(\mathbf{r}, t)\big). \tag{12.172}$$

The impurity charge is known, but the screening density is the unknown response of the jellium to it. In the T-F approach it is assumed that the Fermi wave vector is a function of space and time and is related to the Fermi energy relative to the potential V by

$$\frac{k_F^2(\mathbf{r}, t)}{2m} = E_F - V(\mathbf{r}, t). \tag{12.173}$$

The electron density is then found from equation (12.100)

$$\rho(\mathbf{r}, t) = \frac{k_F^3(\mathbf{r}, t)}{3\pi^2} = \frac{1}{3\pi^2}[2m(E_F - V(\mathbf{r}, t))]^{3/2}. \tag{12.174}$$

The Fermi energy and electron density of unperturbed jellium are E_F and ρ_0, so

$$\rho(\mathbf{r}, t) = \rho_0\left(1 - \frac{V(\mathbf{r}, t)}{E_F}\right)^{3/2}. \tag{12.175}$$

The screening density is $\rho_s(\mathbf{r}, t) = \rho(\mathbf{r}, t) - \rho_0$. If $V(\mathbf{r}, t) \ll E_F$, the preceding equation can be expanded in a Taylor series, leading to

$$\rho_s(\mathbf{r}, t) = -\frac{3\rho_0 V(\mathbf{r}, t)}{2E_F}. \tag{12.176}$$

The impurity charge is positive and thus the electrons must pile up around it, which means that $\rho_s(\mathbf{r}, t) > 0$.

From this expression for the particle density it follows that

$$\left(-\nabla^2 + q_{TF}^2\right)V(\mathbf{r}, t) = 4\pi e\rho_i(\mathbf{r}, t), \tag{12.177}$$

with

$$q_{TF}^2 = \frac{6\pi e\rho_0}{E_F}. \tag{12.178}$$

The Fourier transform of $V(\mathbf{r}, t)$ is

$$V(\mathbf{r}, t) = \int \frac{d^3q}{(2\pi)^3} V(\mathbf{q}, t)e^{i\mathbf{q}\cdot\mathbf{r}} = 4\pi e \int \frac{d^3q}{(2\pi)^3} \frac{\rho_i}{q^2 + q_{TF}^2} e^{i\mathbf{q}\cdot\mathbf{r}}. \tag{12.179}$$

For the case that the impurity is a time-independent point charge

$$\hat{\rho}_i = Q_i \delta(\mathbf{r}),\tag{12.180}$$

the potential in q space is

$$V(\mathbf{q}) = \frac{4\pi Q_i}{q^2 + q_{TF}^2}.\tag{12.181}$$

Comparing this with the formulae in the section on the Born approximation in chapter 6 shows that $V(\mathbf{q})$ is the Fourier transform of the screened Coulomb potential

$$V(\mathbf{r}) = \frac{Q_i e^{-q_{TF} r}}{r}.\tag{12.182}$$

A potential for the displacement field, $\mathbf{D} = -\nabla V_i$, satisfies the equation

$$-\nabla^2 V_i(\mathbf{r}, t) = 4\pi e \rho_i(\mathbf{r}, t).\tag{12.183}$$

From the relation between $\mathbf{D}$ and $\mathbf{E}$ in equation (12.169) and the linear screening approximation

$$\varepsilon(\mathbf{q}) = \frac{V_i(\mathbf{q})}{V(\mathbf{q})}.\tag{12.184}$$

With the same assumptions as in the preceding paragraph,

$$V_i(\mathbf{r}) = \frac{Q_i}{r};\tag{12.185}$$

and, as in equation (12.64)

$$V(\mathbf{q}) = \frac{4\pi Q_i}{q^2}.\tag{12.186}$$

Then, from equation (12.184),

$$\varepsilon(\mathbf{q}) = 1 + \frac{q_{TF}^2}{q^2}.\tag{12.187}$$

The only dependence on $\mathbf{q}$ in the dielectric function is from the explicit appearance of $\mathbf{q}$ in the denominator. The screened Coulomb form for $V(\mathbf{r})$ in equation (12.182) is physically appealing, but the dielectric function in the preceding equation is not particularly good compared to the one that will be found.

Historically, the physical picture of the dielectric function is that a field is applied to a solid that contains polar molecules which are oriented to form a polarization field $\mathbf{P}$. It is assumed that $\mathbf{P}$ is proportional to $\mathbf{E}$, $\mathbf{P} = \chi \mathbf{E}$, and $\mathbf{D} = \mathbf{E} + \mathbf{P} = (1 + \chi)\mathbf{E}$. There are no polar molecules in jellium, so it is not clear how anything can be oriented to give a polarization field. The answer to this conundrum is that particle–hole pairs can be created by exciting electrons out of the Fermi sea. The infinite-order perturbation theory below describes this process.

The first step is to rewrite the formula for time-dependent perturbation theory in chapter 5 in the form

$$i\hbar \dot{c}_f = V_{fi} e^{i\omega_{fi}t} c_i, \tag{12.188}$$

where $|c_i|^2$ is the probability that the system is initially in state i, and $|c_f|^2$ is the probability that it winds up in the final state f. The phase constant ω_{fi} is the difference in energy between the initial and final states divided by $\hbar$. It appears in the formula because the interaction representation is being used.

The time-dependent perturbation in this formula can be taken to be

$$V = U_q e^{-i\omega t} e^{\alpha t} + U_{-q} e^{+i\omega t} e^{\alpha t}, \tag{12.189}$$

which oscillates with frequency $\nu = \omega/2\pi$. The function $e^{\alpha t}$ assures that there is no perturbation at $t = -\infty$, and it is turned on adiabatically. The potential also varies in space

$$\langle \mathbf{r}| U_q |\mathbf{r}'\rangle = U e^{i\mathbf{q}\cdot\mathbf{r}}\delta(\mathbf{r}-\mathbf{r}')\langle \mathbf{r}| U_{-q} |\mathbf{r}'\rangle = U e^{-i\mathbf{q}\cdot\mathbf{r}}\delta(\mathbf{r}-\mathbf{r}'). \tag{12.190}$$

For jellium, the initial and final states are plane waves $|i\rangle = |\mathbf{k}\rangle$ and $|f\rangle = |\mathbf{k}'\rangle$ where

$$\langle \mathbf{r}|\mathbf{k}\rangle = \frac{1}{\sqrt{\Omega}} e^{i\mathbf{k}\cdot\mathbf{r}}\langle \mathbf{r}|\mathbf{k}'\rangle = \frac{1}{\sqrt{\Omega}} e^{i\mathbf{k}'\cdot\mathbf{r}}. \tag{12.191}$$

It follows that

$$\langle \mathbf{k}'| U_q |\mathbf{k}\rangle = \frac{U}{\Omega}\int_\Omega d\mathbf{r} e^{i(-\mathbf{k}'\cdot\mathbf{r}+\mathbf{q}\cdot\mathbf{r}+\mathbf{k}\cdot\mathbf{r})} = U\delta(\mathbf{k}'-\mathbf{k}-\mathbf{q})$$
$$\langle \mathbf{k}'| U_{-q} |\mathbf{k}\rangle = \frac{U}{\Omega}\int_\Omega d\mathbf{r} e^{i(-\mathbf{k}'\cdot\mathbf{r}-\mathbf{q}\cdot\mathbf{r}+\mathbf{k}\cdot\mathbf{r})} = U\delta(\mathbf{k}'-\mathbf{k}+\mathbf{q}) \tag{12.192}$$

and integrating equation (12.188) from t equals minus infinity to zero leads to

$$c_f(0) - c_f(-\infty) = -\frac{1}{\hbar}\left[\frac{U\delta(\mathbf{k}'-\mathbf{k}-\mathbf{q})}{(\omega_{fi}-\omega-i\alpha)} + \frac{U\delta(\mathbf{k}'-\mathbf{k}+\mathbf{q})}{(\omega_{fi}+\omega-i\alpha)}\right]c_\mathbf{k}. \tag{12.193}$$

The initial values of $\mathbf{k}$ can take on any value as long as the state is occupied, so the final values if $c_{\mathbf{k}'}$ are

$$c_{\mathbf{k}'} = \sum_{k(occ)}\left[\delta_{\mathbf{k}'\mathbf{k}} + \frac{U}{(E_k - E_{k+q} + \hbar\omega + i\alpha)} + \frac{U}{(E_k - E_{k-q} - \hbar\omega + i\alpha)}\right]c_\mathbf{k}, \tag{12.194}$$

so the final wave function is, from time-dependent perturbation theory

$$\psi_f(\mathbf{r}) = \frac{1}{\sqrt{\Omega}} \sum_{\mathbf{k}(occ)} \left[e^{i\mathbf{k}\cdot\mathbf{r}} + \frac{U}{(E_\mathbf{k} - E_{\mathbf{k}+\mathbf{q}} + \hbar\omega + i\alpha)} e^{i(\mathbf{k}+\mathbf{q})\cdot\mathbf{r}} \right.$$
$$\left. + \frac{U}{(E_\mathbf{k} - E_{\mathbf{k}-\mathbf{q}} - \hbar\omega + i\alpha)} e^{i(\mathbf{k}-\mathbf{q})\cdot\mathbf{r}} \right] \qquad (12.195)$$

which may be abbreviated to

$$\psi_f(\mathbf{r}) = \sum_{\mathbf{k}(occ)} \frac{e^{i\mathbf{k}\cdot\mathbf{r}}}{\sqrt{\Omega}} [1 + b_{\mathbf{k}+\mathbf{q}} U e^{i\mathbf{q}\cdot\mathbf{r}} + b_{\mathbf{k}-\mathbf{q}} U e^{-i\mathbf{q}\cdot\mathbf{r}}]. \qquad (12.196)$$

To proceed, it is necessary to expand the absolute square of $\psi_f(\mathbf{r})$ to first order in U, which leads to

$$|\psi_f(\mathbf{r})|^2 = \sum_{\mathbf{k}(occ)} \frac{1}{\Omega} \left[1 + \left(b_{\mathbf{k}+\mathbf{q}} + b_{\mathbf{k}-\mathbf{q}}^* \right) U e^{i\mathbf{q}\cdot\mathbf{r}} + \left(b_{\mathbf{k}-\mathbf{q}} + b_{\mathbf{k}+\mathbf{q}}^* \right) U e^{-i\mathbf{q}\cdot\mathbf{r}} \right]. \qquad (12.197)$$

The function $\theta(x)$ was introduced in equation (12.162), and it can be used to extend the sums in the preceding equation over all $\mathbf{k}$

$$\sum_{\mathbf{k}(occ)} \left(b_{\mathbf{k}+\mathbf{q}} + b_{\mathbf{k}-\mathbf{q}}^* \right) = \sum_{\mathbf{k}} \frac{\theta(k_F - k)}{(E_\mathbf{k} - E_{\mathbf{k}+\mathbf{q}} + \hbar\omega + i\alpha)}$$
$$+ \sum_{\mathbf{k}} \frac{\theta(k_F - k)}{(E_\mathbf{k} - E_{\mathbf{k}-\mathbf{q}} - \hbar\omega - i\alpha)} \qquad (12.198)$$
$$= \sum_{\mathbf{k}} \frac{\theta(k_F - k) - \theta(k_F - |\mathbf{k}+\mathbf{q}|)}{E_\mathbf{k} - E_{\mathbf{k}+\mathbf{q}} + \hbar\omega + i\alpha}$$

The advantage in using $\theta(x)$ is that in the second term the $\mathbf{k}$ can be interchanged with $\mathbf{k}+\mathbf{q}$. The screening charge in jellium to first order in the perturbing potential is

$$\rho_s(\mathbf{r}, \omega) = e \left| \psi_f(\mathbf{r}, \omega) \right|^2 - \frac{e}{\Omega}, \qquad (12.199)$$

or

$$\rho_s(\mathbf{r}, \omega) = \frac{eV}{\Omega} \left\{ \sum_{\mathbf{k}} \frac{\theta(k_F - k) - \theta(k_F - |\mathbf{k}+\mathbf{q}|)}{E_\mathbf{k} - E_{\mathbf{k}+\mathbf{q}} + \hbar\omega + i\alpha} e^{i\mathbf{q}\cdot\mathbf{r}} + c. c. \right\}, \qquad (12.200)$$

reverting to the previous notation in which $U \equiv V$. The $c.c.$ represents the complex conjugate of the preceding term.

As explained above, the total potential V is the sum of the applied potential V_i and the screening potential V_s. From Maxwell's equations, the screening potential is related to the screening charge density by the equation

$$\nabla^2 V_s(\mathbf{r}, \omega) = -4\pi e \rho_s(\mathbf{r}, \omega). \tag{12.201}$$

Focusing on one Fourier component

$$V_s(\mathbf{r}, \omega) = \phi(\mathbf{q}, \omega)e^{i\mathbf{q}\cdot\mathbf{r}} + c.\, c.\,, \tag{12.202}$$

leads to

$$\phi(\mathbf{q}, \omega) = C(\mathbf{q}, \omega)V, \tag{12.203}$$

where

$$C(\mathbf{q}, \omega) = \frac{4\pi e^2}{q^2\Omega} \sum_{\mathbf{k}} \frac{\theta(k_F - k) - \theta(k_F - |\mathbf{k} + \mathbf{q}|)}{E_{\mathbf{k}} - E_{\mathbf{k}+\mathbf{q}} + \hbar\omega + i\alpha}. \tag{12.204}$$

The first Born approximation to the electron–electron Coulomb potential has been identified previously as

$$v_q = \frac{4\pi e^2}{q^2}. \tag{12.205}$$

Still focusing on one Fourier component

$$\begin{aligned} V(\mathbf{q}, \omega) &= V_i(\mathbf{q}, \omega) + \phi(\mathbf{q}, \omega) \\ &= V_i(\mathbf{q}, \omega) + C(\mathbf{q}, \omega)V' \end{aligned} \tag{12.206}$$

so

$$V_i(\mathbf{q}, \omega) = [1 - C(\mathbf{q}, \omega)]V. \tag{12.207}$$

From equation (12.184), the dielectric function for pure jellium within the approximations that have been used is

$$\varepsilon_{\text{RPA}}(\mathbf{q}, \omega) = 1 - v_q P^{(1)}(\mathbf{q}, \omega), \tag{12.208}$$

with

$$P^{(1)}(\mathbf{q}, \omega) = \frac{1}{\Omega} \sum_{\mathbf{k}} \frac{\theta(k_F - k) - \theta(k_F - |\mathbf{k} + \mathbf{q}|)}{E_{\mathbf{k}} - E_{\mathbf{k}+\mathbf{q}} + \hbar\omega + i\alpha}. \tag{12.209}$$

The last two equations define the very famous random phase approximation (RPA) to the dielectric constant. It normally appears nearly half way through a text on many-body theory and is derived using diagrams. It was derived analytically earlier by the Danish professor Jens Lindhard, and is also known as the Lindhard dielectric function. It differs from all previous theories of the dielectric function for jellium, or a metal conductor in the real world, in that it depends on $\mathbf{q}$ and ω. If it is Fourier transformed, the arguments will be $\mathbf{r}$ and t. Analysis of $\varepsilon_{\text{RPA}}(\mathbf{q}, \omega)$ gives an explanation for the plasma oscillations, the many-body excitation in a metal described earlier. There is an entire field of experimental physics known as 'plasmonics' based on the study of particle description of these oscillations known as plasmons. For example, energy saving coatings for the glass panes that make up the exterior walls of many office buildings are obtained by synthesizing materials that have plasma frequencies near the transition point between the frequencies of the electromagnetic waves that describe visible light and infrared radiation. The visible light is allowed to pass through the coated glass, but the infrared radiation is reflected away.

In a more scientific vein, the screening formula from T-F theory in equation (12.182) is incorrect for large r. Experimentalists have shown that, rather than approaching zero smoothly, the screened charge shows oscillations in the tail. These oscillations have a wavelength given by $\pi(2mE_F)^{-1/2}$, where E_F is the Fermi energy of the jellium or the related metal. They are called Friedel oscillations, and are predicted by $\varepsilon_{\mathrm{RPA}}(\mathbf{q}, \omega)$.

The original Feynman diagrams had a time axis that pointed up, and a space axis that pointed to the right. The diagrams used in many-body theory have evolved away from this construction, but they are still frequently referred to as Feynman diagrams. It is a simple analytical fact that the Lindhard expression of $P^{(1)}(\mathbf{q}, \omega)$ is exactly equivalent to

$$P^{(1)}(\mathbf{q}, \omega) = 2\int \frac{dE}{2\pi} \int \frac{d\mathbf{p}}{(2\pi)^3} G^{(0)}(\mathbf{p}, E)G^{(0)}(\mathbf{p} + \mathbf{q}, E + \omega), \qquad (12.210)$$

where the Green's functions are

$$G^{(0)}(\mathbf{p}, E) = \frac{1}{E - \xi_{\mathbf{p}} + i\delta\, \mathrm{sgn}(\xi_{\mathbf{p}})}, \qquad (12.211)$$

and

$$\xi_{\mathbf{p}} = \frac{p^2}{2m} - E_F. \qquad (12.212)$$

The signum function is

$$\mathrm{sgn}\,(x) = \frac{x}{|x|}. \qquad (12.213)$$

These Green's functions are very similar to the ones introduced in chapter 6.

If $\xi_{\mathbf{p}} > 0$ the particle is an electron, and Green's function (or propagator) is represented by a line pointing to the right. If $\xi_{\mathbf{p}} < 0$ the particle is a hole, and Green's function is represented by a line pointing to the left. The momentum $\mathbf{q}$ added to the pair comes from the electron–electron interaction v_q, and the frequency is such that

$$\left|\xi_{\mathbf{p}+\mathbf{q}} - \xi_{\mathbf{p}}\right| = \hbar\omega. \qquad (12.214)$$

The diagram that represents $P^{(1)}(\mathbf{q}, \omega)$ is shown in figure 12.2. One of the rules of diagrams is that they imply integration over internal momentum and energy. Clearly, a Feynman diagram is an abstraction, in the spirit of Pablo Picasso, rather than an effort at photographic realistic draftsmanship, as with Rembrandt van Rijn.

$$\left| \xi_{\mathbf{p+q}} - \xi_{\mathbf{p}} \right| = \hbar \omega .$$

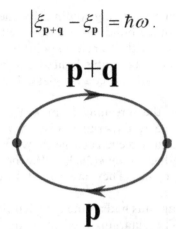

Figure 12.2. A Feynman diagram representing the integral $P^{(1)}(\mathbf{q}, \omega)$ in equation (12.210).

It can be imagined that time runs to the right in the diagram. The upper part of the diagram represents an electron moving forward in time, and the lower part is a hole moving forward in time or an electron moving backward in time. The diagram implies the negative charge on the electron and the positive charge on the hole. The electron–hole pair is a charge dipole that can be oriented by a field. It is for this reason that the integral in equation (12.210) describes a closed fermion loop or a polarization diagram.

The Coulomb interaction between the electrons is shown by a vertical dotted line. The line is vertical because the interaction v_q is instantaneous.

The physical quantity of interest is the inverse of $\varepsilon_{RPA}(\mathbf{q}, \omega)$, and it is a geometric power series in $v_q P^{(1)}(\mathbf{q}, \omega)$

$$\frac{1}{\varepsilon_{RPA}(\mathbf{q}, \omega)} = 1 + v_q P^{(1)}(\mathbf{q}, \omega) + v_q P^{(1)}(\mathbf{q}, \omega) v_q P^{(1)}(\mathbf{q}, \omega)$$
$$+ v_q P^{(1)}(\mathbf{q}, \omega) v_q P^{(1)}(\mathbf{q}, \omega) v_q P^{(1)}(\mathbf{q}, \omega) + ...$$

(12.215)

In terms of diagrams, the above series looks like figure 12.3. It is a superposition of states in which zero, one, two, ...infinity of particle–hole pairs have been formed and then disappeared. The ground state of the degenerate electron gas is very much like the quantum vacuum state. QED led to the understanding that the vacuum is not just geometry between massive objects, but is a very active place with electron–positron pairs blinking on and off. According to equation (12.215), electron–hole pairs blink on and off in the jellium ground state. Feynman diagrams have provided the tool to sum certain terms in perturbation theory of $\varepsilon_{RPA}(\mathbf{q}, \omega)^{-1}$ to infinite order, and also give an understanding of the physics that is involved that could not be obtained by simply staring at mathematical expressions.

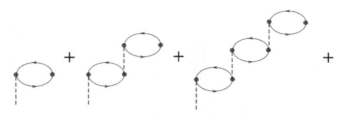

Figure 12.3. A series of linked closed fermion loops.

The equations for $\varepsilon_{\mathrm{RPA}}(\mathbf{q}, \omega)$ have been evaluated and lead to expressions for the real and imaginary part, $\mathrm{Re}[\varepsilon_{\mathrm{RPA}}(\mathbf{q}, \omega)]$ and $\mathrm{Im}[\varepsilon_{\mathrm{RPA}}(\mathbf{q}, \omega)]$, as a function of the wave vector and frequency of the perturbation. The formula for $\mathrm{Im}[\varepsilon_{\mathrm{RPA}}(\mathbf{q}, \omega)]$ is not particularly interesting except to notice that

$$\mathrm{Im}[\varepsilon_{\mathrm{RPA}}(\mathbf{q}, \omega)] \neq 0 \text{ when } \omega \leqslant \varepsilon_q + qv_F. \tag{12.216}$$

In this formula

$$\varepsilon_q = \frac{q^2}{2m} v_F = \frac{k_F}{m}. \tag{12.217}$$

The equation for $\mathrm{Re}[\varepsilon_{\mathrm{RPA}}(\mathbf{q}, \omega)]$ is

$$\mathrm{Re}\left(\varepsilon_{\mathrm{RPA}}(q, \omega)\right)$$
$$= 1 + \frac{q_{TF}^2}{2q^2}\left\{1 + \frac{m^2}{2k_Fq^3}\left[4E_F\varepsilon_q - (\varepsilon_q + \omega)^2\right]\ln\left|\frac{\varepsilon_q + qv_F + \omega}{\varepsilon_q - qv_F + \omega}\right|.\right. \tag{12.218}$$
$$\left. + \frac{m^2}{2k_Fq^3}\left[4E_F\varepsilon_q - (\varepsilon_q - \omega)^2\right]\ln\left|\frac{\varepsilon_q + qv_F - \omega}{\varepsilon_q - qv_F - \omega}\right|\right\}$$

As shown above, q_{TF} gives the screening factor for an impurity in the T-F theory of the electron gas. It is related to the electron density n by

$$q_{TF}^2 = \frac{6\pi ne^2}{E_F}. \tag{12.219}$$

It can also be expressed as

$$q_{TF}^2 = \frac{3m\omega_p^2}{2E_F}, \tag{12.220}$$

where

$$\omega_p^2 = \frac{4\pi ne^2}{m}. \tag{12.221}$$

This ω_p is sometimes referred to as the classical plasma frequency, although that a misnomer. The 'classical' derivation assumes the answer. The many-body theory described here provides the only rigorous theory for plasma oscillations.

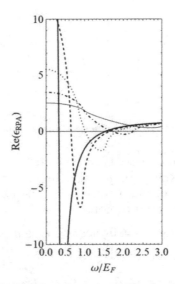

Figure 12.4. Five calculations of the real part of $\varepsilon_{RPA}(\mathbf{q}, \omega)$. The density of the degenerate electron gas is defined by choosing $\omega_p/E_F = 1.5$ for all cases. The wavelength of the perturbations in the calculations shown by the solid, dashed, dotted, and dot-dashed lines are $q/k_F = 0.2, 0.4, 0.6, 0.8$. The thin solid line that does not intersect zero is for $q/k_F = 1.0$.

The formula for $\text{Re}[\varepsilon_{RPA}(\mathbf{q}, \omega)]$ in equation (12.218) is plotted as a function of frequency ω/E_F in figure 12.4 for five values of potential wave number q/k_F. The density of electrons in the system is such that the plasma frequency is $\omega_p = 1.5E_F$. The functions cross zero twice for $q/k_F \leqslant 0.8$. At these points the function $\varepsilon_{RPA}(\mathbf{q}, \omega)^{-1}$ passes through infinity like c/x. According to Sohotsky's theorem

$$\lim_{\varepsilon \to 0^+} \frac{c}{x \pm i\varepsilon} = \text{Principal Value}\left(\frac{c}{x}\right) \mp i\pi c\delta(x), \qquad (12.222)$$

so an operator with $\varepsilon_{RPA}(\mathbf{q}, \omega)$ in the denominator will have a delta-function singularity in its spectrum. The lower frequency zero of $\varepsilon_{RPA}(\mathbf{q}, \omega)$ will not lead to a singularity because $\text{Im}[\varepsilon_{RPA}(\mathbf{q}, \omega)] \neq 0$ for those values of ω/E_F, but there will be a singularity in the spectrum at the higher frequency zero. This identifies the frequency of the plasma oscillation for the given value of ω_p/E_F and q/k_F. The fact that there is no zero in $\text{Re}[\varepsilon_{RPA}(\mathbf{q}, \omega)]$ for $q/k_F = 1.0$ predicts correctly that there will be no plasma oscillation for that case. The prediction of plasmons (or their absence) in good agreement with experiment is one of the great successes of the RPA.

Figure 12.5 shows the positions of plasmon peaks for jellium with $\omega_p = 1.5E_F$ and various values of q/k_F. The curve approaches 1.5 as $q/k_F \to 0$, which demonstrates that the plasmon peaks are given by equation (12.221) in the limit as the jellium becomes uniform in space. As mentioned in the discussion of figure 12.4, there are no plasma peaks for $q/k_F > 0.87034$.

The agreement with experiment of the data in figure 12.4 and figure 12.5, along with other predictions based on $\text{Re}[\varepsilon_{RPA}(\mathbf{q}, \omega)]$ and $\text{Im}[\varepsilon_{RPA}(\mathbf{q}, \omega)]$ give great

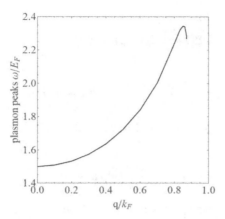

Figure 12.5. Positions of plasmon peaks of jellium with an electron density such that $\omega_p = 1.5 E_F$.

confidence that the theory is quite good. Researchers are attempting to improve on it by including diagrams that do not appear in figure 12.3.

The effect of the H-F term on the self-energy of an electron in jellium was discussed in connection with equation (12.66). The self-energy is a quantity that must be added to the kinetic energy in order to obtain a more exact one-particle energy

$$\varepsilon(\mathbf{k}) = \varepsilon_0(\mathbf{k}) + \Sigma(\mathbf{k}) = \frac{k^2}{2m^*} = \left(\frac{m}{m^*}\right)\varepsilon_0(\mathbf{k}). \qquad (12.223)$$

Given the success of the dielectric function calculated using the RPA, it is reasonable to use the RPA to calculate the electron self-energy. The diagrams for this process are shown in figure 12.6.

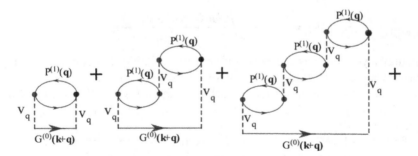

Figure 12.6. Diagrams for the electron self-energy based on the RPA.

The algebraic equation for this self-energy is

$$\hat{\Sigma}(\mathbf{k}) = \int \frac{d\mathbf{q}}{(2\pi)^3} \int \frac{d\omega}{2\pi} G^{(0)}(\mathbf{k} + \mathbf{q}, \omega) \frac{v_q P^{(1)}(\mathbf{q}, \omega)}{1 - v_q P^{(1)}(\mathbf{q}, \omega)} v_q$$

$$= \int \frac{d\mathbf{q}}{(2\pi)^3} \int \frac{d\omega}{2\pi} G^{(0)}(\mathbf{k} + \mathbf{q}, \omega) \left[\frac{1}{\varepsilon_{RPA}(\mathbf{q}, \omega)} - 1\right] v_q . \qquad (12.224)$$

This breaks up into two terms

$$\hat{\Sigma}(\mathbf{k}) = \Sigma_{RPA}(\mathbf{k}) - \Sigma_x(\mathbf{k}).$$ (12.225)

The portion

$$\Sigma_x(\mathbf{k}) = \int \frac{d\mathbf{q}}{(2\pi)^3} \int \frac{d\omega}{2\pi} G^{(0)}(\mathbf{k} + \mathbf{q}, \omega) v_{\mathbf{q}},$$ (12.226)

is just the H-F exchange self-energy discussed above. The connection with equation (12.67) is made more clear by noting

$$n(\mathbf{k} + \mathbf{q}) = \int \frac{d\omega}{2\pi} G^{(0)}(\mathbf{k} + \mathbf{q}, \omega),$$ (12.227)

which can be done because $v_{\mathbf{q}}$ does not depend on the frequency. The diagram that corresponds to $\Sigma_x(\mathbf{k})$ is shown in figure 12.7.

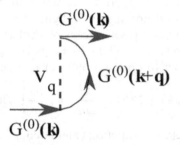

Figure 12.7. The diagrammatic representation of the Hartree–Fock exchange self-energy $\Sigma_x(\mathbf{k})$.

The RPA self-energy is

$$\Sigma_{RPA}(\mathbf{k}) = \int \frac{d\mathbf{q}}{(2\pi)^3} \int \frac{d\omega}{2\pi} \frac{G^{(0)}(\mathbf{k} + \mathbf{q}, \omega)}{\varepsilon_{RPA}(\mathbf{q}, \omega)} v_{\mathbf{q}},$$ (12.228)

and for many years it was the best self-energy available. It behaves in a physically reasonable way, and the effective mass m^* does not go to zero at k_F.

A great success of infinite-order perturbation theory is that the contributions from the diagrams in figure 12.6 cancel the badly behaving $\Sigma_x(\mathbf{k})$ and replace it with the much better $\Sigma_{RPA}(\mathbf{k})$.

The first high-level formula for the exchange-correlation potential, given in equation (12.156), was derived by Lars Hedin and Stig Lundquist by fitting numerically to the RPA self-energy. The search for improved exchange-correlation potentials is a field unto itself. As pointed out in reference [8], over 400 functionals are listed in LIBXC. They include local density approximations that depend only on the density and generalized gradient approximations that include derivatives of the density.

There are an infinity of diagrams that are left out of the RPA. An example is shown in figure 12.8.

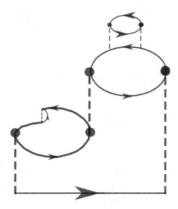

Figure 12.8. A diagram that is left out of the RPA.

Further developments in infinite-order perturbation requires the inclusion of more diagrams, but this has to be done very carefully.

12.15 Progress requires cooperation

Progress in modern quantum mechanics and quantum information requires the cooperation of specialists in many fields. In the early days, physicists who focused on one aspect of their field cast aspersions on those who focused on others. For example, at the 1969 Electronic Density of States Conference at Gaithersburg, Maryland, Professor John Ziman compared main frame computers to elephants and those who used them to mahouts riding on their backs. Professor Jacques Friedel had a preference for back-of-the-envelope calculations, saying that, once numbers were put in a computer, god only knows what could happen to them. On the other hand, the brilliant Nobel laureate Kenneth Wilson spoke glowingly in his acceptance speech in 1982 about the usefulness of computers in his research.

At one time, there was a chasm between scientists using density functional theory and many-body physicists using diagrams. Today, it is natural for theorists to use the most sophisticated versions of many-body theory to improve the density functional theory, while many-body theorists who want to explain the behavior of a specific material ask first what the energy bands look like.

The evolution towards cooperation is to be celebrated.

Problems

P12.1 Explain the Born–Oppenheimer approximation.

P12.2 A list of functionals are given below. Find the indicated functional derivatives.

$$F[f] = \int_a^b w(x')f(x')dx' \quad \frac{\delta F[f]}{\delta f(x)} =$$

$$F[f] = \int_a^b w(x')f^n(x')dx' \quad \frac{\delta F[f]}{\delta f(x)} =$$

$$F[g, f] = \int_a^b g(x'')w(x', x'')f(x')dx' \quad \frac{\delta F[g, f]}{\delta f(x)} =$$

$$F[g, f] = \int_a^b g(x'')w(x', x'')f(x')dx' \quad \frac{\delta^2 F[g, f]}{\delta f(x)\delta g(x)} =$$

$$F[f^*, f] = \int_a^b f^*(x'')w(x', x'')f(x')dx' \quad \frac{\delta F[f^*, f]}{\delta f^*(x)} =$$

P12.3 Insert the Slater determinants in equation (12.34) to obtain equation (12.39).

P12.5 The functional $K[\rho, \rho]$ is defined in equation (12.87). Evaluate the functional derivatives $\frac{\delta K[\rho, \rho]}{\delta \rho(\mathbf{r})}$ and $\frac{\delta^2 K[\rho, \rho]}{\delta \rho(\mathbf{r})\delta \rho(\mathbf{r}')}$.

P12.6 Derive an expression for $h(\mathbf{r}, \mathbf{r}')$ in the single configuration approximation.

P12.7 Redo the hydrogen molecule problem of chapter 7 using the Thomas–Fermi theory. Include the proton–proton interaction.

P12.8 From equations (12.84)–(12.87) what precisely is being replaced by $\varepsilon_{xc}[\rho(\mathbf{r})]$?

P12.9 From equation (12.92) what precisely is being replaced by $\varepsilon_{xc}[\rho(\mathbf{r})]$ in the single configuration approximation?

P12.10 How does the Green function in equation (12.165) compare with the one in chapter 6?

P12.11 What feature of the Lindhard function in equation (12.218) is responsible for Friedel oscillations?

References

[1] https://en.wikipedia.org/wiki/Lagrange_multiplier
[2] https://en.wikipedia.org/wiki/Antisymmetrizer
[3] https://en.wikipedia.org/wiki/Effective_mass_(solid-state_physics)#Experimental
[4] Mahan G D 1981 *Many-Particle Physics* (New York: Plenum)
[5] Kohn W and Hohenberg P C 1964 *Phys. Rev.* **136** B864–71
[6] Kohn W and Sham L J 1965 *Phys. Rev.* **140** A1133
[7] Hedin L and Lundqvist S 1969 *Solid State Phys.* **23** 1–181 (New York: Academic)
[8] Lehtola S, Steigemann C, Oliveira M J T and Marques M A L 2018 Recent developments in Libxc—a comprehensive library of functionals for density functional theory *Software X* **7** 1
[9] Feynman R P 1949 *Phys. Rev.* **76** 769

CPSIA information can be obtained
at www.ICGtesting.com
Printed in the USA
BVHW012246130322
631286BV00003B/15